Essential Topics on Air Pollution

Essential Topics on Air Pollution

Edited by **Raven Brennan**

R CALLISTO REFERENCE

New York

Published by Callisto Reference,
106 Park Avenue, Suite 200,
New York, NY 10016, USA
www.callistoreference.com

Essential Topics on Air Pollution
Edited by Raven Brennan

International Standard Book Number: 978-1-63239-324-1 (Hardback)

Printed in the United States of America.

Contents

Preface

This book provides a comprehensive overview of various facets of air pollution and its effects. There are many complex health issues caused by air pollutants. The ecological health community is being challenged to take some crucial actions to maintain the health of the population. Identifying, observing and evaluating the matter is a favorable way forward, but also raises numerous challenges including what such approaches are, how they can be materialized, and how they are protecting human health. This book gives a summary of key issues in air pollution. Reviews and researches included in this book describe air pollution in a range of situations and explain the available response management regarding this form of pollution.

Various studies have approached the subject by analyzing it with a single perspective, but the present book provides diverse methodologies and techniques to address this field. This book contains theories and applications needed for understanding the subject from different perspectives. The aim is to keep the readers informed about the progresses in the field; therefore, the contributions were carefully examined to compile novel researches by specialists from across the globe.

Indeed, the job of the editor is the most crucial and challenging in compiling all chapters into a single book. In the end, I would extend my sincere thanks to the chapter authors for their profound work. I am also thankful for the support provided by my family and colleagues during the compilation of this book.

Editor

Current Characteristic of Air Pollutants

Old and New Air Pollutants:
An Evaluation on Thirty Years Experiences

Margherita Ferrante, Maria Fiore, Gea Oliveri Conti, Caterina Ledda, Roberto Fallico and Salvatore Sciacca

Additional information is available at the end of the chapter

1. Introduction

Air pollutants are generally defined as those substances which alter the composition of the natural atmosphere.

Emissions of air pollutants derive from almost all economic and societal activities but also by natural disaster (eg.: particulate matter or gaseus emitted by volcanic activities or forest fires, dust by desert winds, pollen scattering, sea aereosol, etc..). The energy production and the general industry activity, all types of transport and agriculture are key emission sources of air pollutants. They result in clear risks to human health and the ecosystems integrity. Air pollution is not only a local phenomenon but also a transboundary issue, in fact, the air pollutants emitted in one Country may be transported in the atmosphere and they harming human health and the environment elsewhere .

In Europe, policies and actions at all levels have greatly reduced the anthropogenic emissions and exposure but some air pollutants still harm human health. Air pollutants are divided into primary pollutants like carbon monoxide, sulphur dioxide, hydrocarbon species, dust and soot, which are emitted directly by air pollutant sources, and secondary pollutants like nitrogen dioxide, photochemical ozone, and aerosol, which are created by chemical changes which occur in the atmospheric environment (WHO, 2005).

After the Meuse Valley fog in 1930 (Firket, 1936) or the London smog in 1952 (Ministry of health, 1954), the air pollution is considered today an important research driver for a global public health protection (WHO, 2005). In fact a high-level exposure to these pollutants at the long-term and short-term can lead to some important adverse health effects, ranging from irritation of the respiratory system to contributing to increased prevalence and incidence of respiratory and cardiovascular diseases and premature death in people of all ages. Particularly children are very susceptible for their very fast metabolism (WHO, 2005).

Emissions of the main air pollutants in Europe have declined significantly in recent decades, greatly reducing exposure to substances such as sulphur dioxide (SO_2) and lead (Pb). Nevertheless, poor air quality remains an important public health issue.

Many EU Member States do not comply with legally binding air quality limits protecting human health. Exposure of vegetation to ground level ozone (O_3) will continue to exceed long-term EU objectives.

In terms of controlling emissions, only 14 European countries expect to comply with all four pollutant-specific emission ceilings set under EU and international legislation for 2010.

The upper limit for nitrogen oxides (NO_x) is the most challenging 12 countries expect to exceed it, some by as much as 50 % (SOER, 2010).

The Thematic Strategy on Air Pollution from the European Commission (2005) set the objectives for the improvement of human health and the environment through the improvement of air quality to the year 2020 (see table 1). At present, airborne PM, tropospheric O_3, and NO_2 are Europe's most problematic pollutants in terms of causing harm to health (EEA, 2010).

The main air pollutants before human's exposure are subject to a range of atmospheric processes including atmospheric transport, mixing and chemical transformation.

Air pollutants, also, depending on their physical-chemical characteristics and on the basis of factors such as atmospheric conditions or characteristics of receiving surfaces, may be deposited after either short (local, regional) or long-range (European, inter-continental) transport. Pollutants can be washed out of the atmosphere by precipitation rain, snow, fog, dew, frost and hail or deposited dry as gases or particulate matter.

2. "Old" and "new" pollutants trend

In most cities air quality has improved over the past decades. In particular, emissions of the main old air pollutants such as sulphur dioxide (SO_2) and lead (Pb) together with other hazardous pollutants including persistent organic pollutants (POPs) and heavy metals, in Europe have declined significantly in recent decades. Nitrogen (N) and nitrogen dioxide (NO_2), on the other hand, has not been dealt with as successfully. Between 1990 and 2008 emissions of polycyclic aromatic hydrocarbons (PAHs) decreased by 60% overall; emissions of polychlorinated biphenyls, dioxins and furans decreased too. While the majority of Countries report that emissions of these substances have fallen during that period, some Countries report that emissions have increased. Emissions of primary particulate matter, $PM_{2.5}$ and PM_{10}, have both decreased by about 13% since 2000. At present, airborne particulate matter (PM), tropospheric (ground-level) ozone (O_3) and polycyclic aromatic hydrocarbons (PAHs) are the new problematic pollutants in Europe in terms of causing harm to health. Moreover, there is an increasing recognition of the importance of long-range hemispheric transport of air pollutants to and from Europe and other continents. VOC (Volatile Organic Compounds) and small dust particles are examples of large-scale air pollutants. At the end the wide-scale use of catalytic converters for automotive traction in most industrialised Countries has led, over the years, to a substantial increase in environmental concentrations of palladium, platinum, and rhodium, also known as the platinum group elements (PGEs). The detection of PGEs, even in remote areas of the planet, provides evidence of the global nature of the problem.

The following paragraphs describe the old and new contaminants in relation to their characteristics, emission sources, health effects and trends through 30 years exeperiences.

Human health	Limit or target (*) value				Time extension (***)	Long-term objective		Information (**) and alert thresholds	
Pollutant	Averaging period	Value	Maximum number of allowed occurrences	Date applic-able	New date applicable	Value	Date	Period	Threshold value
SO$_2$	Hour	350 µg/m³	24	2005				3 hours	500 µg/m³
	Day	125 µg/m³	3	2005					
NO$_2$	Hour	200 µg/m³	18	2010	2015			3 hours	400 µg/m³
	Year	40 µg/m³	0	2010					
Benzene (C$_6$H$_6$)	Year	5 µg/m³	0	2010	2015				
CO	Maximum daily 8-hour mean	10 mg/m³	0	2005					
PM$_{10}$	Day	50 µg/m³	35	2005	2011				
	Year	40 µg/m³	0	2005 ·	2011				
PM$_{2.5}$	Year	25 µg/m² (*)	0	2010		8.5 to 18 µg/m³	2020		
		20 µg/m³ (ECO)		2015					
Pb	Year	0.5 mg/m³ (*)	0	2005					
As	Year	6 ng/m³ (*)	0	2013					
Cd	Year	5 ng/m³ (*)	0	2013					
Ni	Year	20 ng/m³ (*)	0	2013					
BaP	Year	1 ng/m³ (*)	0	2013					
O$_3$	Maximum daily 8-hour mean averaged over 3 years	120 µg/m³ (*)	25	2010		120 µg/m³	Not defined	1 hour	180 µg/m³ (**)
								3 hours	240 µg/m³

Note: The majority of EU Member States (MS) have not attained the PM10 limit values required by the Air Quality Directive by 2005 (EC, 2008a). In most urban environments, exceedance of the daily mean PM10 limit is the biggest PM compliance problem. 2010 is the attainment year for NO2 and C₆H₆ limit values. A further important issue in European urban areas is also exceedance of the annual NO2 limit value, particularly at urban traffic stations.

(#) Signifies that this is a target value and not a legally binding limit value; see EC, 2008a for definition of legal terms (Article 2).

(*) Exceptions are Bulgaria and Romania, where the date applicable was 2007.

(**) Signifies that this is an information threshold and not an alert threshold; see EC, 2008a for definition of legal terms (Article 2).

(***) For countries that sought and qualified for time extension.

Source: SOER 2010.

Table 1. Summary of air-quality directive limit values, target values, assessment thresholds, long-term objectives, information thresholds and alert threshold values for the protection of human health.

3. Carbon monoxide (CO)

Carbon monoxide is a tasteless, odorless, colorless and toxic gaseous pollutant ubiquitous in the outdoor atmosphere that is generated by combustion (Bell M, et All. 2009). EPA initially established NAAQS (National Ambient Air Quality Standard) for CO on April 30, 1971. The standards were set at 9 ppm, as an 8-hour average, and 35 ppm, as a 1-hour average, neither to be exceeded more than once per year. On January 28, 2011, EPA proposed to retain the existing NAAQS for carbon monoxide. After careful review of the available health science, EPA concludes that the current standards provide the required level of public health protection, including protection for people with heart disease, who are especially susceptible to health problems associated with exposures to CO in ambient air.

4. Sulphur dioxide (SO₂)

Historically, SO_2 derived from the combustion of fossil fuels have been the main components of air pollution in many parts of the world. The most serious problems have been experienced in large urban areas where coal has been used for domestic heating purposes, or for poorly controlled combustion in industrial installations (WHO, 2000). In recent years the use of high-sulfur coal for domestic heating has declined in many western European countries, and powder generation is now the predominant source. These changes in pattern of usage have led to urban and rural concentrations becoming similar; indeed in some areas rural concentrations now exceed those in urban areas (WHO, 2000). The city of Catania (Sicily, Italy) has established a network of air quality monitoring stations. The analysis of data show a clear and significant decline since 1993 to 2000 (First Report on the state of the environment of the City of Catania). The significant reduction in emissions of sulphur dioxide achieved since the 1970s is one of the great success stories of Europe's past air pollution policy (EEA, 2010).

5. Particulate Matter (PM)

Over the past decade, 20–50 % of the urban population was exposed to PM_{10} concentrations in excess of the EU daily limit values set for the protection of human health— a daily mean of 50 µg/m³ that should not be exceeded on more than 35 days in a calendar year. The same situation happened in Siracusa (Sicily, Italy), so our research group analyzed the phisico-chemical characteristics of PM_{10} and $PM_{2,5}$ fractions in order to determine the major aerosol contributions to these two granulometric size fractions of the urban aerosol. We found that vehicular traffic is only one cause of the daily elevation of PM in Siracusa City and we exclude industrial derivation of particulate (Sciacca et al., 2007).

The Air Quality Guideline level for PM_{10} set by the WHO is 20 µg/m³. Exceedances of this level can be observed all over Europe, also in rural background environments. In many European urban agglomerations, PM_{10} concentrations have not changed since about 2000. One of the reasons is the only minor decreases in emissions from urban road traffic. Increasing vehicle-km and dieselisation of the vehicle fleet jeopardise achievements from other PM reduction measures. Further, in several places emissions from the industry and domestic sectors — for example, from wood burning — may even have increased slightly. The EU Air Quality Directive of 2008 includes standards for fine PM (PM2.5): a yearly limit value that has to be attained in two stages, by 1 January 2015 (25 µg/ m³) and by 1 January 2020 (20 µg/m³). Further, the directive defines an average exposure indicator (AEI) for each Member State, based on measurements at urban background stations. The required and absolute reduction targets for the AEI have to be attained by 2020. Focusing on PM mass concentration limit values and exposure indicators does not address the complex physical and chemical characteristics of PM. While mass concentrations can be similar, people may be exposed to PM cocktails of very different chemical composition (WHO, 2007).

6. Ammonia (NH₃) emission

Ammonia has become the most abundant gas-phase alkaline species in the atmosphere. Most of the ammonia released into the atmosphere is converted into particulate ammonium sulfate and nitrate. Gaseous ammonia and ammonium compounds in particles are deposited from the air by wet deposition and dry deposition. NH₃ is mainly emitted from livestock and production and application of fertilizers. Natural sources including soil, vegetation and wild animal might also be contributors to the total amount of ammonia emission.

According to the United Nations Food and Agriculture Organization recent research findings, livestock are responsible for almost two thirds of anthropogenic NH₃ emissions that contribute significantly to acid rain and acidification of ecosystems (Sidiropoulos & Tsilingiridis, 2009; EMEP/CORINAIR, 2007). Following deposition, soil microbes can convert ammonia into acidic compounds by nitrification. Through these processes, NH₃ can contribute to acidic compounds on natural ecosystems and also cause eutrophication.

7. Nitrogen oxides (NO₂)

Nitrogen dioxide a combustion-generated oxidant gas, is widely present in indoor and outdoor environments. Outdoors, where it comes primarily from high temperature fuel combustion of engines, industry, and power generation, it is a precursor to particles and ozone (G. Viegi,2004).

Using a nationwide network of monitoring sites, EPA has developed ambient air quality trends for nitrogen dioxide. Nationally, average NO₂ concentrations have decreased substantially over the years. In January 2010, EPA set the primary NO₂ standard at a level of 100 parts per billion and the secondary NAAQS remains to 0.053 ppm (EPA, 2012). The European air quality guidelines suggest a daily maximum concentration of 200 mg/m³ (1 h) for NO₂, while the WHO recommends a limit of 40 mg/m³ (annual average) for long-term exposure. In general, the levels reported for Europe, Canada and the United States (except for New Mexico) are below this exposure threshold, whereas higher levels have been measured in Asiatic countries and in Mexico (100 mg/m³).

There is still no robust basis for setting an annual average guideline value for NO_2 through any direct toxic effect. Evidence has emerged, however, that increases the concern over health effects associated with outdoor air pollution mixtures that include NO_2. A number of recently published studies have demonstrated that NO_2 can have a higher spatial variation than other traffic-related air pollutants, for example, particle mass. These studies also found adverse effects on the health of children living in metropolitan areas characterized by higher levels of NO_2 even in cases where the overall city-wide NO_2 level was fairly low. A number of short-term experimental human toxicology studies have reported acute health effects following exposure to 1-hour NO_2 concentrations in excess of 500 $\mu g/m^3$. Although the lowest level of NO_2 exposure to show a direct effect on pulmonary function in asthmatics in more than one laboratory is 560 $\mu g/m^3$, studies of bronchial responsiveness among asthmatics suggest an increase in responsiveness at levels upwards from 200 $\mu g/m^3$.

Since the existing WHO AQG short-term NO_2 guideline value of 200 $\mu g/m^3$(1-hour) has not been challenged by more recent studies, it is retained. In conclusion, the guideline values for NO_2 remain unchanged in comparison to the existing WHO AQG levels, i.e. 40 $\mu g/m^3$ for annual mean and 200 $\mu g/m^3$ for 1-hour mean.

8. Ozone (O_3)

It is the principal component of smog, which is caused primarily by automobile emissions, predominantly in urban areas. Normal levels of ozone in the air are between 20 and 80 mg/m^3. Ozone concentrations in urban areas rise in the morning, peak in the afternoon, and decrease at night. Ozone has become a significant pollutant as a result of increased population growth, industrial activities, and use of the automobile. Ozone is at present the primary air pollution problem in the United States. A trend analysis covering the years from 1993 to 2005 showed that the average number of hours with an ozone concentration above 180 $\mu g/m^3$ (the EU information threshold) for any given monitoring site was higher in the summer of 2003 than in any of the previous years (Park JW et al., 2004). On the February 2002 European Parliament approved a guideline (2002/3/CE) that indicates the "information threshold" (180 $\mu g/m^3$) and "alert threshold" (360 $\mu g/m^3$) for ozone and imposes urgent obbligation of the population's information. In Italy, legislative decree 155/2010 sets long term aims for human health protection (120 $\mu g/m^3$ during 8 hours). Ozone is monitored in Catania (Sicily, Italy) since 1997 by two control units but since then attention levels for this compound have never been exceeded (Comune di Catania, 2001). Respiratory tract responses induced by ozone include reduction in lung function, aggravation of preexisting respiratory disease (such as asthma), increased daily hospital admissions and emergency department visits for respiratory causes, and excess mortality (Corsmeier et al., 2002). Controlled human exposure studies have demonstrated that short-term exposure - up to 8 hours - causes lung function decrements such as reductions in forced expiratory volume in one second (FEV1), and the following respiratory symptoms: cough, throat irritation, pain, burning, or discomfort in the chest when taking a deep breath, chest tightness, wheezing, or shortness of breath. The effects are reversible, with improvement and recovery to baseline varying from a few hours to 48 hours after an elevated ozone exposure.

9. Heavy metals

Heavy metals are a class of pollutants extremely widespread in the various environmental matrices. They are natural components of the earth's crust. Their presence in air, water and soil erosion resulting from natural phenomena and human activities. To a small extent they enter human bodies where, as trace elements, they are essential to maintain the normal metabolic reactions. They cannot be degraded or destroyed, and can be transported by air, and enter water and human food supply. Respect to air pollution, the metals that are generally more concerned are: As, Cd, Co, Cr, Mn, Ni, Pb as conveyed by particulate air pollution. Their origin is different: Cd, Cr and As are mostly from mining and steel industries; Cu and Ni from combustion processes; Co, Cu, Zn and Cr from cementitious materials obtained by recycling scrap steel industries and incinerators. The effect of heavy metals on the human health depends on the mode of assumption, as well as the amount absorbed. Heavy metals (such as cadmium, mercury and lead) are recognised as being directly toxic to biota. All have the quality of being progressively accumulated higher up the food chain, such that chronic exposure of lower organisms to much lower concentrations can expose predatory organisms, including humans, to potentially harmful concentrations. In humans they are also of concern for human health because of their toxicity, their potential to cause cancer and their ability to cause harmful effects at low concentrations. Their relative toxic/carcinogenic potencies are compound specific. Specifically, exposure to heavy metals has been linked with developmental retardation, various cancers, kidney damage, and even death in some instances of exposure to very high concentrations. Heavy metals can reside in or be attached to PM. For several metals there are the standard reference, in particular for lead the limit is intended as an average annual value of 0.5 ug/m^3.

Urban sources of lead are fossil fuels, mining and manufacturing. The use of lead as an additive to gasoline was banned in 1996 in the United States and since then the cases of acute intoxication are notably decreased. In the Municipality of Catania lead is measured through control programmes long for 15- 20 days. Since 1999 to 2000 it can be note a decrease of mean concentrations (from 0,38 µg/m^3 to 0,15 µg/m^3) with values lower than those required in the European Rule 99/30/CE. This result is correlated with the abolishment of use of lead in gasoline (Comune di Catania, 2001).

In our experience high-level exposure to metals of men can damage sperm production and motility, as are suggest by an our study that shows adverse impact of heavy metals on male reproductive health. We have conducted a case-control study to investigate the exposition to lead, arsenic, nickel and male fertility. The results show a sperm motility reduction of 50 % (Ferrante M et al., 2011). Another study in the industrial triangle of Priolo-Melilli-Augusta shows that males living in these towns show a sperm progressive motility decrease from 45% to 23% whereas density and morphology were into the reference limit of WHO parameters (Ferrante M et al., 2011).

Cadmium is widely spread in the environment. Its consumption is growing, as a result cadmium contamination of soil, water and air increases. Cadmium enters soil, water, and air from mining, industry, and burning coal and household wastes and its particles in air can travel long distances before falling to the ground or water. Cadmium is accumulated in fish, plants, animal and human body. People can be exposed to cadmium eating contaminated foods, smoking cigarettes or breathing cigarette smoke, drinking contaminated water, living or working near industrial facilities which release cadmium into the air. Following the European Law 155/2010, Sicily adopted a plane to value and manage air quality, aiming to not exceed levels of 5,0 ng/m^3 of cadmium in the air (ARPA Sicilia). The form of cadmium that is of most interest for health effects from inhalation exposure is cadmium oxide because that is the main form of airborne cadmium. Our research group has conducted a case–control study to examine relationships between environmental exposures, particularly to Cd, and male infertility. Cd showed concentrations in seminal plasma of the cases (1.67 µg/l) higher than controls (0.55 µg/l) and cases showed a motility reduction from 45% to 23 % (50% reduction). Our results indicate that the males exposed to environmental Cd showed a deleterious effect on fertility (Ferrante M et al., 2011). Another study, carried out also by our research group and not yet published, has shown the toxic effect of Cd even on male reproductive organs because this metal cause a blood-testis barrier disruption and consequently an impairment in sperm production. Finally we have to consider the effect of cadmium on cancer development. Cd was classified as a cancer-causing agent in humans by the WHO (1993), based on consistent reports of an association between Cd exposure and lung cancer (ATSDR 2008).

Heavy metals can reside in or be attached to PM.

10. Organic compound

10.1. Persistent Organic Pollutants (POPs)

Persistent organic pollutants are a group of chemicals which have been intentionally or inadvertently produced and introduced into the environment and because of theirs resistance to degradation, they persist in the environment.

Due to their stability and transport properties, they are now widely distributed around the world, and are even found in places where they had never been used, such as the arctic regions. Given their long half-lives and their fat solubility, POPs tend to bioaccumulate in the food-chain including fish, meat, eggs and milk. POPs are also present in the human body and traces can be found in human milk (WHO, 2007).

The United Nations Environment Programme Governing Council (GC) originally created a have listed 12 POPs, known as the "dirty dozen." Nine of these are old organochlorine pesticides, including including aldrin, dichlorodiphenyltrichloroethane (DDT), chlordane, dieldrin, endrin, heptachlor, hexachlorobenzene, mirex and toxaphene, whose production and use have been banned or strictly regulated by most countries for some time. In addition other three POPs of concern are industrial chemicals, including the widely used polychlorinated biphenyls (PCBs) as well as two groups of industrial by-products, polychlorinated dibenzodioxins (PCDDs or dioxin) and polychlorinated dibenzofurans (PCDFs or furans). In recent years, this list has been expanded to include polycyclic aromatic hydrocarbons (PAHs), polybrominated diphenyl ethers (PBDE), and tributyltin (TBT). The groups of compounds that make up POPs are also classed as persistent, bioaccumulative, and toxic (PBTs) or toxic organic micro pollutants (TOMPs). These terms are essentially synonyms for POPs (Crinnion, 2011).

While production of PCBs has been largely banned for many years, the electrical transformers and other equipment containing still these chemicals are still in use and they present, today, serious disposal problems.

Regard to PCDDs and PCDFs, the better manufacturing controls and reduction of emissions from industrial combustion processes, e.g. power generation and waste incineration plants, have made a measurable impact on decreasing levels of these chemicals in human milk, particularly in Europe (WHO, 2007).

Humans can be exposed to POPs through the direct exposure, e.g. occupational accidents or by environment exposure (including indoor). Short-term exposures to high concentrations of POPs may result in severe illness and death. Chronic exposure to POPs may also be associated with a wide range of adverse health effect as a the endocrine disruption, reproductive and immune dysfunction, neurobehavioral and developmental disorders and cancer (Ritter et al, 1995).

Polycyclic Aromatic Hydrocarbons (PAHs) and PCDD/Fs are perhaps the most obvious example. However, because an extensive array of POPs occur and accumulate simultaneously in biota it is very difficult to say conclusively that an effect is due to one particular chemical or a family of chemicals, in fact several chemicals act synergistically (Jones, 1999).

10.2. Dioxins and furans (PCDDs and PCDFs)

Polychlorinated dibenzo-p-dioxins and dibenzofurans are two similar classes of chlorinated aromatic chemicals that are produced as contaminants or by products.

Most dioxins and furans are not man-made or produced intentionally, but are created when other chemicals or products are made. They are formed as unwanted byproducts of certain chemical processes during the manufacture of chlorinated intermediates and in the combustion of chlorinated materials. The chlorinated precursors include polychlorinated biphenyls (PCB), polychlorinated phenols, and polyvinyl chloride (PVC) (Faroon M. et. al., 2003). Of all of the dioxins and furans, one, 2,3,7,8-tetrachloro-p-dibenzo-dioxin (2,3,7,8 TCDD) is considered the most toxicand the most extensively studied. Like the other POPs are chemically stable and highly lipophilic; in the environment, are persistent, undergo transport, and preferentially bioconcentrate in higher trophic levels of the food chain (Safe SH, 1998). In terms of dioxin release into the environment, uncontrolled waste incinerators (solid waste and hospital waste) are often the worst culprits, due to incomplete burning.

Technology is available that allows for controlled waste incineration with low emissions. Dioxins also have been detected at low concentrations in cigarette smoke, home-heating systems, and exhaust from cars running on leaded gasoline or unleaded gasoline, and diesel fuel. The larger particles will be deposited close to the emission source, while very small particles may be transported longer distances will be deposited on land or water, contaminating the food of animal origin, as they are persistent in the environment and accumulate in animal fat and finding himself well in dairy, meat, fish and shellfish (Alcock R, 2003). Excluding occupational or accidental exposures the most common way is by eating food contaminated with dioxins particularly important is, also, the exposure of infants through breast-feeding because of the high content of fat in human milk and may exceed the exposure of adults by one or two orders of magnitude (Gies A, 2007).

Releases from industrial sources have decreased approximately 80% since the 1980s (Consonni, 2012) and have been the subject of a number of federal and state regulations and clean-up actions; however, current exposures levels still remain a concern. (EPA, http://www.epa.gov/pbt/pubs/dioxins.htm). TDI (tolerable daily intake) values recommended by WHO is 1-4 pg TEQ/kg/day. However, several nations have performed their own reassessment of the available toxicity data for dioxin to derive a TDI (ASTDR, 2011). Today, the largest release of these chemicals occurs as a result of the open burning of house- hold and municipal trash, landfill fires, and agricultural and forest fires. Breast milk is a substantial source of exposure for infants (Lundqvist et al., 2006), though breast milk levels have been decreasing in recent years (Arisawa et al., 2005).

10.3. Polycyclic AromaticH (PAHs)

Polycyclic aromatic hydrocarbons are ubiquitous pollutants formed from the combustion of fossil fuels, industrial powder generation, incineration, production of asphalt, coal tar and coke, petroleum catalytic cracking and primary aluminium production (WHO, 2003). The specific emissions of PAHs from modern cars were observed to be 5 times higher from diesel engines than from gasoline cars during transient driving conditions. Older diesel cars and gasoline cars with a catalytic converter of outmoded design have 5–10 times higher PAH emissions than modern cars. PAHs can react with pollutants such as ozone, nitrogen oxides and sulfur dioxide, yielding diones, nitro- and dinitro-PAHs, and sulfonic acids, respectively (WHO, 2000). PAHs have a tendency to be associated with particulate matter and may be subject to direct photolysis (WHO, 2010). A number of PAHs are mutagenic and genotoxic, and induce DNA adduct formation in vitro and in vivo. IARC considers several purified PAHs and PAH derivatives to be probable (group 2A) or possible (group 2B) human carcinogens. Benzo[a]Pyrene has decreased fertility and caused embryotoxicity (WHO 2010) Our research group has conducted a study aimed to evaluate concentration of priority PHAs in seminal plasma samples of healthy men (age 20-45 years) living in a polluted area of Sicily (Priolo-Augusta-Melilli triangle), declarated "area at elevated environmental crisis". Our results show that the semen quality could not be affect by PAHs concentration air (Oliveri Conti et al., 2011).

Emissions of PAHs decreased by 60 % overall between 1990 and 2008 in the EEA-32 but increased in a small number of countries (EEA 2010). In Europe and the United States, urban traffic contributes 46–90% of the total PAHs in ambient air. Recent initiatives included the conversion to ultralow-sulfur diesel fuel, aimed to be 97% cleaner (Narváez, 2008)

10.4. Non-Methane Volatile Organic Compounds (NMVOCs or TNMHC)

Non-methane volatile organic compounds are a collection of organic compounds that differ widely in their chemical composition but display similar behaviour in the atmosphere. Essentially, NMVOCs are identical to VOCs, but with methane excluded. NMVOCs are emitted into the atmosphere from a large number of sources including combustion activities, paint application, road transport, dry-cleaning and other solvent uses and production processes. NMVOCs contribute to the formation of ground level (tropospheric) ozone. In fact, the hydrocarbons have a strong tendency to react, in the presence of light, with the oxides of nitrogen and oxygen. In addition, certain NMVOC species or species groups such as benzene and 1,3 butadiene are hazardous to human health. Quantifying the emissions of total NMVOCs provides an indicator of the emission trends of the most hazardous NMVOCs. Biogenic NMVOCs are emitted by vegetation, with amounts dependent on the species and on temperature (EEA, 2010; EEA, 2011). There are thousands of organic compounds known ascribable to NMVOC, both of natural origin that is released into the air from plants (biogenic), and anthropogenic (anthropogenic), both species we can find them in the air or in the form of gas or in the form steam. The main TNMCH are generally: aliphatic hydrocarbons or carbon chain with a linear structure, or with aromatic ring structure (benzene, toluene, xylenes, etc..), oxygenated (aldehydes, ketones, etc.)., etc. Their concentration in the atmosphere in urban areas and industrial centers are directly related to vehicular traffic, domestic heating, to phenomena of evaporation of gasoline (engine compartments and tanks), the exhaust gas vehicle (incomplete combustion of fuels), the emissions from petrol stations fuel and many industrial activities (eg, oil refining, storage and handling of fuels, production of paints and solvents, etc. ...) (ARPAT, 2010; Broderick & Marnane, 2002). Solvent use and road transport are the two most significant sources of NMVOC emissions in urban environments. (Sidiropoulos C, 2009) but Today the major contributions of NMVOCs anthropogenic emissions are ascribed primarily to vehicular traffic. Adverse operating conditions of the vehicle (low speed, repeated gear changes, and frequent stops to a minimum) as those due to heavy traffic have resulted in greater emission of unburned hydrocarbons. The evaporative emissions mainly stem from the volatility of fuel and are therefore made up only of hydrocarbons. They occur when walking, both with the engine off at stops (ARPAV, 2004). The effects on human health are very different depending on the types of compounds present in the mixture thus depend solely on the type of hydrocarbons present and their concentrations (WHO, 1989; WHO, 2005). Hydrocarbons Alkanes are absolutely non-toxic. Are toxic and carcinogenic in some cases a part of the aromatic hydrocarbons ("Air Quality Guideline for Europe" WHO, 1989; Sciacca S. & Oliveri Conti G, 2009).

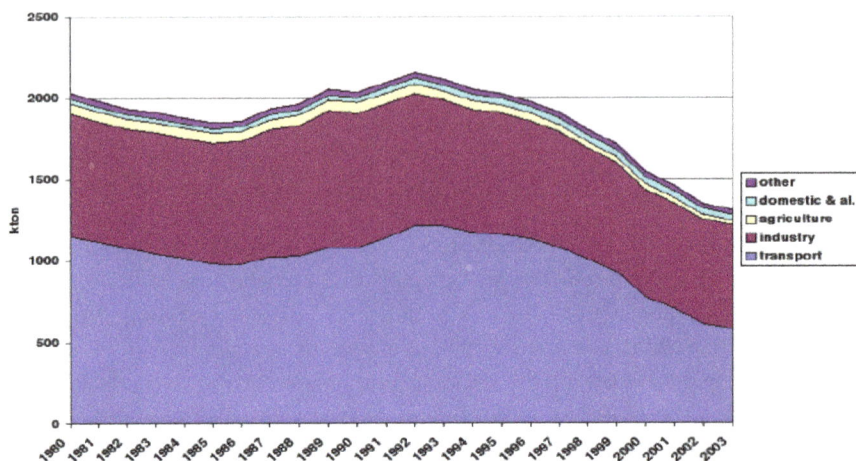

Figure 1. TNMCH Anthropogenic emission in Italy (da APAT)

Improved loading technology in the oil industry in the last 10 years has contributed to the decline, and this trend continued in 2010. However, the emission of solvents from products increased steeply in 2010, after a drop in the previous year, and made up for the decline in the oil industry. The NMVOC emission still ended 28% below the Gothenburg Protocol target. The emissions of non-methane volatile organic compounds have decreased by 51% since 1990. In 2009, the most significant sources of NMVOC emissions were 'Solvent and product use' (36%) (comprising activities such as paint application, dry-cleaning and other use of solvents), followed by 'Commercial, institutional and households' (15%). The decline in emissions since 1990 has primarily been due to reductions achieved in the road transport sector due to the introduction of vehicle catalytic converters and carbon canisters on petrol cars, for evaporative emission control driven by tighter vehicle emission standards, combined with limits on the maximum volatility of petrol that can be sold in EU Member States, as specified in fuel quality directives. The reductions in NMVOC emissions have been enhanced by the switching from petrol to diesel cars in some EU countries, and changes in the 'Solvents and product use' sector (a result of the introduction of legislative measures limiting for example the use and emissions of solvents.

10.5. Polychlorinated biphenyls (PCBs)

Polychlorinated biphenyls are mixtures of chlorinated hydrocarbons that have been used extensively since 1930 in a variety of industrial uses and are another major contaminant of concern in communities (Johnson BL, 1999). PCB production in most countries was banned in the 1970s and 1980s (Vallack, 1998).

According to the position of the chlorine atoms in the molecule of biphenyl may be obtained 209 congeners, 12 of which have characteristics similar to the dioxins and therefore defined dioxin-like (ASTDR, 2000). All PCB congeners are lipophilic (lipophilicity increases with increasing degree of chlorination) and have very low water solubilities. Once in the environment, PCBs do not readily break down and therefore may remain for very long periods of time. They can easily cycle between air, water, and soil and can be carried long distances. Air is probably the most significant compartment for environmental distribution. Their presence is ubiquitous in the environment, and residues have even been detected in the Arctic air, water and organisms (Alcock et al, 2003).

PCBs were formerly used as dielectric fluids in transformers and large capacitors, as pesticide extenders, plasticisers in sealants, as heat exchange fluids, hydraulic lubricants, cutting oils, flame retardants, dedusting agents, and in plastics, paints, adhesives and carbonless copy paper. Although PCBs are no longer made people can still be exposed to them. Old fluorescent lighting fixtures and old electrical devices and appliances, such as television sets and refrigerators, therefore may contain PCBs if they were made before PCB use was stopped. When these electric devices get hot during operation, small amounts of PCBs may get into the air and raise the level of PCBs in indoor air (ASTDR, 2000).

Food is the main source of exposure for the general population. PCBs enter the food chain by a variety of routes, including migration into food from external sources, contamination of animal feeds, and accumulation in the fatty tissues of animals. PCBs are found at higher concentrations in fatty foods (e.g., dairy products and fish). Other sources of exposure in the general population include the release of these chemicals from PCB-containing waste sites and from fires involving transformers and capacitors. The transfer of PCBs from mother to infant via breast milk is another important source of exposure. The lesser-chlorinated PCBs are more volatile and indoor inhalational exposure from buildings containing caulking made with these PCBs prior to 1979 can increase background serum levels.

Today, PCBs can still be released into the environment from poorly maintained hazardous waste sites that contain PCBs; illegal or improper dumping of PCB wastes, such as old transformer fluids; leaks or releases from electrical transformers containing PCBs; and disposal of PCB-containing consumer products into municipal or other landfills not designed to handle hazardous waste. PCBs may be released into the environment by the burning of some wastes in municipal and industrial incinerators (ASTDR, 2000).

Human health effects that have been reported after investigations of occupational and accidental exposures to high levels of PCBs include elevations of serum hepatic enzymes, dermal changes (such as chloracne and rashes), inconsistent associations with serum lipid levels, and some types of cancer in particular of the gastrointestinal tract (e.g., liver, biliary). They are classified as probable human carcinogens by IARC (ATSDR, 2000; Carpenter, 2006). PCBs weakly interact with estrogen and thyroid receptors and with transport proteins (Purkey et al., 2004). Developmental and fetotoxic effects may also be observed in humans.

Our research group has conducted a study with the aim to evaluate the presence of possible alterations of sperm parameters of male exposed to PCB. We studied a group of 96 volunteers (aged 20-46 years) resident in the Priolo-Augusta-Melilli (SR) triangle, which has been declared "area at elevated environmental crisis" (Italian Government note). Of all congeners analyzed the 74 appears to be the predominant (Fig.1). Our results show an alteration of the morphology and motility of spermatozoa. (Ferrante et al., 2006; Altomare et al, 2012).

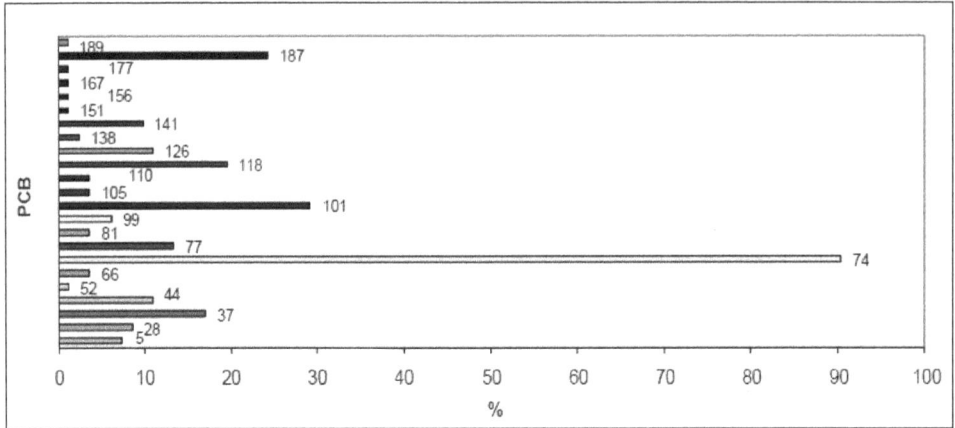

Figure 2. Percentage of subjects with detectable PCB levels in seminal plasma.

In general, published emission estimates for PCBs are difficult to compare as the methodological and empirical basis are different (Breivik et al., 2002; Breivik et al., 2004).

For decades, many countries and intergovernmental organizations have banned or severely restricted the production, usage, handling, transport and disposal of PCBs so, since the early 1980s, PCB concentrations in the air have shown a significantly decreasing trend for urban, rural, and marine/coastal areas. PCBs are also significantly decreased the concentrations in blood and human milk (Porta et al, 2012; Alivernini et al, 2011).

Concentrations of PCB in human blood decreased about 34-56% from 2002 to 2006 with difference for age, body mass index, weight (Porta et al., 2012). The comparison between PCBs measured in human milk samples collected in Rome between 2005 and 2007 with two previous studies performed in Rome in 1984 and in 2000-2001 indicates a 64% decrease of PCB levels, still in progress data are in good agreement with recent European studies (Alivernini et al., 2011).

10.6. Volatile Organic Compounds (VOCs)

Volatile organic compounds, major air pollutants in the indoor environment, are molecules typically containing 1–18 carbon atoms that readily volatilize from the solid or liquid state and are easily released into indoor air. All organic chemical compounds that can volatize under normal indoor atmospheric conditions of temperature and pressure are VOCs. These substances are classified in Very Volatile Organic Compound (VVOC), Volatile Organic Compound (VOC), and semivolatile Organic Compound (SVOC) to show the wide range of volatility among organic compounds. VOCs are emitted from many household products, including paints and lacquers, paint strippers, cleaning supplies, combustion appliances, aerosol sprays, glues, adhesives, dry-cleaned clothing, and environmental tobacco smoke. VOCs, which predominantly exist in the vapor phase in the atmosphere, and SVOCs, which exist in both vapor and condensed phase, redistribute to indoor surfaces and may persist from several months to years (Weschler et al., 2008). VOCs are of concern as both indoor air pollutants and as outdoor air pollutants but several studies have reported a two to fivefold increase in indoor concentrations of VOCs as compared to outdoors (Sexton et al. 2004). VOCs emission in Italy are regulated by the legislative decree 152/2006 that establishes threshold values between 50 mg C/Nm^3 - 150 mg C/Nm^3. Moreover both adults and children spend an estimated 90% of daily hours in indoor setting and energy conservation measures for buildings have led to reduced air exchange rates and promotion of indoor moisture buildup (Bornehag CG et al. 2005).

Exposure to VOCs can lead to acute and chronic health effects. The risk of health effects from inhaling any chemical depends on how much is in the air, how long and how often a person breathes it in (MDH, 2010). The major potential health effects include acute and chronic respiratory effects, such as bronchitis and dyspnea. Over the past few decades concern has increased about respiratory health effects from exposure to indoor air pollution.

Several studies have shown that the major effect of VOC's exposure is the development of asthma and allergic symptoms. Global trend in prevalence of allergic airway disease and other types of allergies in children and young adults appears to be increasing in Western countries, not only as seasonal decreases or in the context of particular ambients (Green et al., 2003). Between 2008 and 2010, our research group has conducted a study to evaluate the exposure to VOCs of secondary school students of first and second degree of Melilli Augusta and Priolo, area whit high environmental impact. The exposition was assessed by the passive samplers know, the Radiello, worn by student. The results show that the concentrations of major VOCs such as benzene, toluene and ethylbenzene, with a few exceptions, are within the limits of the law (Acerbi et al, 2010). Another study of our Department shows that the exposure to various environmental pollutants, including VOC, can cause development and exacerbation of asthma symptoms especially in children (Oliveri Conti G et a.,l 2011). Causal factors underlying these diseases and other contributors of global trend in prevalence since the 1970s remain unknown. Global secular trend in asthma and the allergy disease prevalence draw a parallel with vast shift in diet, lifestyle, and consumer product uses within the western societies since the World War II (Weschler et al., 2009). Enormous quantity and array of chemical compounds have been introduced in the societies which adopted western lifestyles. Consumer products, such as computer, TV, and synthetic building materials, including artificial carpets, composite wood, polyvinyl chloride (PVC) flooring, foam cushions, and PVC pipes emit an array of volatile organic compounds (VOCs), semi-volatile organic compounds (sVOCs) and nonorganic compounds. In a study, conducted as part of the European Community Respiratory Health Survey (ECRHS), authors reported higher concentration of total VOCs in recently painted homes, which was significantly associated with increased odds of asthma. Similarly, in a recent population based case-control study of children in Western Australia, two to threefold increased odds of asthma was reported among children exposed to benzene, ethylbenzene, and toluene (Rumchev et al., 2004). Indoor residential chemicals, emitted from particle board, plastic materials, recent painting, home cleaning agents, air freshener, pesticide, and insecticide, consistently increase the risks of multiple allergic symptoms and asthma-like symptoms (Henderson et al., 2008; Mendell et al ., 2007). As far as respiratory deseases, VOCs are responsible for allergic skin reaction, neurological toxicity, lung cancer, and eye and throat irritation, fatigue, headaches, dizziness, nausea and neurological symptoms such as lethargy and depression (Guo et al., 2004).

One of the most important VOCs is benzene. Benzene is an aromatic volatile hydrocarbon, having a characteristic smell. It come mainly into air from vehicles emissions, and from refueling losses; smoke from tobacco contains benzene and, in closed spaces, it constitutes the greater source of such polluter. Short term effects on man act on nervous system while long term ones produce progressive reduction of blood plateles and effect on leucocytes. Due to its toxicity benzene has been inserted from IARC in group I. In an our study as far as the city Catania is concerned, the benzene concentration has been maintained under the objective values of 10 microgrammi/m^3 (Ferrante et al., 2004)

We made a critical analysis of recent literature on biomarkers used in the assessment of exposure to benzene in order to identify, on the basis of personal research, and reliable biological indicators appropriate for monitoring exposure to low levels of solvent. We concluded identifying the urinary compartment, the site of elimination of benzene and its metabolites as such, as the most suitable for biological monitoring of benzene. The urinary benzene is certainly a valid biomarkers in the estimation of the internal dose of benzene, even if the measure prevail in precautions. The trans, trans-muconic acid, metabolite of benzene, as there is of toxicological interest, can be considered the most important biomarker in the assessment and of exposure and individual susceptibility to adverse effects of benzene. (Vivoli et al., 2002)

11. Platinum Group Elements (PGEs or PGMs)

The platinum group metals sometimes referred as the platinum group elements, comprise the rare metals platinum (Pt), palladium (Pd), rhodium (Rh), ruthenium (Ru), iridium (Ir) and osmium (Os).

Accumulation of PGEs is increasing in the environment over the time. The detection of PGEs, even in remote areas of the planet, provides evidence of the global nature of the problem. Catalytic converters of modern vehicles are considered to be the main sources of PGE contamination in addition to some other application (e.g. industrial, jewelry, anticancer drugs, etc.). The wide-scale use of catalytic converters for automotive traction in most industrialised Countries has led, over the years, to a substantial increase in environmental concentrations of PGEs. Along with PGEs the vehicle exhaust catalysts contain also a number of stabilizers, commonly oxides of rare earth elements and alkaline earth elements such as cerium (Ce), lantanium (La) and zirconium (Zr). Platinum content of road dusts, however, can be soluble, consequently, it enters to the waters, sediments, soil, and finally, the food chain. The effect of chronic occupational exposure to Pt compounds is well-documented, and certain Pt species are known to exhibit allergenic potential, but PGEs have also been found to be related to asthma, nausea, increased hair loss, increased spontaneous abortion, dermatitis, and other serious health problems in humans. Some researchers have shown that the Rh and Pd have a role on the emergence of certain tumors of the blood in rat. The vast majority of studies on airborne PGMs have however been carried out in urban areas characterised by high traffic density or in areas adjoining the aforementioned zones.

Analytical difficulties restrict the number of studies carried out for PGE concentration estimation in air and airborne particles.

In fact the low PGE concentration in the environmental samples combined with numerous interferences in the most sensitive analytical techniques are considered to be the major difficulties by many technicians.

Ravindra et al. (2004) says that : "the Pt concentration in air was reported to be lower than 0.05 pg/m^3 near a freeway in California. However, other studies in Germany have shown that the total Pt concentration in air along a highway ranged from 0.02 to 5.1 pg/m^3 (0.6 to 130 ng/g) with the Pt mainly present in the small particle size fraction (from 0.5 to 8 μm), whilst the larger airborne particles had a lower Pt content. The proportion of soluble platinum in air particles varied from 30 % to 43 %. A mean Pt concentration of 7.3 pg/m^3 has been measured inside Munich city buses and tramways during regular rides, with a strong correlation with traffic density. Bocca et al. (2003) reported a significant difference for the PGE content of air in urban and remote sites of Rome. The PGE concentration in urban airborne particulate matter ranged at 21.2-85.7 pg/m^3 for Pd, 7.8-38.8 pg/m^3 for Pt, and 2.2-5.8 pg/m^3 for Rh. In Madrid, the Pt and Rh concentrations in airborne particulate matter ranged from 3.1 to 15.5 pg/m^3, and from not detectable to 9.32 pg/m^3, respectively".

The present literature survey shows that the concentrations of these metals have increased significantly in the last decades in diverse environmental matrices; like airborne particulate matter, soil, roadside dust and vegetation, river, coastal and oceanic environment. Generally, PGEs are referred to behave in an inert manner and to be immobile.

Our research group has carried out a study on this issue entitled: "First data about Pt, Pd, and Rh in air, foods and biological samples in the district of Catania" (Ferrante et al., 1998) in order to acquire data about Rh, Pt and Pd concentrations in air, food, blood and urine samples of the territory of Catania in order to establish a set of values to make an initial bibliography. Metals investigated have been found in the samples assayed, although discontinuously and in trace concentrations.

Another study carried out on the Italian territory by Spaziani et al. (2008) investigated the Pt distribution in urban matrices (soils and dusts) in five cities, from north (Padova), central

(Rome and Viterbo), and south (Naples and Palermo) Italy in order to obtain a large set of data concerning pollution from autocatalysts. Analyses show a beginning of Pt enrichment in urban soils, with concentration ranges of 0.1–5.7 ng/g (Padova), 7–19.4 ng/g (Rome), 4.9–20 ng/g (Viterbo), 4.7–14.3 ng/g (Napoli), and 0.2–3.9 ng/g (Palermo).

Platinum group elements from automotive catalytic converters are continuously increasing in environmental matrices over the time. It is still under discussion, whether the emitted PGEs are toxic for human beings. The potential health risk from these elements would have to be taken in consideration for the possible risk of exposure for those living in urban environments, or along major highways.

12. Conclusion

The movement of atmospheric pollution between continents attracts increasing political attention. In a context dominated by the struggle against the emission of greenhouse gases, problems of air quality should not be underestimated and policies relating to climate protection must be taken into account.

All the above topics need further investigation (both experimental and model), partly on the base of health studies for novel air pollutants, to reach a better understanding of the behaviour of these in the environment. Greater international cooperation, also focusing on links between climate and air pollution policies, is required more than ever to address the phenomenon of air pollution.

However, the most important thing that emerges from our forty years of experience on the environmental topics is the need for a more precise and careful risk management (identification, esteem, evaluation and interventions on the risk). Too often technologies that pose a serious risk to the human health are replaced with technology just as risky or even riskier for the health of the population with enormous burdens also of the social and health costs.

Author details

Margherita Ferrante, Maria Fiore, Gea Oliveri Conti, Caterina Ledda,
Roberto Fallico and Salvatore Sciacca
Department "G.F. Ingrassia", Sector of Hygiene and Public Health, Catania University, Italy

13. References

Acerbi G, Zuccarello M, Toscano, Cipresso R, Mazzarino A, Bandini L, Ferrante M, Sciacca S. Valutazione dell'esposizione ai COV negli studenti delle scuole di Augusta-Priolo-Melilli, Congresso nazionale SiTI ottobre 2010.

Agency for Toxic Substances and Disease Registry (ATSDR). Toxicological Profile for Lead. 2007.

Agency for Toxic Substances and Disease Registry (ATSDR). Toxicological profile for heptachlor and heptachlor epoxide [online]. August 2007. Available at URL: http://www.atsdr.cdc. gov/toxprofiles/tp12.html. 4/21/09

Agency for Toxic Substances and Disease Registry (ATSDR). Toxicological profile for polycyclic aromatic hydrocarbons 1995 [online]. Available at URL: http://www.atsdr.cdc.gov/toxprofiles/ tp69.html. 5/26/09

Agency for Toxic Substances and Disease Registry (ATSDR). Toxicological profile for polychlorinated biphenyls. 2000 [online]. Available from URL: http://www.atsdr.cdc.gov/toxprofiles/tp17. html. 03/17/05.

Agency for Toxic Substances and Disease Registry (ATSDR). Toxicological profile for polycyclic aromatic hydrocarbons 1995 [online]. Available at URL: http://www.atsdr.cdc.gov/toxprofiles/ tp69.html. 5/26/09

Agency for Toxic Substances and Disease Registry (ATSDR). Toxicological Profile for Mercury. 1999.

Agency for Toxic Substances and Disease Registry (ATSDR). Toxicological Profile for Cadmium. 2008.

Alcock et. Al. Health risks of persistent organic pollutants from long-range transboundary air pollution, World Health Organization 2003.

Alcock R. et. al. (2003). Health risk of persistent organic pollutants from long range transboundary air pollution. World Health Organization

Alivernini S, Battistelli CL, Turrio-Baldassarri L. Human milk as a vector and an indicator of exposure to PCBs and PBDEs: temporal trend of samples collected in Rome. Bull Environ Contam Toxicol. 2011; 87(1):21-5.

Altomare M, Vicari LO, Oliveri Conti G, Condorelli RA, La Vignera S, Asero P, Giuffrida MC, Manag A, Arena G, Fallico R, Calogero C, D'Agata R, Calogero AE, Vicari E. PCB contamination in an industrial area with high environmental risk of the south-eastern Sicily, Riproduzione e Sessualità dalla sperimentazione alla clinica, cleup 2012.

Arisawa K, Takeda H, Mikasa H. Background exposure to PCDDs/PCDFs/PCBs and its potential health effects: a review of epidemiologic studies. J Med Invest. 2005;52(1-2):10-21.

ASTDR- U.S. Environmental Protection Agency Research, Locating and Estimating Air Emissions from Sources of Dioxins and Furans, May 1997

ASTDR Public Health Service Agency for Toxic Substances and Disease Registry U.S. Department of Health and Human Services Toxicological Profile for Polychlorinated Biphenyls (PCBs) November 2000).

ASTDR- U.S. Department of Health and Human Services Public Health Service Agency for Toxic Substances and Disease Registry , Addendum to the Toxicological Profile for Chlorinated Dibenzo-p-Dioxins (CDDs), April 2011

ASTDR- U.S. Department of Health and Human Services Public Health Service Agency for Toxic Substances and Disease Registry, Toxicological Profile for Chlorinated Dibenzo-p-Dioxins, December 1998

ASTDR-U.S. Department of Health and Human Services Public Health Service Agency for Toxic Substances and Disease Registry, Toxicological Profile for Sulfur Dioxide, 1998

ASTDR-U.S. Environmental Protection Agency Research. (1997). Locating and Estimating air Emissions from Sources of Dioxins and Furans, May.

Bell M, PhD; Roger D. Peng, PhD; Francesca Dominici, PhD; Jonathan M. Samet, MD Emergency Hospital Admissions for Cardiovascular Diseases and Ambient Levels of Carbon Monoxide Results for 126 United States Urban Counties, 1999–2005 Circulation 2009, 120:949-955.

Bocca B, Petrucci F, Alimonti A, Caroli S. Traffic-related platinum and rhodium concentrations in the atmosphere of Rome. J Environ Monit. 2003 Aug;5(4):563-8

Breivik K, Alcock R, Li YF, Bailey RE, Fiedler H, Pacyna JM. Primary sources of selected POPs: regional and global scale emission inventories. Environ Pollut. 2004;128(1-2):3-16.

Breivik K, Sweetman A, Pacyna JM, Jones KC. Towards a global historical emission inventory for selected PCB congeners--a mass balance approach. 2. Emissions. Sci Total Environ. 2002a May 6;290(1-3):199-224.

Carpenter DO. Polychlorinated biphenyls (PCBs): routes of exposure and effects on human health. Rev Environ Health 2006;21(1):1-23

Consonni D, Sindaco R, Bertazzi PA. Blood levels of dioxins, furans, dioxin-like PCBs, and TEQs in general populations: A review, 1989-2010. Environ Int. 2012 Feb 23.

Corsmeier, U., Kalthoff, N., Kottmeier, Ch., Vogel, B.,Hammer, M., Volz-Thomas, A., Konrad, S., Glaser, K.,Neininger, B., Lehning, M., Jaeschke, W., Memmesheimer, M., Rappenglu¨ ck, B., Jakobi, G. Ozone budget and PAN formation inside and outside of the Berlin plume process analysis and numerical process simulation. Journal of Atmospheric Chemistry 2002; 42: 289–321.

Crinnion WJ. Polychlorinated biphenyls: persistent pollutants with immunological, neurological, and endocrinological consequences. Altern Med Rev. 2011 Mar;16(1):5-13

Department of Health and Human Services Centers for Disease Control and Prevention. (2009), Fourth National Report on Human Exposure to Environmental Chemicals.

EEA, The European environment state and outlook 2010.

EMEP/CORINAIR. (2007). Atmospheric Emissions Inventory Guidebook Group 10: Agriculture (3rd ed.). Copenhagen: European Environment Agency.

Environmental Protection Agency 40 CFR Part 50 [EPA-HQ-OAR-2007-1145] RIN: 2060-AO72 "Secondary National Ambient Air Quality Standards for Oxides of Nitrogen and

Sulfur Final Revisions to Nitrogen and Sulfur Oxides Secondary National Air Quality Standards" 3/20/2012

Environmental Protection Agency . Sources of indoor air pollution— organic gases (Volatile Organic Compounds—VOCs). Available at http://www.epa.gov/iaq/voc.html. Site last updated

Environmental Protection Agency. Sources of indoor air pollution—respirable particles, 2005 a Available at http://www.epa.gov/iaq/rpart.html.

EPA, http://www.epa.gov/pbt/pubs/dioxins.htm

European Environmental Agency (EEA). The European Environment state and outlook, 2010.

Faroon O.M.et. al. (2003), Polychlorinated Biphenyls: Human Health Aspects. WHO.

Ferrante M, Fallico R, Fiore M, Barbagallo M, Castagno R, Salemi M, Caltavituro G, Lombardo C, Manciagli E, Sciacca S. Evaluation of xenobiotics presence in maternal milk. Epidemiology 2006;17:298.

Ferrante M, Fallico R, Fiore M, Brundo MV, Sciacca S. Sustainable development and air quality in the Catania city within. Atti 13th World Clean Air and Environmental Protection. IUAPPA, NSCAEP and ISEEQS. London, 22-27 Agosto 2004.

Ferrante M., Fallico R., Smecca G., Fiore M., Sciacca S. First information about Pt, Pd and Rh in air, foods and biological samples in the district of Catania. 11th World Clean Air e Environment Congress "The interface between developing and developed countries" IUAPPA Durban, South Africa 14 – 18 September 1998.

Ferrante Margherita, Aldo Calogero, Gea Oliveri Conti, Enzo Vicari, Caterina Ledda Paola Asero, Salvatore Sciacca, Rosario D'Agata. Cadmium toxicity: a possible cause of male infertility., ISEE Barcellona 2011?)

Ferrante Margherita, Aldo Calogero, Gea Oliveri Conti, Vincenzo Vicari, Maria Fiore, Giovanni Arena, Salvatore Sciacca, Rosario D'Agata. As, Hg, and Ni levels in serum and seminal plasma and male fertility. 2011. Abstracts of the 23rd Annual Conference of the International Society of Environmental Epidemiology (ISEE). September 13 - 16, 2011, Barcelona, Spain. Environ Health Perspect.

Firket, J. Fog along the Meuse Valley, Trans. Faraday Soc. 1936;32:1192-1197.

Gies A, Neumeier G, Rappolder M, Konietzka R. (2007). Risk assessment of dioxins and dioxin-like PCBs in food--comments by the German Federal Environmental Agency. Chemosphere. Apr;67(9):S344-9

Green RJ. Inflammatory airway disease. Current Allergy and Clinical Immunology, 2003; 16:181.

Guo H., S.C. Lee, L.Y. Chan, andW.M. Li. Risk assessment of exposure to volatile organic compounds in different indoor environments. Environmental Research 2004;94: 57–66.

Henderson J, Sherriff A, Farrow A, Ayres JG. Household chemicals, persistent wheezing and lung function: effect modification by atopy? Eur Respir J 2008; 31: 547–554.

Johnson DE, Braeckman RA, Wolfgang GH. Practical aspects of assessing toxicokinetics and toxicodynamics. Curr Opin Drug Discov Devel. 1999 Jan;2(1):49-57.

Jones, K.C. & De Voogt, P. (1999). Persistent organic pollutants (POPs): state of the science. Environmental Pollution (100): 209±221.

Lundqvist C, Zuurbier M, Leijs M, Johansson C, Ceccatelli S, Saunders M, et al. (2006). The effects of PCBs and dioxins on child health. Acta Paediatr Suppl;95(453):55-64.

Mendell MJ. Indoor residential chemical emissions as risk factors for respiratory and allergic effects in children: a review. Indoor Air 2007; 17: 259–277.

Minnesota Department of Health (MDH). Volatile Organic Compounds (VOCs) in Your Home. 2010.

Narváez, Lori Hoepner, Steven N. Chillrud, Beizhan Yan, Robin , Garfinkel, Robin Whyatt, David Camann, Frederica P. Perera, Patrick L.Kinney, And Rachel L. Miller. Spatial and Temporal Trends of Polycyclic Aromatic Hydrocarbons and Other Traffic-Related Airborne Pollutants in New York City, Environ Sci Technol. 2008; 42(19): 7330–7335.

Oliveri Conti G, Ledda C, Bonanno, Romeo, Fiore M, Ferrante M. (2011). Evalutation of PHAs in seminal plasma of volunteers from an Sicilian industrialized area. Proceedings of Environmental health Conference 6-9 February 2011 Salvador, Brazil.

Oliveri Conti G, Ledda C, Fiore M, Mauceri C, Sciacca S, Ferrante M. Rinite allergica e asma in età pediatrica e inquinamento dell'ambiente Indoor. Ig. Sanità Pubbl. 2011; 67: 467-480.

Park JW et al. Interleukin-1 receptor antagonist attenuates airway hyperresponsiveness following exposure to ozone. American Journal of Respiratory Cell and Molecular Biology 2004; 30:830–836.

Porta M, López T, Gasull M, Rodríguez-Sanz M, Garí M, Pumarega J, Borrell C, Grimalt JO. Distribution of blood concentrations of persistent organic pollutants in a representative sample of the population of Barcelona in 2006, and comparison with levels in 2002. Sci Total Environ. 2012 Mar

Primo rapporto sullo stato dell'ambiente della Città di Catania, 2001. http://www.liceogalileict.it/Helianthus/catania.pdf

Purkey HE, Palaninathan SK, Kent KC, Smith C, Safe SH, Sacchettini JC, et al. Hydroxylated polychlorinated biphenyls selectively bind transthyretin in blood and inhibit amyloidogenesis: rationalizing rodent PCB toxicity. Chem Biol 2004;11(12):1719-1728.

Ravindra K, Bencs L, Van Grieken R. (2004). Platinum group elements in the environment and their health risk. Science of The Total Environment. 318(1-3):1-43.

Ritter et al, 1995. A review of selected persistent organic pollutants, WHO.

Rumchev K, Spickett J, Bulsara M, Phillips M, Stick S. Association of domestic exposure to volatile organic compounds with asthma in young children. Thorax 2004; 59:746–751.

Safe SH. (1998). Development Validation and Problems with theToxic Equivalency Factor Approach for Risk Assessment of Dioxins and Related Compounds. J ANIM SCI, 76:134-141.

Sciacca Salvatore and Gea Oliveri Conti. Mutagens and carcinogens in drinking water. Mediterranean Journal Of Nutrition And Metabolism. 2009; 2(3):157-162.

Sciacca S, Fallico R, Brundo V.M, Fiore M, Oliveri Conti G, Sinatra M. L, Bella F, Galata' R, Castagno R, Caltavituro P, Cirrone Cipolla A, Ferrante M. Characterization of inhalable particulate matter in Siracusa city. 14th International Union of Air Polllution Prevention and Environmental. Protection Associations (IUAPPA) World Congress 2007 incorporating 18th Clean Air Society of Australia and New Zealand (CASANZ) Conference. Brisbane 9-13 Settembre 2007.

Sexton K, Adgate JL, Ramachandran G, Pratt GC, Mongin SJ, Stock TH, Morandi MT. Comparison of personal, indoor, and outdoor exposures to hazardous air pollutants in three urban communities. Environ Sci Technol 2004; 38:423–430.

Sidiropoulos C. & Tsilingiridis G.(2009). Trends of Livestock-related NH_3, CH_4, N_2O and PM Emissions in Greece. Water Air Soil Pollut , 199:277–289.

SOER 2010. The European Environment. State and outlook 2010. Air pollution. European Environment Agency (EEA). EEA, Copenhagen, 2010. Luxembourg: Publications Office of the European Union, 2010. ISBN 978-92-9213-152-4.

Spaziani F, Angelone M, Coletta A, Salluzzo A, Cremisini C. (2008). Determination of Platinum Group Elements and Evaluation of Their Traffic-Related Distribution in Italian Urban Environments. Analytical Letters; 41(14):2658-2683. DOI:10.1080/00032710802363503.

Vallaci H W, D J Bakker, I Brandt, E Broström-Lundén, A Brouwer, K R Bull, C Gough, R Guardans, I Holoubek, B Jansson,R Koch, J Kuylenstierna, A Lecloux, D Mackay, P McCutcheon, P Mocarelli, R D Taalman. Controlling persistent organic pollutants-what next? Environmental Toxicology and Pharmacology (1998) Volume: 6, Issue: 3, Pages: 143-175

Viegi G, M. Simoni, A. Scognamiglio, S. Baldacci, F. Pistelli, L. Carrozzi, I. Annesi-Maesano "Indoor air pollution and airway disease" INT J TUBERC LUNG DIS 8(12):1401–1415 2004 IUATLD

Vivoli G, Fallico R, Gilli G, Bergomo M, Rovesti S, Vivoli R, Ferrante M, Fiore M, Bono R, Amodio Cocchieri R, Cirillo T. Il monitoraggio biologico nella valutazione dell'esposizione ambientale al benzene. Atti 40° Congresso Nazionale S.It.I., Como 8-11 Settembre 2002, Relazioni: 126-130.

Weschler CJ. Changes in indoor pollutants since the 1950s. Atmospheric Environment 2009; 43: 153–169.

WHO, 2007. Health relevance of particulate matter from various sources. Report on a WHO Workshop Bonn, Germany, 26–27 March 2007. World Health Organization Regional Office for Europe

WHO, Air quality guidelines for particulate matter, ozone, nitrogen dioxide and sulfur dioxide, Global Update 2005 , 2006

WHO, Polynuclear aromatic hydrocarbons in Drinking-water, 2003

WHO, Sulfur dioxide Air Quality Guidelines – Second Edition, 2000

WHO, WHO guidelines for indoor air quality: selected pollutants, 2010

Chapter 2

Particulate Matter Exposure in Agriculture

Selçuk Arslan and Ali Aybek

Additional information is available at the end of the chapter

1. Introduction

World Health Organization (WHO) defines agriculture as all kinds of activity concerning growing, harvesting, and primarily processing of all kinds of crops; with breeding, raising and caring for animals; and with tending gardens and nurseries (Jager, 2005). Agriculture is estimated to have the greatest labor force in the world with over one billion people and employs about 450 million waged woman and men workers (FAO-ILO-IUF, 2005). Agriculture requires a wide variety of operations in order to meet the food, feed, and fiber demands of mankind, requiring specific tasks in the fields, orchards, greenhouses, animal production facilities, and in the agriculture based industry. The methods of production, mechanization levels and labor needs differ significantly in each work setting. Post-harvest operations including grain processing, fruit and vegetable sorting, packaging, and meat processing add different types of operations to the conventional agricultural production practices. In agri-industry, feed mills, flour mills, cotton ginners, textile industry, etc have different nature in processes involved.

Particulates are generated during the agricultural operations and processes. The "particulates" (or particles) is a term referring to fine solid matter dispersed and spread by air movement (Förstner, 1998). Particulate matter (PM) may be either primary or secondary in origin and is generated naturally (pollen, spores, salt spray, and soil erosion) and by human activities (soot, fly ash, and cement dust) occurring in a wide range of particle sizes (Krupa, 1997). The human health is affected as the PM penetrates into the respiratory system. Size of particulates may range from less than 0.01 to 1000 microns and are generally smaller than 50 microns. As a principle, PM can be characterized as discrete particles spanning several orders of magnitude in size and the inhalable particles fall into the following general size fractions (EPA, 2012a):

- PM_{10} (generally defined as all particles equal to and less than 10 microns in aerodynamic diameter; particles larger than 10 microns are not generally deposited in the lung);

- $PM_{2.5}$, also known as fine fraction particles (generally defined as those particles with an aerodynamic diameter of 2.5 microns or less)
- $PM_{10-2.5}$, also known as coarse fraction particles (generally defined as those particles with an aerodynamic diameter greater than 2.5 microns, but equal to or less than a nominal 10 microns); and
- Ultrafine particles generally defined as the particles less than 0.1 microns.

The fresh air at sea level is composed of a variety of gases, including nitrogen (70.09%), oxygen (20.94%), argon (0.93%) and more than ten other gases at small proportions (Salvato et al., 2003). But natural events and human activities change the composition slightly across the world. Additionally, industrial production, forest fires, dust storms, acid rains, agricultural operations, etc. pollute the fresh air with gases and solid particulate matter. Pollution may be described as "the undesirable change in the physical, chemical, or biological characteristics of air, land, and water ...." or "the presence of solids, liquids, or gases in the outdoor air in amounts that are injurious or detrimental to humans, animals, plants, or property or that unreasonably interfere with the comfortable enjoyment of life and property" (Salvato et al., 2003). The humans, animals, and plants are exposed to different concentrations of PM (or dust) due to polluted air depending on the environment and PM exposure has health implications of living organisms, including humans (Salvato et al., 2003). Therefore, it is of utmost importance to deal with issues associated with air pollution. Agricultural field operations, animal production, and agri-industry are the sources of indoor or outdoor air pollution, resulting in personal exposure to different concentrations of dusts from different sources at different size fractions described above.

Although agriculture is thought of as a single sector, it is extremely diverse with substantial respiratory hazards resulting from organic and inorganic particulates, chemicals, gases, and infectious agents (Jager, 2005). In some industries there may be one or two predominant respiratory hazards or categories of hazards. The nature of agricultural practice, however, also varies with climate, season, geographic location, moisture content and other properties related to growing practices, and with the degree of industrialisation of the region. The contents of particulates also depend on where, when and how the dust is produced (The Swedish National Board of Occupational Safety and Health, 1994). Consequently, the permutations of potential exposures in agricultural work environments are virtually infinite (Jager, 2005).

The sources of air pollution either as a single source or as combination can be field and orchard operations, unpaved roads, farm equipment exhaust, agricultural burning, processing and handling facilities, pesticides, livestock, and windblown dust (HSE, 2007). In the work environment mineral dusts, such as those containing free crystalline silica (e.g., as quartz); organic and vegetable dusts, such as flour, wood, cotton and tea dusts, and pollens may be found (WHO, 1999). Grain dust is the dust caused by harvesting, drying, handling, storage or processing of barley, wheat, oats, maize and rye. And this definition includes any contaminants or additives within the dust (HSE, 1998). The grain dust includes bacteria, fungi, insects and possibly pesticide residues as well as dry plant particles. Organic dust may contain not only the grain and hay contents but pollen, fungal spores, fungal hyphae, mycotoxins, bacteria and endotoxins and dust from livestock pens may contain skin, hair, feathers and excrement particles (The Swedish National Board of Occupational Safety and Health, 1994).

As cited by Jager (2005), the University of Iowa's Environmental Health Sciences Research Centre exclaims that it is organic dust that accounts for the most widely exposure leading to agricultural respiratory diseases and that virtually everyone working in agriculture is exposed to some level of organic dust. It was also noted that, in general, the studies of respiratory hazards in agriculture lags the investigation of hazards in mining and other heavy industries.

2. PM exposure in agriculture

Agricultural field operations causing dust production in conventional crop production includes soil tillage and seed bed preparation, planting, fertilizer and pesticide application, harvesting and post-harvest processes. In most countries, awareness in sustainability has been increasing so as to accomplish soil and water conservation. Minimum soil tillage reduces tillage operations resulting in less soil manipulation and direct or zero planting methods eliminate tillage operations that are conventionally applied before planting. These differences in soil tillage, seedbed preparation and planting methods create significant variations in the level of soil perturbation. Thus the amount of mineral dust generated as a result of soil tillage and planting are likely to vary significantly not just due to the differences in these field operations but to different soil types and climate variations. Determining personal PM exposure is important during these operations because of the health hazards to be explained in sub-section 2.6. Another important task is to determine the total amount of dust generated during agricultural operations because the impact of agriculture on air quality is not well-known. Some information on the air pollution in agriculture might be useful before introducing the topic of personal exposure.

An eight-year extensive field study conducted at University of California, along with previous research results obtained in the same university, allowed development of PM10 fugitive dust emission factors for discing, ripping, planing, and weeding, and harvesting of cotton, almonds, and wheat (Gaffney and Yu, 2003). As a result of more than ten-year of studies the researchers developed activity specific and crop specific emission estimates for all agricultural land preparation and harvesting activities within California. In the San Joaquin Valley, PM10 emissions estimates for land preparation and all harvest operations were 13 000 tons year^{-1} and 13 300 tons year^{-1}, respectively. The researchers considered this step as a critical one since they can seek cost-effective means to reduce fugitive dust emissions from agricultural field operations, and determine future research needs associated with air quality.

Since air pollution is known to be the result of industrialization and mechanization, agriculture is not considered a major cause of air pollution. The emission trends (Figure 1) estimated from all sources and from agriculture in European Union shows that agriculture is a major source of emission and should be studied further to increase the health and welfare of rural community. PM2.5 and PM10 contribute to emissions by 5% and 25% in

Europe, respectively, however recent studies imply that agricultural PM emissions in intensive emission areas might be more (Erisman et al., 2007).

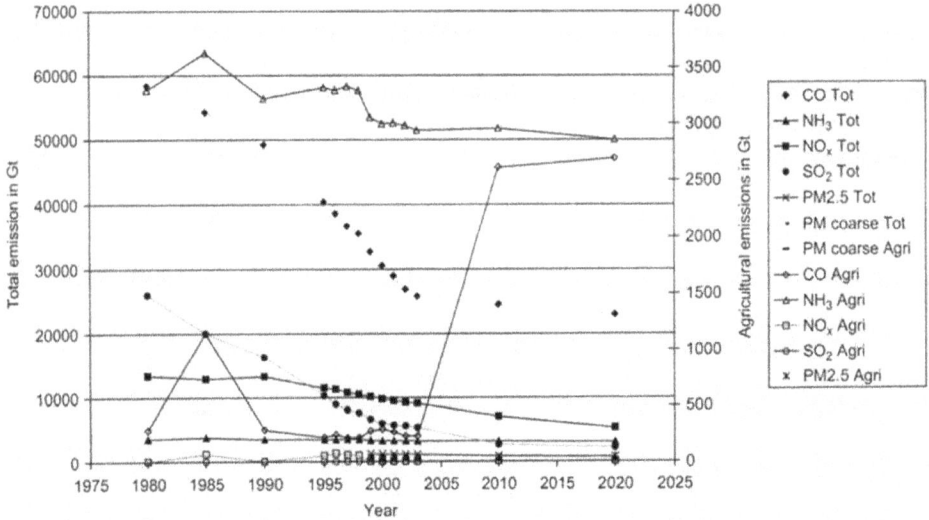

Figure 1. Trend in best estimate emissions from all sources and from agriculture in EU-15 (Erisman et al., 2007)

Bogman et al. (2007) assessed the particle emission from farming operations in Belgium and depending on the conversion factor used, they assessed 10.1 kton or 7.5 kg ha^{-1} total suspended particle (TSP) emission per year, 2.0 to 3.1 kton PM10, or 1.5 to 2.3 kg ha^{-1}, respectively. It was estimated that agriculture generates 35% of total TSP emission and 24% of total PM10 emission. In a study conducted over one year they also found that mineral dust was approximately 8 times higher than organic dust. Particles smaller than 20 μm made up of more than 50% and the particles smaller than 40 μm were more than 80%, suggesting that harmful PM10 is not negligible within the total suspended particles. These findings are informative in that the agriculture is one of the major sectors polluting the air.

Air quality standards were violated at certain times of the year, particularly during row crop agricultural operations, implying that row crop agriculture could be a major contributor to PM10 (Madden et al., 2008).

The stubble burning is a common practice in agriculture in many parts of the world. It is often preferred to remove the harvest leftovers from the field to reduce the draft force needed in soil tillage equipment, to prevent tillage and planting machines from being clogged by the stubble, to achieve a smoother seed bed, etc. In areas where second crop production is a practice, stubble burning saves time before planting because the time available for planting is limited after the first crop is harvested. Smoke from field burning, however, may be distruptive or hazardous to the people in the rural and neighboring urban areas. Rural communities living in eastern Washington and northern Idaho in the United States were worried about health hazards posed by smoke exposure resulting from stubble burning, but research showed that air quality standards were not violated by pollution from field burning (Jimenes, 2006). The contributions of PM2.5 from soil, vegetative burning, and sulfate aerosol, vehicles and cooking were 38%, 35%, 20%, 2%, and 1%, respectively in the Pullman airshed.

2.1. Agricultural field operations

Conservation tillage accomplished approximately 85% and 52% reduction in PM10 emissions on two different farms (Madden et al., 2008). In this study conservation tillage systems required zero or one operation whereas conventional tillage required six operations. Furthermore, conservation tillage could be done at higher soil moisture contents resulting in even less dust compared to dry soil conditions. Other studies also found significantly less amount of dust in conservation tillage compared to standard tillage applications because of decreased number of field operations (Baker et al., 2005). In a two-year study, the cumulative dust production in no-till was one third of traditional tillage. The reduction in dust production was due to the elimination of the two dustiest operations, which were disking and rotary tilling (Baker et al., 2005). However, Schenker (2000) discusses that in conservation tillage, some benefits of reduced dust may be offset to some extent by an increased organic fraction resulting from the cover crop treatment. In the cover crop treatments high organic constituent was found in the respirable dust, suggesting that there may be the potential for increased allergic responses in agricultural workers due to organic dust, but the potential health effect of increased organic matter is not known (Schenker, 2000).

The exposure of tractor operators to dust depends on the availability of a cabin on the tractor and the ventilation system. Early studies found dust exposure levels to be much higher than the ACGIH's threshold limit value (10 mg m^{-3}) for inhalable dust, however the exposure levels were considerably lower in the case of a tractor with an enclosed cabin (Nieuwenhuijsen et al., 1998). The tractor operators were subjected to personal respirable quartz concentrations of 2 mg m^{-3} in an open cabin and 0.05 mg m^{-3} in a closed cabin. Pull type soil tillage equipment may generate great amount of dust clouds but the scientific data is not sufficient to determine to what extent quartz exposure creates a risk in agriculture (Swanopoel et al., 2010). This is probably due to the difficulties in exposure assessment posed by varied and cyclic nature of the farmers' work and the diverse locations of the farms (Nieuwenhuijsen et al., 1998).

The field work may require different durations to be completed, resulting in exposure times more or less than 8 hours. It is important to correct measurements and assess the exposure based on actual working time for a task since the threshold limits for occupational exposure are based on 8 h working duration. For instance, dust levels would have been found well below the ACGIH's TLV of respirable nuisance dust (3 mg m^{-3}), but well above the TLV of inhalable dust (10 mg m^{-3}) if the exposure had lasted for an 8-hour period for many operations (Nieuwenhuijsen et al., 1998).

Mean dust concentrations in field operations, transportation and conveying of materials, and indoor tasks showed that dust concentrations were higher than 10 mg m^{-3} in soil tillage, plant harvesting, and confinements, as shown in Figure 2 (Molocznik and Zagorski, 1998). Annual work cycle of operators and farmers show that the dust exposure vary significantly during the year depending on the tasks in different seasons (Figure 3). The dust levels were more variable for the tractor/harvester operators considering the work cycle throughout the year while both drivers and farmers were subjected to high dust levels from July to October.

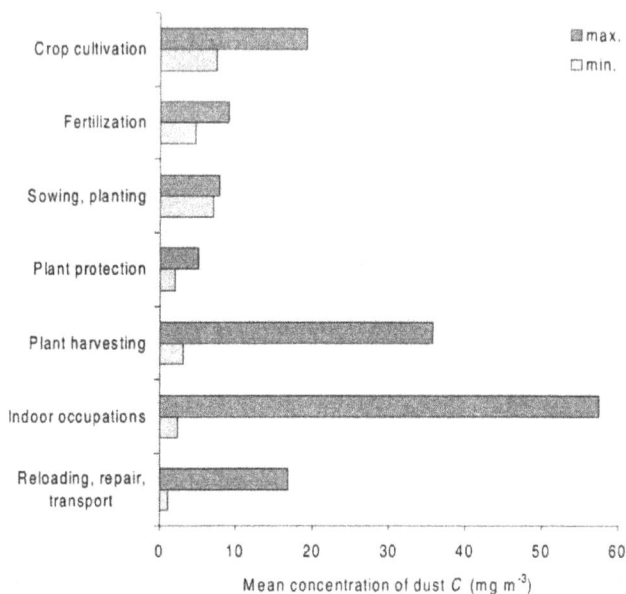

Figure 2. Dust levels in individual groups of farming activities (Molocznik and Zagorski, 1998)

The soil tilling with rotary tiller, disc harrow, soil packer, fertilizer and the planter predominantly generate inorganic dusts (Figure 4) while harvesting wheat and corn, hay making and baling produce predominantly organic dusts (Figure 5). PM10 concentrations measured gravimetrically were greater than the OSHA threshold (15000 µg m^{-3}) in rotary tilling (25770 µg m^{-3}), wheat harvesting (29300 µg m^{-3}), and hay making (24640 µg m^{-3}).

Also, PM2.5 concentration levels were higher than the TLV (5000 µg m^{-3}) in these operations, respectively with 5888, 10560, 8470 µg m^{-3}. PM1.0 concentration was too high particularly during wheat harvest (3130 µg m^{-3}) and hay making (6026 µg m^{-3}). It may be striking that PM1.0 concentrations measured during hay making were higher than the TLV set for PM2.5. PM10 and PM2.5 concentrations were below the threshold limits in all other field applications (Arslan et al., 2010).

Smokers (63% of operators) had complains about coughing with 60% and phlegm with 83% (Arslan et al., 2010). The operators' complaints about chest tightness were 31% and breathlessness about 29%. But, coughing rate decreased to 47% and chest tightness reduced to 13% when smokers and non-smokers were evaluated separately (Figure 6). The operators should use personal preventions to avoid adverse health effects when operating tractors and combine harvesters without cabins.

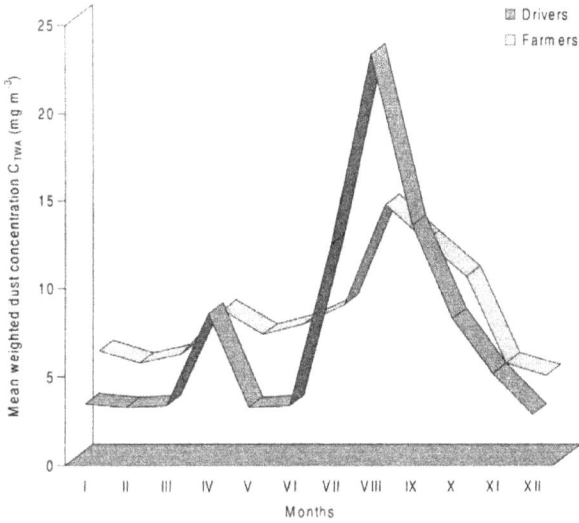

Figure 3. Distribution of exposure to dust in annual work cycle among drivers and farmers (Molocznik and Zagorski, 1998)

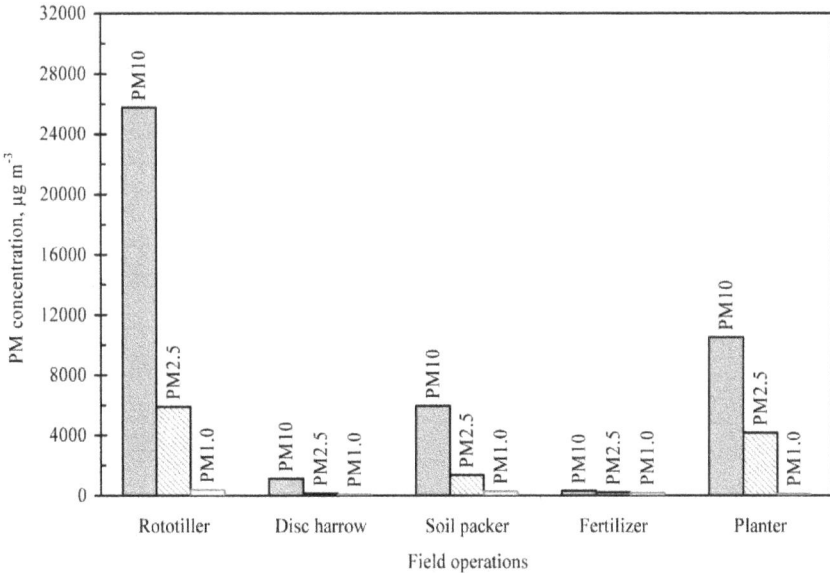

Figure 4. Particulate matter exposure in agricultural operations – predominantly inorganic PM sources (Arslan et al., 2010)

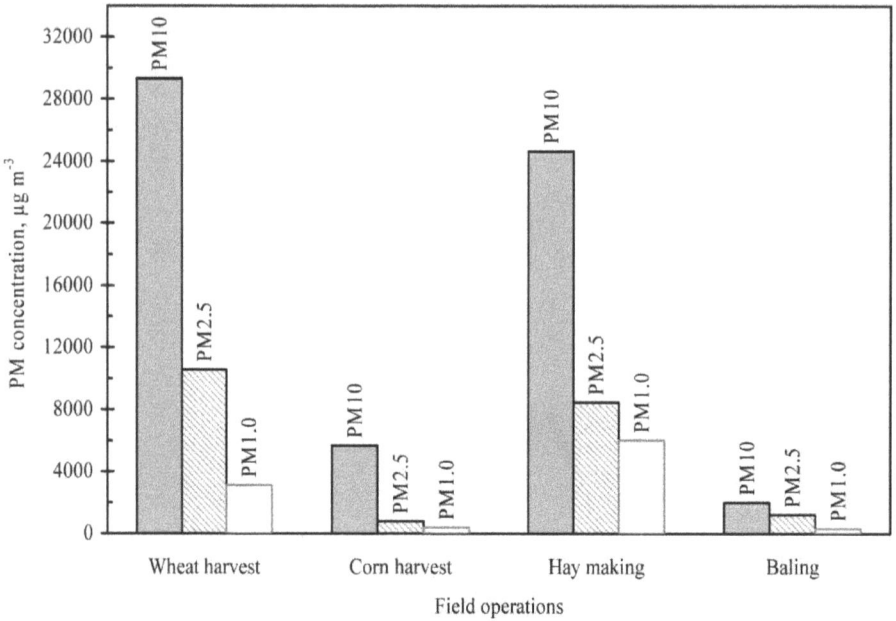

Figure 5. Particulate matter exposure in agricultural operations – predominantly organic PM sources (Arslan et al., 2010)

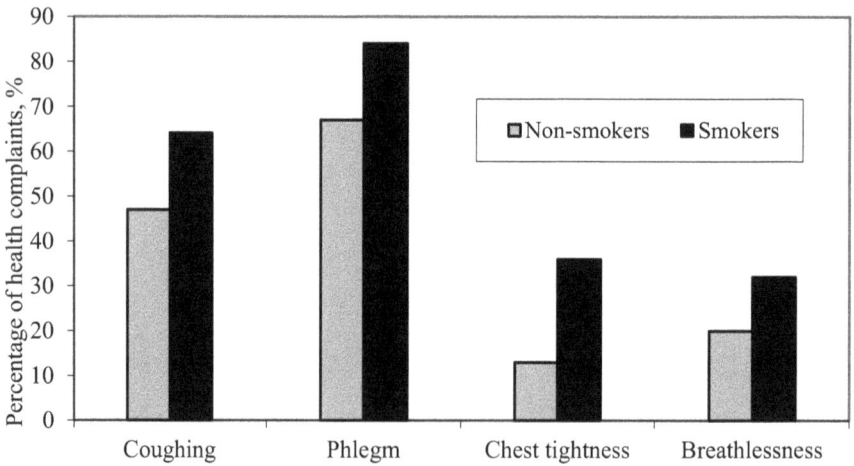

Figure 6. Effect of smoking on health complaints of operators (Arslan et al., 2010)

Literature review and South African Survey was conducted on quartz exposure and quartz related diseases in order to conduct a comprehensive exposure assessment of respirable dust and quartz during farming on a central South African farm (Swanepoel et al., 2010).

Respirable quartz was measured to be not detectable to 626 μg m⁻³. The maximum time weighted average concentration was found during wheat planting. They found that twelve of 138 respirable dust measurements (9%) and 18 of 138 respirable quartz measurements (13%) were greather than the occupational exposure limits of 2 mg m³ and 100 μg m⁻³, respectively.

The ACGIH threshold limit value (25 μg m⁻³) was exceeded in 57% of respirable quartz measurements and quartz percentages of the fine dust were between 0.3 and 94.4% with a median value of 13.4%. Swanepoel et al. (2010) concluded that the published literature regarding quartz exposure in agriculture is not sufficient and that the quartz risk in agriculture should be quantified systematically and further discussed that the public health could be seriously threatened especially in poor and middle-income countries since tuberculosis and HIV rates might be high in these countries employing large numbers of people in agriculture.

The dust concentration at operator's breathing zone ranged in a wide interval during grain harvesting and handling (HSE, 2007). On the other hand, the measured concentration range was very small in combine harvesters equipped with cabins or with filtration (Table 1). The dust concentrations were above the workplace exposure limit (WEL) of 10 mg m⁻³ for grain dust in combining without the cabins, grain carting work, and grain drying. The harvesters without cabins had extremely higher concentrations than the permissible level. Thus HSE (2007) emphasized that dust exposure needs to be reduced below the WEL and should be kept as low as possible in practice.

Process	Dust level measurement averaged over 8 hours	Comments
Combining (no cabin)	18 to 41 mg m⁻³	2.4 times daily legal amount
Combining (with cabin and air filtration)	0.2 to 2.5 mg m⁻³	1/4th of daily legal amount
Grain carting work	1 to 40 mg m⁻³	Up to 4 times daily legal amount
Grain drying	4 to 57 mg m⁻³	Almost 6 times daily legal amount
Milling and mixing	0.1 to 11 mg m⁻³	Can exceed daily legal amount

Table 1. Dust concentration levels in an operator's breathing zone during grain harvesting and handling equipment (HSE, 2007)

Spankie and Cherrie (2012) exclaimed that employers should be aware of both short-term peaks in exposure and longer term time-weighted averages. They explained thay the grain workers were often exposed to high levels of inhalable dust concentration in Britain. The exposed levels in 1990s were usually about 20 mg m⁻³ and were greater than the general guidance level of inhalable dust (10 mg m⁻³) expecially at import and export grain terminals. Therefore the authors conducted a small survey of industry representatives to determine updated exposure levels. According to the newer data, long-term average exposure to inhalable dust was generally estimated to be less than 3 mg m^{-3} (Figure 7). It was also calculated that 15–20% of personal exposures was greater than 10 mg m^{-3}. The British experience clearly shows that improved engineering systems are very effective in reducing dust concentrations and dust exposure. Improved technologies to reduce dust exposure enable the governments and institutions to set newer and stricker limits to protect public and worker health. For instance, The Dutch Expert Committee on Occupational Safety (DECOS) has set a long-term limit of 1.5 mg m^{-3} for inhalable grain dust (DECOS, 2012). This exposure limit, based on Dutch experiences, shows that there is a need to lower the dust concentrations as much as possible in the work environments, also as recommended by HSE (2007).

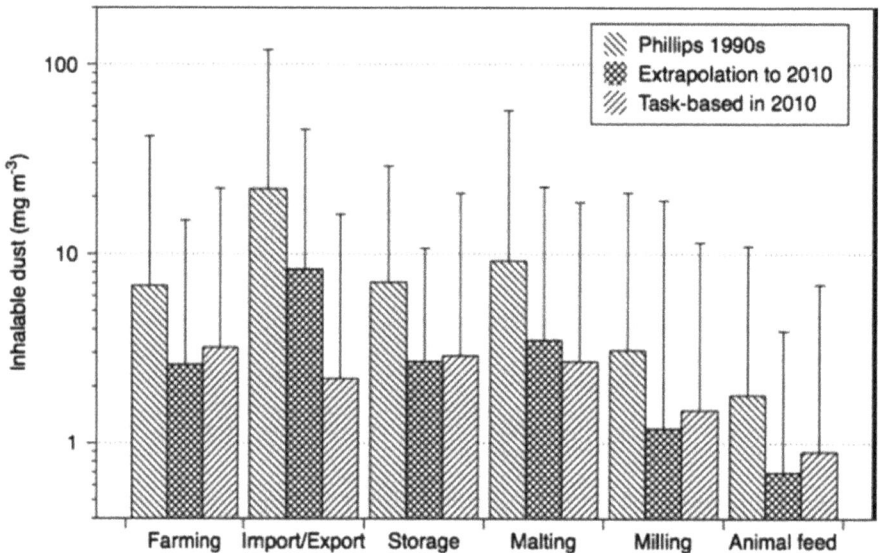

Figure 7. Inhalable dust levels measured in the British grain industry in the early 1990s and estimated levels in 2010—long-term (8 h) average data (Spankie and Cherrie, 2012)

Based on previous research on dust emission and exposure in crop production in agriculture, several conclusions may be drawn. First, the dust generated in agriculture should not be underestimated since the contribution of agricultural activities to air pollution is not negligible. Second, the soil perturbation mostly causes mineral dusts whereas harvesting, hay making, and conveying grain mostly generate organic dusts. Third, the personal exposure is more likely to be a problem for operators when tractor and combine harvesters are used without an enclosed cabin. The operator exposure depends heavily on the presence of an enclosed cabin with a proper filtering system, rather than the PM concentration in the air casued by field applications. Similarly, the farm worker exposure to size fractionated PM particles depends on the personal preventions taken, but farm workers rarely use personal preventions. Forth, the mineral dust generation can be significantly reduced through conservation agriculture, but the potential in organic dust increase due to cover crops during all applications should be monitored through scientific studies.

2.2. Orchards and vegetable growing

In one of the earliest comprehensive studies 103 cascade impactor measurements and 108 cyclone measurements were done to determine personal dust exposure and particle size distribution during field crop, fruit and nut farming, and dairy operations at three farms in California (Nieuwenhuijsen et al., 1998). Personal dust exposure levels were high, especially during land planing with 57.3 mg m⁻³, and discing with 98.6 mg m⁻³. The great majority of the collected dust particles were large and belonged to the extrathoracic fraction. Measured exposure levels were significantly lower when the tractor was equipped with a closed cabin, resulting in sixtyfold reduction in large particles and more than fourfold reduction for the respirable dust fraction. In peach harvesting total and respirable dust levels were 13 mg m⁻³ and 0.50 mg m⁻³, respectively during hand harvesting (Poppendorf et al., 1982).

Among other agricultural machines, tractors are also used to operate sprayers that apply agro-chemicals. When applied in open spaces pesticide droplets smaller than 100 microns (according to CIGR Handbook of Agricultural Engineering (1999), 100-200 micron range is considered fine particles in agricultural spraying) are more susceptible to drift and relatively small droplets are preferred for a better coverage and biological efficiency in spraying. Average droplet diameter, for instance, may be 200-300 microns (medium size droplets in spraying) when herbicides are applied and smaller mean diameter is needed for fungicides and insecticides. However, the ratio of particles smaller than 100 microns could be substantial if the pressure settings are not appropriate resulting in excess drift due to smaller particles. Larger droplets would settle with or without drift if they cannot reach the target. Kline et al. (2003) determined the surface levels of pesticides and herbicides at different interior and exterior locations using five tractors, one of which was without a cabin and four with cabin using carbon–bed air–filtering systems and three commercial spray rigs. The equipment was used during fruit and vegetable growing. They found the greatest chemical concentrations on steering wheels and gauges, and in the dust obtained from the fabric seats. It was suggested that the contamination observed in the steering wheel and seat might cause pesticide exposure if the operator uses the tractor without personal protection in other field applications. It was also noted that the outlet louvers of the air filtering systems on the enclosed cabins frequently had more compounds at higher levels compared to the samples taken from the inlet louvers. Thus the carbon-bed may release the chemical compounds back into the cabin later on. While the efficiency of enclosed cabins in protecting against pollutants is known, the efficiency of carbon-bed systems in removing chemicals may be an important topic to study in the future. And the operators should be careful in thorough and regular cleaning of the tractors (Kline et al., 2003).

The researchers found that the workers were exposed to a complex mixture of organic and inorganic particles during manual harvest of citrus and table grapes (Lee et al., 2004). Geometric means for inhalable dust and respirable dust were 39.7 mg m⁻³ and 1.14 mg m⁻³, respectively during citrus harvest, which were significantly higher compared to the levels that were determined for table grape operations and exceeded the TLVs for inhalable dust and respirable quartz. The exposure levels did not exceed the TLVs in table grape operations with the exception of inhalable dust exposure during leaf pulling. It was determined that the degree of contact with foliage was significant in determining exposure factors. It was concluded that inhalable dust and respirable quartz exposures may be sufficiently high to result in respiratory health effects.

During mechanical harvest of almonds, dust levels in the dust plume were 26513 mg m^{-3} and 154 mg m^{-3}, respectively for inhalable dust and respirable dust (Lee et al., 2004). Mechanical harvesting of tree crops caused average personal exposure of 52.7 mg m^{-3} of inhalable dust and 4.5 mg m^{-3} of respirable dust. During manual harvest the workers are closer to the plant compared to mechanical harvest, but emissions are less and the plume size is smaller during manual harvest. Mean dust exposures was 1.82 mg m^{-3} during manual tree crops harvest and 0.73 mg m^{-3} during vegetable harvest whereas during peach harvest the respirable dust exposure was 0.5 mg m^{-3}.

2.3. Animal production

Class of animal, animal activity levels, type of bedding material, cleanliness of the buildings, temperature, relative humidity, ventilation rate, stocking density, and feeding method are among the factors affecting the dust concentrations in animal production (Jager, 2005). The components of the particulate matter found in concentrated animal production systems may include soil particles, bedding materials, fecal matter, litter, and feed, as well as bacteria, fungi, and viruses (EPA, 2004; Guarino et al., 2007).

Compared to non farmers, pig, poultry or cattle farmers have greater prevalence of work related and chronic respiratory symptoms and these farmers may have non specific respiratory symptoms or specific syndromes including organic dust toxic syndrome (ODTS) (Reed et al., 2006). Early studies showed excess amount of PM concentrations from all sources in animal production buildings (Mitloehner and Calvo, 2008). Although the dust generated from soil or crops may tend to have large particles, the particles found in animal confinement facilities belong to respirable fraction (Lee et al., 2006).

Exposure levels were 0.02 and 81 mg m^{-3} for inhalable dust and between 0.01 and 6.5 mg m^{-3} for respirable dust for poultry houses (Jager, 2005). On the other hand, heavy endotoxin concentrations were observed in poultry houses and swine confinement buildings in Sweden. Relatively low amount of endotoxin (0.1 µg m^{-3} air) can cause acute feverish reaction and airways inflammation while concentrations were as high as 1.5 µg endotoxin m^{-3} air in poultry houses (The Swedish National Board of Occupational Safety and Health, 1994).

Both composition and distribution of feed in a barn are important key factors affecting the release of dust from animal feed (Guarino et al., 2007). Thus researchers have studied the effect of adding different oils to the feed in order to reduce the feed dust. An early study by Xiwei et al. (1993) showed that the fine fraction of the airborne dust could be reduced as much as 85% through pelleting and coating the feed. The effect of adding food grade soybean oil or two commercial feed additives to animal feed at 1% or 3% levels was studied under laboratory conditions (Guarino et al., 2007). Reductions of 80% to 95% was achieved in inhalable fraction using soybean oil and were much higher than either of the two commercial additives whereas commercial additives were better in reducing fine particles (<4 m in diameter), particularly with 1% oil treatment. The respirable dust reductions (70% to 90%) were noticeable but the Guarino et al. (2007) notes that feed additives would not affect pig skin or dander and in actual pig growing conditions the dust reductions could be less. The respirable dust concentrations in poultry barns were found to be much higher during winter than during summer in Canada, making it difficult to keep respirable dust to an acceptable level in the winter months (Jager, 2005). According to the latter study, high animal activity propels dust into the air causing dust concentration fluctuations and peaks and at such rates that the ventilation usually fails to remove the dusts whereas in the summer respirable dusts are generally lower because of high ventilation rates.

No occupational hazard was found in terms of lung function for the workers in poultry farms in the Potchefsroom district, South Africa due to exposure to ammonia, particulate matter and microorganisms in the short term, but the long term effects are not known (Jager, 2005). The measured concentrations did not exceed the limits of OSHA, NIOSH and the Regulations for Hazardous Chemical Substances of 1995. It was concluded that the current legal limits provide sufficient protection in the short term for poultry farm workers.

The European Farmers' Project showed that animal farmers had significantly lower prevalence of allergic diseases while they had higher prevalence of chronic phlegm than the general population (Radon et al., 2003). A major predictor of chronic bronchitis was the ODTS indicating that the allergens could be carried to the living environment of the farmer. Additionally, the ventilation was poor and the temperatures were high inside the animal buildings, causing a negative impact on respiratory symptoms and lung function parameters. Radon et al. (2003) concluded that animal producers were at high risk of chronic bronchitis and ODTS and should be studied.

The accurate determination of personal dust exposure depends on time schedule of a worker. A farmer or a worker may be involved in a variety of tasks daily, seasonally or annually. Based on the duration of each task, the worker may be exposed to different environments consisting of a variety of biological or other pollutants that might affect health. The duration of farmers' exposure to various factors were studied on 30 farms by grouping farmers based on their major interests, which were Group A–plant production, Group B–animal production and Group C-mixed production (Moloznik, 2004). Working time ranged from 106–163% of the legal working time in plant production, 75–147% in animal production, and 136–167% in mixed production. Thus working time on private farms usually exceeded the legal working time in all types of operations. Other conclusions from the work cycle study were as follows (Moloznik, 2004): 1) Agricultural tasks are frequently accompanied by a variety of hazards simultaneously, 2) Among the factors most frequently occurring and creating risks are dusts, thermal elements, and biological agents, 3) Sixty percent of farming operations are accompanied by biological agents, 4) Workers were exposed to biological agents 51% of the total time in plant production, 80% in animal production, and 77% in mixed production systems, 5) Work cycle data constitutes a basis for biasing prophylactic actions.

Animal feeding operations (AFOs) are "agricultural enterprises where animals are kept and raised in confined situations" (EPA, 2012b). Information on PM2.5 concentrations and the spatio-temporal variations of PM2.5 in AFOs is insufficient (Li et al., 2011). In a high-rise layer egg production house the ambient PM2.5 levels were greater than 35 µg m^{-3} (24 h) and 15 µg m^{-3} (annual) PM2.5 National Ambient Air Quality Standards (NAAQS). The ambient and in-house measurements showed the effect of season on PM2.5 concentrations. Highest levels in ambient PM2.5 occurred in the summer whereas the greatest in-house concentration levels were measured in winter (Li et al., 2011). Also, PM2.5 levels were negatively correlated with ambient relative humidity, egg production, and ventilation rate.

Lee et al. (2006) discussed that more information is needed on the combined effect of organic and inorganic dust exposures considering size fractions of sampled particulates on different types of farming operations since the effect of biological dust in the total dust exposure is not well known in agriculture. Therefore they collected data on six farms (three types of animal confinements (swine, poultry, and dairy), and three grain farms) on personal exposure to dust and bioaerosols in size range of 0.7 to 10 µm to cover the range of most bacteria and fungi. The number concentrations of small particles (0.7 µm to 3 µm) were greater than those of large particles (3 µm to 10 µm) in all animal confinements. Particle concentrations were higher on the swine farm in winter. The concentrations at the workers' breathing zone were 1.7×10^6 to 2.9×10^7 particles m^{-3} for total dust in animal confinements and 4.4×10^6 to 5.8×10^7 particles m^{-3} during grain harvesting (Lee et al., 2006). The total particles were composed mainly of large particles (3–10 µm) during grain harvesting whereas in animal confinement facilities the total dust was composed mostly of smaller particles (<3 µm). It was noted that about 37% of the particles were fungal spores in the size of 2–10 µm, implying that predominantly large particles during grain harvesting were partly due to the increased fungal spores. However, the overall combined effect of dust and microorganism exposure was more severe in harvesting compared to confined animal production.

Organic dust and endotoxin exposures are widely described for agricultural industries, however a detailed overview of concentration levels of airborne exposure to endotoxins and a systematic comparison using the same exposure measurement methods to compare different sub sectors of agriculture are needed (Spaan et al., 2006). Therefore the researchers collected 601 personal inhalable dust samples in 46 companies of three agricultural industrial sectors: grains, seeds and legumes sector (GSL), horticulture sector (HC) and animal production sector (AP), with 350 participating employees. Figure 8 shows the means and the variations in measured average concentrations. The greatest dust and endotoxin levels were found in the GSL sector while smallest levels were observed in HC sector. The exposure was higher in the primary production section compared to the parts of all sectors. Occupational exposure limit (50 EU m^{-3}) and the temporary legal limit (200 EU m^{-3}) of the Dutch for exdotoxin were exceeded frequently. Spaan et al. (2006) concluded that a 10–1000 fold reduction is required in endotoxin exposure to accomplish reduction in health related hazards. The authors also noted that the wet processes resulted in reduced exposure to endotoxin and less dusty environment.

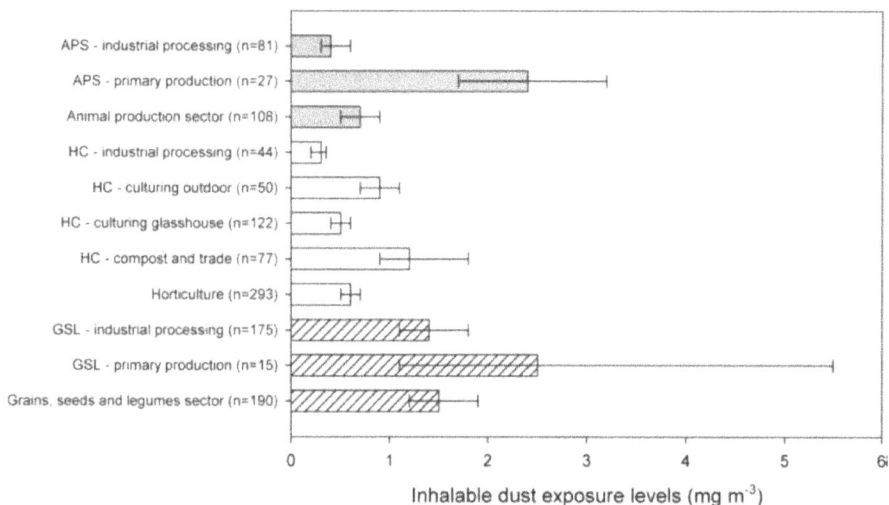

Figure 8. Inhalable dust exposure (Geometrik Mean and 95% CI) levels in three sectors and subsectors of the agricultural industry (Spaan et al., 2006)

2.4. Agriculture based industry

Another important area of interest in occupational exposure to dusts is the agri-industry where agricultural products are processed to be consumed by humans, animals or plants. Agriculture based industry may include a variety of different facilities, including storages, feed mills, flour mills, cotton ginners, hullers and shellers of nuts, etc.

Respirable dust (PM2.5) and very fine particle (PM1.0) concentrations in Turkey were higher than the OSHA TLV (1000 μg m^{-3}) in the ginner, press, and storage areas of two cotton ginners, except for PM1.0 in the storages (Arslan and Aybek, 2011). There is no threshold limit value for very fine particles generated from raw cotton, but the concentrations of PM1.0 were even greater than the TLV set for PM2.5 (Figure 9). The range of coefficient of variation was 0.33-0.54 for PM10 and was the narrowest range among the three fractions, implying large variations in measured quantities for all PM fractions. The technology used is not advanced and engineering controls are very weak in these facilities, resulting in excess amount of dust in all fractions. Therefore, the workers should use personal preventions to minimize the potential adverse health effects of personal PM exposure (Arslan and Aybek, 2011).

Most ginners are in operation for only several months following the cotton harvest in autumn in eastern Mediterranean, Turkey. Thus the workers in cotton ginners are usually exposed to cotton dust seasonally. The workers are employed in other agricultural and non agricultural jobs for the rest of the year or may be unemployed for some time. Thus it may be difficult to assess the long-term health effects of personal exposure with such a work cycle.

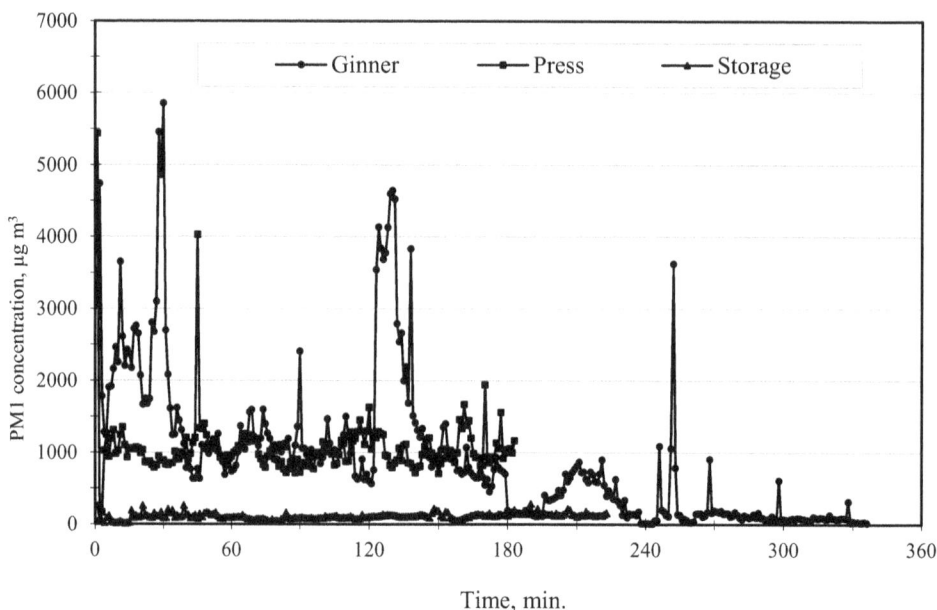

Figure 9. Continuous PM1.0 concentration measurements in three different working areas of ginners (Arslan and Aybek, 2011)

Pure endotoxin causing adverse pulmonary effect can be as low as 9 ng m^{-3} if the subjects are sensitive to cotton dust (Omland, 2002). However, in the cotton industry, healthy subjects may experience a cross shift decline in forced expiratory volume FEV1 when exposed to concentration levels of about 100–200 ng m^{-3}, chest tightness with 300–500 ng m^{-3} and fever with 500–1000 ng m^{-3} (Rylander, 1987). Therefore the latter study concluded that exposure to endotoxin pose a possible risk for workers and also endotoxin content in cotton dust may have a different effect compared to the effect of endotoxin in dust from various farming operations. According to personal PM2.5 exposures measured gravimetrically in four different working areas of three feed mills (bulk storage, dosing, mill, and bagging units) in Turkey, the highest PM2.5 concentration was in the mill section (3033 µg m^{-3}), and the smallest in weighing section (782 µg m^{-3}) (Aybek et al., 2009). The measured concentrations were lower than OSHA TLV for PM2.5. A health survey administered on the workers revealed that the workers did not have serious health complaints, including coughing, phlegm, chest tightness, and breathlessness. However, smokers had more complaints about coughing and phlegm.

Textile industry may employ large numbers of workers depending on the capacity and the level of technology used. The workers in spinning factories were exposed to concentration levels greater than the occupational TLV (200 µg m^{-3}) for respirable dust (PM2.5) in eastern Mediterranean, Turkey (Aybek et al., 2010). In the textile industry, weaving caused higher PM concentrations compared to spinning. TLV for PM2.5 (750 µg m^{-3}) was exceeded in weaving with rapier weaving machines. Air jet weaving machines generated finer dusts due to agitation around the machine caused by the air stream. PM2.5 concentration was higher when air jet weaving machines were used whereas rapier types caused more coarse particles.

Organic dusts are generated in hemp processing plants and the measured dust concentrations were usually ten times more or higher than that of cotton processing (spinning) with 1580, 3730 ve 360 µg m^{-3} in spinning, as cited from other studies (Fishwick et al., 2001).

2.5. Standards and threshold limit values

The regulations and related terminologies to improve air quality may differ in different countries. For instance, national standards in the USA (NAAQS) were set for a variety of air pollutants "to protect public health and welfare", Canada uses air-quality objectives, in Germany air-quality guidelines are effective and World Health Organization recommends desirable air-quality levels (Krupa, 1997). EU has established directives to reduce gas and particle pollution in the air (EU Council Directive, 1999). These regulations are aimed at improving public health and try to limit concentration levels of a specified pollutant in the air. Air Quality Index (EPA, 2009) classifies air quality into several categories within certain limits such as good (0-50 µg m^{-3}), moderate (51-100 µg m^{-3}), unhealthy for sensitive groups (101-150 µg m^{-3}), unhealthy (151-200), very unhealthy (201-300 µg m^{-3}), and hazardous (301-500 µg m^{-3}). Occupational exposures require different regulations to limit PM in specific work related environments. Threshold Limit Value (TLV) is the average 8-hr occupational exposure limit and is calculated to be safe exposure for a working lifetime (Salvato et al., 2003). Occupational Safety and Health Organization (OSHA) in the United States determined the threshold limit values for several agricultural sources (Table 2).

Feature	Limit values (µg m^3)	Particle size
Lower respiratory system nuisance limit	5000	PM2.5
Total nuisance limit	15000	PM10
Grain dust (wheat, barley, rye)	15000	PM10
	5000	PM2.5
Raw cotton	1000	PM10
Spinning	200	PM2.5
Weaving	750	PM2.5

Table 2. TLVs for some agricultural pollutants for PM10 and PM2.5 (OSHA, 2010)

The criteria to evaluate occupational environment have been established by national institutions in different countries. The National Institute for Occupational Safety and Health (NIOSH) in the United States uses recommended exposure limits (RELs); the American Conference of Governmental Industrial Hygienists (ACGIH) uses threshold limit values (TLVs); American Thoracic Society uses permissible exposure limits (PELs); and Occupational Exposure Limits (OELs) are used in some European countries. Table 3 shows the current permissible exposure levels currently set by organizations in the United States.

Hazard	OSHA PEL	NIOSH REL	ACGIH TLV	Animal Confinement Research
Nuisance total dust	15 mg m^{-3}	NE	10 mg m^{-3}	
Nuisance hazard respirable dust	5 mg m^{-3}	NE	3 mg m^{-3}	
Grain dust	10 mg m^{-3}	4 mg m^{-3}	4 mg m^{-3}	
Organic dust	NE	NE	NE	2.4-2.5 mg m^{-3}
Respirable organic	NE	NE	NE	0.16-0.23 mg m^{-3}
Endotoxin	NE	NE	NE	640-1000 ng m^{-3}
Ammnonia	50 ppm	25 ppm	25 ppm	7.5 ppm

Table 3. Recommended maximum exposures in agriculture (Kirkhorn and Garry, 2000)

As far as quartz exposure is considered, reference values of respirable dust and quartz concentrations are 2 mg m^{-3} for respirable dust and 100 μg m^{-3} (South African Occupational Exposure Limit), 50 μg m^{-3} (NIOSH REL), and 25 μg m^{-3} (ACGIH TLV) (Swanepoel et al., 2010).

Grain dust, as defined by HSE (1998), has been assigned a maximum exposure limit (MEL) of 10 mg m^{-3}, 8-hour time-weighted average (TWA). Also the exposure should not exceed 30 mg m^{-3} over any 15-minute period. Even then, exposures should still be kept as low as reasonably practicable. Additionally, grain dust has been given a workplace exposure limit (WEL) by Control of Substances Hazardous to Health Regulations (COSHH, 2002). It was noted that WEL is a maximum concentration value, not a target limit. The exposures should be reduced far below the WEL if possible (HSE, 1998). In order to advise on indicative limits in EU, The Health and Safety Directorate of the Directorate-General of Employment, Industrial Relations, and Social Affairs of the Commission of the European Union formed a Scientific Expert Group. PELs are legal standards, buy they do not apply to farms and most agricultural field operations. But TLVs are consensus exposure guidelines.

Scientific data have not been accumulated enough to set threshold limit values for some specific particulate matters. General nuisance levels are known (Table 2) and may vary depending on the institution that sets the standard. Threshold limits for mineral or organic PM concentrations do not exist due to the difficulties to characterize the PMs found in the soil and in agricultural products since they are made up of different sources. Because of such complexities exposure limits for soil-implement interactions for PM10 and PM2.5 are not known yet. Another important difficulty in determining threshold limists for coarse and fine dust exposure comes from the fact that the combined effects of dusts with toxic gases and other microorganisms are not known.

Literature has dealth more with PM10 while PM2.5 exposure studies are gaining more attraction. On the other hand threshold levels for very fine particles (PM1.0) have rarely been published in agriculture.

2.6. PM health effects in agriculture

Exposure to PM has been related to a series of respiratory and cardiovascular health problems (EPA, 2012a): "The key effects associated with exposure to ambient particulate matter include: premature mortality, aggravation of respiratory and cardiovascular disease, aggravated asthma, acute respiratory symptoms, chronic bronchitis, decreased lung function, and increased risk of myocardial infarction. Recent epidemiologic studies estimate that exposures to PM may result in tens of thousands of excess deaths per year, and many more cases of illness among the US population." The exposure to dust in agriculture is a combination of occupational and environmental exposures with widely varying work practices. The specific respiratory hazards related to different commodities and related work practices are given in Table 4 (Kirkhorn and Gary, 2000) and a list of respiratory hazards by American Thoracic Society is given in Table 5 (Schenker, 1998).

Categories	Sources	Environment	Conditions
Organic dusts	Grain, hay, endotoxin, silage, cotton, animal feed, animal byproducts, microorganisms	Animal confinement operations, barns, silos, harvesting and processing operations	Asthma, asthmalike syndrome, ODTS, chronic bronchitis, hypersensitivity pneumonitis (Farmer's Lung)
Inorganic dusts	Silicates	Harvesting/tilling	Pulmonory fibrosis, chronic bronchitis
Gases	Ammonia, hydrogen sulfide, nitrous oxides, methane, CO	Animal confinement facilities, silos, fertilizers	Asthmalike syndrome, tracheobronchities, silo-filler's disease, pulmonary edema
Chemicals Pesticieds Fertilizers Disinfectants	Paraquat, organophospates, fumigatants Anhydrous ammonia Chlorine, quarternary compounds	Applicators, field workers Application in fields, storage containers Dairy barns, hog confinement	Pulmonary fibrosis, pulmonary edema, bronchospasm Mucous membrane irritation, tracheobronchitis Respiratory irritant, bronchospasm
Others Solvents Welding fumes Zoonotic infections	Diesel fuel, pesticed solutions Nitrous oxides, ozone, metals Microorganisms	Storage containers Welding operations Animal husbandry, veterinary services	Mucous membrane irritation Bronchitis, metal-fume fever, emphysema Anthrax, Q fever, psittacosis

Table 4. Agricultural respiratory hazards and diseases (Kirkhorn and Garry, 2000)

Maximum levels of indoor air contaminant levels, including total dust, ammonia, respirable dust, and total microbes were determined for workers in swine buildings (Donham, 1995). The levels given in the last column in Table 3 are for animal confinement research while Table 6 relates maximum indoor levels for swine building, explaining the slight differences in the two tables.

Maynard and Howard (1999) cited several literatures regarding PM effect on human health: "PM10 is currently regarded as the size fraction best representing those particles most likely to cause ill health (DoE, 1995). PM10 is not as long-lived as PM2.5, with a life-time of some 7±3 days, as the latter is less subject to efficient removal by gravitational settling or scavenging by rain (DoE, 1993). However, particles have to be < 2.5 μm in order to penetrate into the gas exchange regions of the lungs. Numerous epidemiological studies have found a relationship between particulate air pollution and increased cardiorespiratory morbidity and mortality (Pope et al., 1995), and hospital admissions for asthma and chronic obstructive pulmonary disease (Schwartz, 1994; Schwartz et al., 1993)".

Respiratory Reqion	Principal Exposures	Diseases/Syndromes
Nose and nasopharynx	Vegetable dusts Aeroallergens Mites Endotoxins Ammonia	Allergic and nonallergic rhinitis Organic dust toxid syndrome (ODTS)
Conducting airways	Vegetable dusts Endotoxins Mites Insect antigens Aeroallergens Ammonia Oxides of nitrogen Hydrogen sulfides	Bronchitis Asthma Asthma-like syndrome ODTS
Terminal bronchioles and alveoli	Vegetable dusts Endotoxins Mycotoxins Bacteria and fungi Hydrogen sulfide Oxides of nitrogen Paraquat Inorganic dusts (silica, silicates)	ODTS Pulmonary edema/adult respiratory distress syndrome Bronchiolitis obliterans Hypersensitivity pneumonitis Interstitial fibrosis

Table 5. Agricultural respiratory disease common exposures and effects (Schenker, 1998)

Air contaminant	Recommended 8-hour TWA for human health
Total dust (mg m^{-3})	2.40
Respirable dust (mg m^{-3})	0.23
Endotoxin (g m^{-3})	0.08
Ammonia (ppm)	7.00
Total microbes (cfu m^{-3})	4.3 x 10^5

Table 6. Maximum air contaminant levels for humans in swine buildings (Donham, 1995)

Air pollution has been thought of as an urban phenomenon, but it has been understood better that urban–rural differences in PM10 are small or even absent in many regions of Europe, implying that PM exposure is widespread (WHO, 2000). Although most studies provided data on PM10 exposure, the data on fine particulate matter (PM2.5) has been increasing and the recent studies show that PM2.5 is a better predictor of health effects compared to PM10 (WHO, 2000).

Agricultural and food industries introduce various dangers because organic dust in these sectors is frequently associated with endotoxins, mycotoxins and microorganisms (Zock et al., 1995). Granular products generate high quantities of particles during conveying, loading and unloading from hoppers, trailers, and grain silos. Grain dust may also contain mould spores and if inhaled they can cause a fatal disease called farmer's lung (HSE, 2006). Farmer's lung caused by intoxicant dusts in the lower respiratory tract may result in labor loss, increased health costs, and even death in severe cases (Sabancı, 1999). The wide variety of different exposures may result in numerous respiratory diseases including bronchitis or asthma that may result from or exacerbated by organic dust (grain dusts, animal dander, plant dusts), toxic gases, and infectious agents whereas inorganic dusts and other irritants may exacerbate, if not cause, asthma (Schenker, 2000). Early studies explained that dust exposure might cause inflammation of the eyes, lungs, and the skin (Matthews and Knight, 1971), poisoning and allergy in the respiratory system (Witney, 1988), and grain fever (HSE, 2007). While allergic responses, such as asthma, are generally linked with exposure to organic dust, nonallergic responses, such as bronchitis and chronic obstructive airways disease, are usually associated with exposure to inorganic dust from agricultural origins (Baker et al., 2005). Fine particles can pollute the blood by reaching air packets in the lung or might start various disturbances and diseases. The disease called byssinosis is due to chronic PM inhalation of high levels of microbial products such as endotoxin from cotton processing (Lane et al., 2004).

Jimenes (2006) reports detrimental health effects due to both chronic and acute exposures to biomass smoke (EPA, 2004), which includes both vapor and mostly particulate phase material with PM2.5 fraction. Reduced lung function, depressed immune system and increased risk of respiratory diseases were listed as the effects of chronic exposure to biomass smoke, as cited from Sutherland (2004), Sutherland and Martin (2003), and acute health effects in susceptible people, including chronic obstructive pulmonary disease (COPD) patients, and asthmatic children, as cited from Romieu et al. (1996); Pekkanen et al. (1997); Peters et al. (1997). It was stated that coughing, wheezing, chest tightness and shortness of breath were among the health effects. Long et al. (1998) conducted a survey in Winnipeg, Canada and reported that straw burning had more effect on individuals with asthma or chronic bronchitis. In another study on children, a relationship was found between PM10 from rice straw burning and increased asthma attacs in Niigata, Japan (Torigoe et al., 2000). Another effect of PM10 emissions is the visibility impairment (e.g., Brown Cloud) (Arizona Air Quality Division, 2008).

According to numerous researchers, as cited by Baker et al. (2005), diseases such as asthma, pulmonary fibrosis, and lung cancer are associated with dust inhalation. Additionally, organic PM exposure is related with allergic responses, including asthma. Inorganic PM is generally generated during agricultural field applications and causes non-allergic diseases such as bronchitis and chronic obstructive airways disease. Fume and dust exposures are predominant in animal buildings and barns and the main elements related to respiratory health are dust, bacteria, moulds, endotoxin and ammonia (Omland, 2002).

Farming characteristics were determined through a questionnaire study in 1468 cattle farmers in Schleswig-Holstein, showing a high correlation between the ventilation systems in the cattle house and respiratory symptoms (Radon et al., 2002). Other important factors affecting the symptoms were climatic factors and the size of the animal house. In their study, the pig farmers were found to be at the highest risk for developing respiratory symptoms associated with asthma-like syndrome whereas an increased risk of wheezing was found in poultry farmers. The dose-response relationship was significant between daily hours inside the animal buildings and symptoms. The study also found that the sheep producers had excess cough with phlegm.

Schenker (2000), by citing numerous research papers, summarized the types of hazards arising from agricultural dust as follows: Scientific work generally dealt with dust exposure as it relates to respiratory diseases in terms of allergic diseases resulting from inorganic dusts, namely occupational asthma and hypersensitivity pneumonitis. However, inorganic (mineral) dust exposure may be substantial in the agricultural work force. The frequency of exposure to mineral dust may be more in dry-climate regions. Soil tillage operations such as plowing, chiseling, and harrowing disturb the soil, causing operators to be exposed to 1-5 mg m^{-3} respirable dust and >20 mg m^{-3} total dust in dry regions. The soil composition is usually reflected by the inorganic dust in the soil. For instance, 20% of particles in the soil are made up of crystalline silica and 80% is of silicates. Such high levels of inorganic concentrations possibly explain some of the increase in chronic bronchitis reported in farmers' studies. A disease called pulmonary fibrosis (mixed dust pneumoconiosis) was found in agricultural workers. Chronic obstructive pulmonary disease morbidity and mortality have also been observed in farmers living in different geographies. It is likely that inorganic dust exposure is to some extent related with chronic bronchitis, interstitial fibrosis, and chronic obstructive pulmonary disease however the individual effect of mineral dusts beyond the effects of organic dusts is unknown. Some cross-sectional surveys showed increased prevalence rates of chronic bronchitis among swine confinement farmers and poultry workers independent of the effects of cigarette smoking. Workers in animal production are predominantly exposed to organic dusts compared to field workers. In addition to the the physical, chemical and biological properties of particles, the factors affecting worker's health as a result of PM exposure include concentration level, duration of exposure, gender, age, weight, smoking habits, etc.

A cross sectional study called the European Farmers' Project was conducted in two stages to determine the prevalence and risk factors of respiratory diseases in farmers in seven centers across the Europe (Radon et al., 2003). About 8000 farmers in Denmark, Germany, Switzerland, the UK, and Spain were administered a standardized questionnaire to determine the characteristics and respiratory symptoms of the farmers in the first stage while the second stage studied the exposure assessment and lung function determination in four of the seven centers. Among animal production farmers, pig farmers were found to be at high risk of asthma-like syndrome compared to others. When plant production was considered, greenhouse workers were found to be at higher risk of asthma-like symptoms. According to the European Community Respiratory Health Survey, animal farmers had lower prevalence of allergy symptoms while the prevalence of chronic bronchitis symptoms was significantly higher in animal farmers. It was found that the ventilation in the animal production facilities and greenhouses was the major risk factor for respiratory symptoms. It was shown that the highest median total dust (7.01 mg m^{-3}) and endotoxin (257.58 ng m^{-3}) concentrations occurred in poultry houses in Switzerland. Also, growing vegetables, tomatoes, fruits or flowers in greenhouses was a secondary risk factor for asthma.

Elci et al. (2002) did not observe any dose-response relationship with asbestos, grain, or wood dust exposure. They found 958 larnyx cancer stories in 6731 patients diagnosed with cancer at Okmeydani Hospital, Istanbul. Patients exposed to silica and cotton dusts had more cancer rates but there was no correlation between larnyx cancer and asbestos, wood or grain dusts. They found "an excess risk of laryngeal cancer among workers exposed to silica and cotton dust in a large study in Turkey".

Müller et al. (2006) determined sociodemographic and farming-production parameters of 1379 poultry farmers from Southern Brazil, and determined that workers were exposed to high levels of organic and mineral dusts. Their findings present the significance of income level, gender, age, and smoking habits. The low income farmers had higher prevalence of respiratory symptoms and chronic respiratory disease symptoms were more in poultry workers. Additionally, woman had significantly ($p<0.01$) higher (15%) asthma symptoms compared to men (10%), but the difference was not statistically significant in chronic respiratory disease symptoms with 24% in women and 20% in men ($p=0.09$). Mülller et al. (2006) also found more prevalence among persons over 40 years of age and children under four years of age. Furthermore, smokers, particularly ex-smokers showed more chronic respiratory disease. Also, smokers consuming more than 182 packs per year had more respiratory symptoms.

3. Future research and perspectives

Dust emissions from agriculture and the personal exposure to generated particulates are two major issues to be addressed for both policy makers and researchers. Dust emission results in outdoor and indoor air pollution threatening public health. Occupational exposure to dust, on the other hand, is associated with respiratory health in the work environments. The researchers and health organizations have focused on both aspects in order to reduce air pollution and occupational health hazards.

One of the most important and challenging issues is differentiating between the effect of dust components. The management choices in a feed operation, for instance, affect the compounds in the mixture of emissions and complex mixes of various particles create difficulties to regulation, given the lack of information on the effect of individual components (Mitloehner and Calvo, 2008). However, it would be difficult to separate the individual health effect of a component; also it may be somewhat artificial to separate inorganic mineral dusts from other respiratory toxins (Schenker, 2000). The combined health effects of dust components (both mineral and organic) and microorganisms such as exdotoxins deserve to be further studied thoroughly in different sectors and subsectors of agriculture. The scientific studies regarding the relationships of gaseous and particulate mixtures in biosystems is still in its infancy (Mitloehner and Calvo, 2008), requiring more exploration in agricultural operations.

The emissions from agriculture may create local and regional problems in terms of air quality in Europe and such problems may include PM exposure, eutrophication and acidification, toxics and contribution to greenhouse gas emissions, causing numerous environmental impacts and hence PM emissions should be investigated not only for PM10 but PM2.5 with NH3 as precursor (Erisman et al., 2007).

Spatial distribution of contaminants is of importance when sampling locations are to be determined in confined buildings in agriculture and the same should apply for sampling to assess the emissions during field operations. The researchers tend to make the sampling near the center of buildings or at the breathing zone of workers or animals however the sampling location might be random only if the distribution of the pollutants is homogenous throughout the sampling volume; otherwise the spatial distribution across the building needs to be measured first to accurately determine the best sampling locations (Jerez et al., 2011).

It was suggested that, in Australia, each major animal production industry (pigs, poultry, dairy, horses and sheep) should be investigated in a range of climate and seasonal conditions to determine worker exposure to a range of contaminants. Other issues to deal with are changes in respiratory function before and after exposure, respiratory symptoms for at least a week after exposure, both exposure and respiratory function and symptoms over time, long term changes in respiratory function and symptoms, species of bacteria and fungi to which workers are exposed in the different animal industries, the toxicity, and if needed developing appropriate approaches to occupational hygiene (Reed et al., 2006). Further studies were recommended to explore the independent effects on symptoms of smoking, gender and farm characteristics in Australia and was concluded that the relatively high prevalence of asthma in Australian pig and poultry farmers compared with overseas farmers also requires further investigation.

Dust exposure in and near farm fields is of increasing concern for human health and may soon be facing new emission regulations. Dust plumes have rarely been documented due to the unpredictable nature of the dust plumes and the difficulties of accurate sampling of the plumes, requiring further research on dust dispersion measurements and simulations to better assess the dust emissions (Wang et al., 2008). Also more focus should be put on the air quality during agricultural burning and related health effects because few studies have been conducted on air quality during stubble burning and even fewer studies on characterization of exposure (Jimenes, 2006).

Kline et al. (2003) discussed that thoroughness and frequency of cleaning enclosed cabins, the work practices, training, and behavior of operators are important variables in future studies because these factors have direct effects on the sources of chemical contamination. The authors emphasized that further studies were needed in carbon bed air filtration systems in cabin to assess efficiency and specificity of chemical removal, particularly in relation to bed size and chemical breakthrough trends.

Crystalline silica may make up to 20% of the soil composition, representing a risk for interstitial fibrosis and other silicates up to 80% may also result in or contribute to mixed-dust pneumoconiosis (pulmonary fibrosis) (Schenker, 2000). Since the prevalence and clinical severity of pulmonary fibrosis is unknown, more research is needed in this area (Schenker, 2000).

It is emphasized that worker exposure studies should be conducted with a link to health outcomes and similarly health studies should be associated with personal exposure measurements (Reed et al., 2006). Additionally, when occupational exposure studies are conducted, personal sampling should be preferred because stationary sampling estimates of airborne allergens are lower (Lee et al., 2006).

Since poultry workers are exposed to high dust concentrations resulting in increased risk of occupational respiratory symptoms, respiratory protection programs should be implemented and should include poulty production workers (Müller et al., 2006).

As previously discussed, different institutions have set different threshold or permissible exposure levels for the same particulates. For instance, ACGIH limits the personal exposure concentration for PM10 and PM2.5 to be 10000 $\mu m \ m^{-3}$ and 3000 $\mu m \ m^{-3}$, respectively whereas NIOSH recommends 4000 $\mu m \ m^{-3}$ concentration for granular dust. However, health effects might be seen at concentrations just above 2400-2500 $\mu m \ m^{-3}$ in pig production whereas health effects may be observed in poultry at concentrations as low as 1600 $\mu m \ m^{-3}$ (Kirkhorn and Garry, 2000). The composition of dust may vary substantially along with accompanying microorganisms in animal production compared to crop production or agro-industry since more toxic gasesous particles might be generated in animal confinements. This could result in adverse health effects in animal production with less concentration compared to other subsectors in agriculture. Therefore research should continue until health organizations set more practicable limits for a wide variety of compounds in order to protect workers from both short term and long term health hazards of occupational dust exposure. The differences in threshold levels determined by different organizations also imply the need for further research in organic and mineral dust exposure.

Much research is required to characterize the nature and pathogenicity of exposure to agricultural dust, particularly on inorganic dusts and actual dose (Schenker, 2000). Numerous questions raised by Schenker (2000) still seeking answers, to varying degrees in different countries/geographies/climates/subsectors of agriculture, are as follows: "What is the composition of mineral dusts to which agricultural workers are exposed? How do climatic conditions, agricultural operations, soil conditions, and personal characteristics affect dust exposure? What are average and extreme cumulative exposures to inorganic dust among agricultural workers, and where do they occur? What cumulative dust exposure occurs with agricultural work? What is the pulmonary response to inorganic dusts, and what is the mechanism of that response? What are the critical components and relative potency of different inorganic dusts? Does chronic exposure cause pulmonary fibrosis? Is airway inflammation a critical component of the response? If so, by what mediators? Is the response similar to that seen for organic dusts? How do personal characteristics such as age, gender, smoking atopy, and genetic factors affect the response to inorganic dusts? Is there a similar pattern of acute cross-shift change in pulmonary function, and is it predictive of long-term pulmonary function decline? What are effective measures to reduce exposure and other control methods suitable for the agricultural setting? This should include educational strategies, engineering controls, and regulatory interventions. Should the occupational silica standard be applied to the agricultural workplace? What is the role of the practicing physician in recognizing, treating, and preventing respiratory disease among agricultural workers?" The research in air quality and personal health seem to be advanced in industries such as mining, but more research in agriculture and related industries is needed to address all the questions above.

Previous research clearly shows that more scientific studies are needed to accumulate sufficient data to determine dose-response relationships so as to improve engineering control systems and personal protective devices or to increase awareness and implementation of prevention techniques in agriculture.

4. Policies and prevention

The implementation of air quality policies in rural areas usually lags urban settings, resulting in poor monitoring and weak inspection in agricultural work environments. The awareness in air quality issues, the effects of personal exposure to dusts, and personal protective measures hence is generally not strong among farmers and farm workers. An important cause should be related to the fact that the jobs requiring less than 11 people are not subjected to routine inspections by OSHA, resulting in poor awareness, lack of engineering controls and irregular personal protection among agricultural workers in the US (Kirkhorn and Garry, 2000) and the situation is probably similar in other countries.

Policies to reduce air pollution or work-related exposure to dust and microorganisms are of utmost importance. Different countries may impose different measures to improve air quality or may recommend methods to reduce exposure levels. For improved public health the emissions from all industries should be kept as low as possible but may not be economical or practicable under given conditions. For instance, the contribution of particulate matter from agriculture was considered a pressing issue in emissions and to accomplish the objectives on acidification, eutrophication and PM concentrations, much greater reductions should be targeted in EU (Erisman et al., 2007).

Environmental Protection Agency (EPA) in the US redesignated the Moderate PM10 Nonattainment Area to Serious in 1996 to include unregulated sources including unpaved roads, unpaved parking lots, vacant lots and agriculture, requiring emission reduction programs for these areas (Arizona Air Quality Division, 2008). Arizona Legislation required all farmers to comply with PM10 program by the end of 2007, imposing that farmers with 4 ha of contiguous land located within the Maricopa PM10 Nonattainment Area and Maricopa County Portion of Area A must comply with the agricultural PM10 general permit. To aid farmers the Arizona Legislature defined Best Management Practices (BMPs) and farmers are required to implement at least one of BMPs for each of the three categories defined, including tillage and harvest, non-cropland and cropland (Arizona Air Quality Division, 2008).

The personal protection is not common in agriculture mainly due to the fact that they are hot and uncomfortable, but in dusty and mouldy conditions two-strap dust and mist respirators approved by NIOSH may be used (Arizona Air Quality Division, 2008).

The use of sixteen types of engineering controls and thirteen types of personal protective equipment (PPE) was studied using the information obtained from 702 certified pesticide applicators (Coffman et al., 2009). The results of this study showed that 8 engineering control devices were used out of 16 by more than 50% of the applicators. The adoption of engineering control was affected by the crop produced, field size, and type of pesticide application equipment. Engineering devices were usually adopted on large farms, when hydraulic sprayers are used. Most respondents used PPE with chemical-resistant gloves resulting in the highest level of compliance. Appropriate headgear use also increased in pesticide applicators.

Mitchell and Schenker (2008) surveyed 588 farmers longitudinally from 1993 to 2004 to determine respiratory protective behaviors and the personal characteristics of farmers. They identified some characteristics related to smoking and farm size, and found that about 75% of the farmers were not "very" concerned about respiratory health risks. Interestingly, the use of a dust mask or respirator decreased significantly from 54% in 1993 to 37% in 2004, whereas 20% was consistent in use of respiratory protection. From 1993 to 2004 closed-cabin useage increased slightly from 14% to 17%. Those who regularly used a dust mask or respirator were ex-smokers or the ones concerned about the health risks. Also, closed-cabin tractors were used in larger areas and were related to higher salary and the farmers preferred using personal protection in small areas. The researchers recommended that farmers be educated about the long term respiratory health risks.

Some practices in the field can be helpful in reducing dust emissions and hence personal exposure levels. Some of these practices include no tillage or soil preparation if wind speed exceeds 40 km h^{-1} at 2 m height, adopting reduced tillage including minimum tillage system, mulch tillage system, and reduced tillage system (Arizona Air Quality Division, 2008). These practices should also incorporate other preventions such as avoiding conditions in which soil is susceptible to produce PM10, reducing vehicle speed, planting vegetative barriers to the wind (tree, shrub, or windbreak planting), managing residues to reduce soil erosion and maintain soil moisture (Arizona Air Quality Division, 2008).

Some of the technical preventive measures may include appropriate technology for drying, storage, conservation and handling of granular materials and hay; vacuum cleaning; water sprinkling; mist sprinkling; addition of vegetable oil to flour feed; sprinkling of fine oil aerosol over the animals; and design and ventilation of the premises however the materials should be carefully selected and the design of the equipment, process, and the work should be done properly to reduce dust generation (The Swedish National Board of Occupational Safety and Health, 1994).

When the dust concentrations are high enough respiratory protective equipment should be used to avoid upper and lower respiratory disturbances and diseases. HSE (2005) warns that the use of nuisance dust masks (NDMs) may prevent only from large particles during handling grain, and are inappropriate since they cannot prevent fine particles from reaching the lungs. Recommended RPSs are disposable filtering face piece respirator to BS EN 149 or a half mask respirator to BS EN 140 with particle filters to BS EN 143 (HSE, 2006). Nevertheless, oxygen sufficiency in the work environment may be an important factor to choose the best type of mask since the use of a mask could be hazardous when oxygen level is less than 17% (Hetzel, 2010). Powered filtering equipment (helmets or hoods) should be used when large quantities of dust are released during agricultural operations. Some examples of such operations include manual weighing of animals, the transfer of poultry to and from battery cages, the cleansing of grain hoppers from mouldy grain, and threshing with a cabinless threshing machine (The Swedish National Board of Occupational Safety and Health, 1994).

The cost effectiveness of engineering solutions might be important in implementing various methods to reduce emissions and work-related exposures. Technical solutions are usually poor in developing countries due to low level of technologies used such as cotton sawgins in Turkey (Aybek et al., 2010). Lahiri et al. (2005) estimated a cost of 5000-10000 $ to eliminate the risks for silicosis in factories working in three shifts with 5 workers in each shift for stone grinders in construction sector. In other areas ventilation cost was only 650 $. Lahiri et al. (2005) suggested that an annual cost of 106 $ would be required to improve ventilation in these sectors both in developing and developed countries. It was noted that these estimates were based on reducing silicosis and did not account for the effect of smoking.

In crop production the most advanced technical solution is the use of enclosed cabin with appropriate filtering sytems. An original cabin might sufficiently filter the air and reduce PM concentrations from 2000-20000 $\mu g\ m^{-3}$ to 100-1100 $\mu g\ m^{-3}$ (Kirkhorn and Garry, 2000). Thus an enclosed cabin is important especially in soil tillage operations generating silica. Concentration of grain dust may widely vary and can be as high as 72500 $\mu m\ m^{-3}$ in threshing and cleaning, implying the need for both technical and personal preventions. However, the masks could feel hot and uncomfortable and are not routinly used in agriculture (Kirkhorn and Garry, 2000).

Author details

Selçuk Arslan and Ali Aybek

Department of Biosystems Engineering, Faculty of Agriculture, Kahramanmaraş Sütçü İmam University, Turkey

5. References

Arizona Air Quality Division, 2008. Guide to Agricultural PM10 Best Management Practices, Governor's Agricultural Best Management Practices Committee, Second Edition.

Arslan, S., Aybek, A., Ekerbiçer, H. 2010. Measurement of personal PM10, PM2.5 and PM1.0 exposures in tractor and combine operations and evaulation of health disturbances of operators. Journal Agricultural Sciences, 16: 104-115.

Arslan, S., Aybek, A. 2011. PM10, PM2.5 and PM1 concentrations in cotton ginners. 13th International Congress and Agricultural Mechanization and Energy, Istanbul.

Aybek, A., Arslan, S., Genç, Ş. 2009. The effect of PM10 and PM2.5 pollution on feed mill workers. (IAEC Ref. 250, available on CD), International Agricultural Engineering Conference, Bangkok, Thailand.

Aybek, A., Arslan, S., Ekerbiçer, H. 2010. Detemination of particulate matter concentrations and health risks in agriculture and agri-industry. Final Report (In Turkish), Turkish Scientific and Technological Council Project No: 107 O 513.

Baker, J.B., Southard, R.J., Mitchell, J.P. 2005. Agricultural dust production in standard and conservation tillage systems in the San Joaquin Valley. Journal of Environmental Quality, 34: 1260–1269.

Bogman, P., Cornelis, W., Rollé, H., Gabriels, D. 2007. Prediction of TSP and PM10 emissions from agricultural operations in Flanders, Belgium. DustConf 2007, Maastricht, Netherlands. Available at
http://www.dustconf.org/CLIENT/DUSTCONF /UPLOAD/S9/BOGMAN_B.PDF

CIGR Handbook of Agricultural Engineering, 1999. Plant Production Engineering, Volume III, Edited by CIGR—The International Commission of Agricultural Engineering, Volume Editor: Bill A. Stout, Co-Editor: Bernard Cheze, Published by ASAE.

Coffman, C.W., Stone, J.F., Slocum, A.C., Landers, A.J., Schwab, C.V., Olsen, L.G., Lee, S. 2009. Use of engineering controls and personal protective equipment by certified pesticide applicators. J Agric Saf Health. 15(4): 311-26.

COSHH, 2002. Approved Code of Practice and Guidance. Control of Substances Hazardous to Health Regulations (Fifth edition). ISBN 978 0 7176 2981 7.

DECOS, 2012. Grain dust. Health-based recommended occupational exposure limit. The Hague, The Netherlands: The Health Council of The Netherlands.

Donham, K.J. 1995. Health hazards of pork producers in livestock confinement buildings: from recognition to control. In: McDuffie, H.H., et al editors. Agricultural health and safety: workplace environment sustainability. Boca Raton Florida: CRC press.

Elci, O.C., Akpinar-Elci, M., Blair, A., Dosemeci, M. 2002. Occupational dust exposure and the risk of laryngeal cancer in Turkey. Scand J Work Environ Health, 28(4): 278–284.

EPA, 2004. Risk assessment evaluation for concentrated animal feeding operations. Available at:

www.epa.gov/nrmrl/pubs/600r04042/600r04042.pdf. (Accessed 06 April 2012).

EPA, 2009. Air Quality Index: A Guide to Air Quality and Your Health. EPA-456/F-09.

EPA, 2012a. Particulate Matter. Available at http://epa.gov/ncer/science/pm/. (Accessed on 03.02.2012).

EPA, 2012b. Agriculture: Animal Feeding Operations. Avaliable at http://epa.gov/ncer/science/pm/. (Accessed on 06.04.2012).

Erisman, J.W., Bleeker, A., Hensen, A., Vermeulen, A. 2007. Agricultural air quality in Europe and the future perspectives. Atmospheric Environment, 42: 3209–3217.

EU Council Directive, 1999. Sulphur dioxide, nitrogen dioxide and oxides of nitrogen, particulate matter and lead in ambient air. 1999/30/EC, Official Journal L 163.

Fishwick, D., Allan, L.J., Wright, A., Curran, A.D. 2001. Assessment of exposure to organic dust in a hemp processing plant. Ann. occup. Hyg., 45(7): 577–583.

FAO-ILO-IUF, 2005. Agricultural Workers and Their Contribution to Sustainable Agriculture and Rural Development. Geneva, Switzerland: International Labour Organization.

Förstner, U. 1998. Integrated Pollution Control. Translated and Edited by A. Weissbach and H. Boeddicker. Springer.

Gaffney, P., Yu, H. 2003. Computing agricultural PM10 fugitive dust emissions using process specific emission rates and GIS. US EPA Annual Emission Inventory Conference, San Diego, California.

Guarino, M., Jacobson, L.D., Janni, K.A. 2007. Dust reduction from oil-based feed additives. Applied Engineering in Agriculture, Vol. 23(3): 329-332.

Hetzel, G.H. 2010. Respiratory Protection in Agriculture. Virginia Cooperative Extenstion.

HSE, 1998. Grain Dust. Guidance Note EH66. Second Edition.

HSE, 2005. Respiratory protective equipment at work: A practical guide. Series Code: HSG53.

HSE, 2006. Farmers' Lung. Web version of Leaflet AS5.

HSE, 2007. Controlling grain dust on farms. Agriculture Information Sheet No 3 (rev).

Jager, A.C. 2005. Exposure of poultry farm workers to ammonia, particulate matter and microorganisms in the Otchefstroom District, South Africa. MSc Dissertation, North-West University, South Africa.

Jerez, S.B., Zhang, Y., Wang, X. 2011. Spatial and temporal distributions of dust and ammonia concentrations in a swine building. Transactions of the ASABE, Vol. 54(5): 1873-1891.

Jimenes, J.R. 2006. Aerosol Characterization for Agricultural Field Burning Smoke. PhD Dissertation. Washington State University.

Kirkhorn, S.R., Garry, V.F. 2000. Agricultural lung diseases. Environmental Health Perspectives, Vol. 108, Supplement 4: 705-712.

Kline, A.A., Landers, A.J., Hedge, A., Lemley, A.T., Obendorf, S.K., Dokuchayeva, T. 2003. Pesticide exposure levels on surfaces within sprayer cabs. Applied Engineering in Agriculture, Vol. 19(4): 397–403.

Krupa, S.V. 1997. Air Pollution, People, and Plants. An Introduction. American Phytopahtological Society Press, St. Paul, Minnesota.

Lahiri, S., Levenstein, C., Nelson, D.I., Rosenbery, B.J. 2005. The cost effectiveness of occupational health interventions: Prevention of silicosis. American Journal of Industrial Medicine.

Lane, S.R., Nicholis, P.J., Sewell, R.D.E. 2004. The measurement and health impact of endotoxin contamination in organic dusts from multiple sources: Focus on the cotton industry. Inhalation Toxicology 16: 217-229.

Lee, K., Lawson, R.J., Olenchock, S.A., Vallyathan, V., Southard, R.J., Thorne, P.S., Saiki, C., Schenker, M.B. 2004. Personal exposures to inorganic and organic dust in manual harvest of California citrus and table grapes. Journal of Occupational and Environmental Hygiene, 1: 505–514.

Lee, S.A., Adhikari, A., Grinshpun, S.A., McKay, R., Shukla, R., Reponen, T., 2006. Personal exposure to airborne dust and microorganisms in agricultural environments. Journal of Occupational and Environmental Hygiene, 3: 118–130.

Li Q, Wang Li L, Shah S B, Jayanty R K M, Bloomfield P. 2011. Fine particulate matter in a high rise layer house and its vicinity. Transactions of the ASABE, Vol. 54(6): 2299-2310

Long, W., Tate, R.B., Neuman, M., Manfreda, J., Becker, A.B., Anthonisen, N.R. 1998. Respiratory Symptoms in a Susceptible Population Due to Burning of Agricultural Residue, Chest; 113:; 351-357.

Madden, N.M., Southard, R.J., Mitchell, J.P. 2008. Conservation tillage reduces PM10 emissions in dairy forage rotations. Atmospheric Environment 42: 3795–3808.

Matthews, J., Knight, A.A., 1971. Ergonomics in Agricultural Equipment Design. National Institute of Agricultural Engineering, Silsoe.

Maynard, R.L., Howard, C.V. 1999. Particulate matter: Properties and effects on health. Garland Publishing.

Mitchell, D.C., Schenker, M.B. 2008. Protection against breathing dust: behavior over time in Californian farmers. J Agric Saf Health, 14(2): 189-203.

Mitloehner, F.M., Calvo, M.S. 2008. Worker health and safety in concentrated animal feeding operations. Journal of Agricultural Safety and Health, 14(2): 163-187.

Molocznik, A. 2004. Time of farmers' exposure to biological factors in agricultural working environment. Ann Agric Environ Med, 11: 85–89.

Molocznik, A., Zagorski, J. 1998. Exposure to dust among agricultural workers. Ann Agric Environ Med 5: 127–130.

Müller, N., Xavier, F., Facchini, L.A., Fassa, A.G., Tomasi, E. 2006. Farm work, dust exposure and respiratory symptoms among farmers. Rev Saúde Pública; 40(5) NAAQS. National Ambient Air Quality Standards, USA.

Nieuwenhuijsen, M.J., Kruize, H., Schenker, M.B. 1998. Exposure to dust and its particle size distribution in California agriculture. American Industrial Hygiene Association Journal 58: 34–38.

Omland, Ø, 2002. Exposure and respiratory health in farming in temperate zones – a review of the literature. Ann Agric Environ Med, 9: 119–136.

OSHA, 2010. Occupational Safety and Health Standards. Available at http://www.osha.gov

Popendorf, W. J., Pryor A., Wenk H.R. 1982. Mineral dust in manual harvest operations. Ann. Am. Conf. Gov. Ind. Hyg., 2: 101-115.

Radon, K., Monso, E., Weber, C., Danuser, B., Iversen, M., Opravil, U., Donham, K., Hartung, J., Pedersen, S., Garz, S. 2002. Prevalence and risk factors for airway diseases in farmers - summary of results of the European Farmers' Project. Ann Agric Environ Med, 9: 207–213.

Radon, K., Garz, S., Riess, A., Koops, F., Monso, E., Weber, C., Danuser, B., Iversen, M., Opravil, U., Donham, K., Hartung, J., Pedersen, S., Nowak, D. 2003. Respiratory diseases in European farmers-II. Part of the European farmers' Project. Pneumologie, 57(9): 510-7.

Reed, S., Quartararo, M., Kift, R., Davidson, M., Mulley, R. 2006. Respiratory illness in farmers dust and bioaerosols exposures in animal handling facilities. A report for the Rural Industries Research and Development Corporation, RIRDC Publication No 06/107.

Rylander, R. 1987. The role of endotoxin for reactions after exposure to cotton dust. Am J Ind Med 12: 687-697.

Sabancı, A., 1999. Ergonomi. Baki Publishig (In Turkish), Adana.

Salvato, J.A, Nemerow, N.L., Agardy, F.J. 2003. Environmental Engineering. Fifth Edition. John Wiley & Sons, Inc.

Schenker, M.B. 1998. Supplement: American Thoracic Society. Respiratory Health Hazards in Agriculture. American Journal of Respiratory and Critical Care Medicine, Volume 158 Number 5, Part 2, pp1-76.

Schenker, M.B. 2000. Exposures and health effects from inorganic agricultural dusts. Environmental Health Perspectives, Vol. 108 Supplement 4: Occupational and Environmental Lung Diseases, 661-664.

Spaan, S., Wouters, I.M., Oosting, I., Doekes, G., Heederik, D. 2006. Exposure to inhalable dust and endotoxins in agricultural industries. J. Environ. Monit., 8: 63–72.

Spankie, S., Cherrie, J.W. 2012. Exposure to grain dust in Great Britain. Ann. Occup. Hyg., Vol. 56, No. 1, pp. 25–36.

Swanepoel, A.J., Rees, D., Renton, R., Swanepoel, C., Kromhout, H., Gardiner, K. 2010. Quartz exposure in agriculture: literature review and South African survey. Ann. Occup. Hyg., Vol. 54, No. 3, pp. 281–292.

The Swedish National Board of Occupational Safety and Health. 1994. Organic Dust in Agriculture. General Recommendations of the Swedish National Board of Occupational Safety and Health on Organic Dust in Agriculture, AFS: 11.

Torigoe, K., Hasegawa, S., Numata, O., Yazaki, S., Matsunaga, M., Boku, N., Hiura, M., Ino, H. 2000. Influence of emission from rice straw burning on bronchial asthma in children. Pediatrics 42, 143-50.

Xiwei, L., Owen, J. E. Murdoch, A.J., Pearson, C.C. 1993. Respirable dust from animal feeds. In Proc. Int. Livestock Env. Sym. IV, eds. E. Collins and C. Boon, 747-753. St. Joseph, Mich.: ASAE.

Wang, J., Hiscox, A.L., Miller, D.R., Meyer, T.H., Sammis, T.W. 2008. A dynamic Lagrangian, field scale model of dust dispersion from agriculture tilling operations. Transactions of the ASABE, Vol. 51(5): 1763-1774.

WHO, 1999. Hazard Prevention and Control in the Work Environment: Airborne Dust, Geneva - WHO/SDE/OEH/99.14.

WHO, 2000. Air Quality Guidelines for Europe, Second Edition. WHO Regional Publications. European series; No. 91.

Witney, B., 1988. Choosing and Using Farm Machines. Copublished in The United States with John Wiley & Sons Inc., Newyork.

Zock, J.P., Heederik, D., Kromhout, H. 1995. Exposure to dust, endotoxin and micro-organisms in the potato processing industry. Annals of Occupational Hygiene, 39(6): 841-854.

Chapter 3

Urban Structure and Air Quality

Helena Martins, Ana Miranda and Carlos Borrego

Additional information is available at the end of the chapter

1. Introduction

It is an inquestionable fact that much has been done in the last decades to improve the quality of the air we breathe and live in. Policies, technology and increasing public awareness have taken us to an unprecedented level of protection. On the other hand, it is also a fact that not only our cities but also our countryside continue to show worrying and troubling signs of environmental stress, of which air pollution is one of many.

In 1900, 14% of the world's population lived in cities; fifty years later, the proportion had risen to 30%, and by 2003 to 48%; today half the world's population lives in cities and predictions are that by 2030, 60% of the population will be urban [1]. In Europe, approximately 75% of the population lives in urban areas [2]. The last two centuries have seen a transformation of cities from being relatively contained, to becoming widespread over kilometres of semi-suburban/semi-rural land with commercial areas, office parks and housing developments. People often live miles from where they work, shop or go for leisure activities. This type of urban development has been named urban sprawl, and has its origins from the rapid low-density outward expansion of the United States of America cities in the beginning of the 20th century [3]. In Europe, cities have traditionally been much more compact; however urban sprawl is now also a European phenomenon [4].

Next the scientific and policy background on the subject of this chapter is presented. Urban planning aspects related to urban structure are briefly addressed, and the issue of urban air pollution is introduced, as well as the main air pollution problems that European cities are facing. The most important research studies covering the relation between urban planning and air pollution during the last decades are then reviewed.

1.1. Urban planning

When the first cities emerged, they were created having defence in mind, resulting in compact forms of settlement. With the advent of industrialization first and transport systems later, urban structures have changed dramatically, with an unprecedented process of urbanization that has persisted so far.

1.1.1. Urban planning perspectives

People have imagined ideal cities since ever; urban planners in particular have directed their attention to the types of urban structure that can provide a greater quality of life and environmental protection. In the 20th century, various architects have proposed radical changes in the form of the city [5]. Le Corbusiers's 'Radiant City' and Frank Lloyd Wright's 'Broadacre City' represent two extremes in a broad spectrum between urban density and dispersal. Le Corbusier (1887-1965) proposed high-density urban areas, where different land uses would be located in separate districts, with distinct functions - residential, commercial areas, churches - forming a geometric pattern with a sophisticated transit system. In opposition, Frank Lloyd Wright (1867-1959) defended the need for a closer contact with nature, and defended decentralized low-density cities, composed of single-family homes on large pieces of land, small farms, light industry, recreation areas, and other urban facilities where travel needs would be almost entirely dependent on the automobile [5].

Twenty years after the mid 1970's oil crisis which incited the first search for urban forms that conserved resources, the idea of sustainability has re-emerged, due to the growing awareness of urban problems related with resources depletion, energy consumption, pollution and waste [6]. The role of urban planning in urban sustainability, namely which urban structure will provide higher environmental protection, is today still under discussion. The scope of the debate can be summarized by classifying positions in two groups: the "decentrists", in favour of urban de-centralization, defending the dispersed city characterized by low population densities and large area requirements; and the "centrists", who believe in the virtues of high density cities with low area requirements, defending the compact city. Defenders of dispersal and low density development claim that low densities can be sustainable and that the quality of life within them is much higher in comparison with contained high density developments. The argument against the dispersed city is that low densities, and the consequent large area needs and land use segregation, result in a high dependence from motorized vehicles. Several authors however have associated sprawling urban development patterns with increased vehicle travel and congestion [7], increased volumes of storm-water runoff [8], loss of agricultural lands [9], and, even, increased rates of obesity in children and adult populations [10].

The compact city is characterized by high density and mixed use development, where growth is encouraged within the boundaries of existing urban areas. Those in its favour defend that urban containment will reduce the need for motorized trips, therefore reducing traffic emissions, and promoting public transport, walking and cycling [11]. It is also claimed that higher densities will help to make the supply of infrastructures and leisure services economically feasible, also increasing social sustainability [12]. Other such as [13] however, claim that the environmental benefits resulting from urban compaction are doubtful and that higher urban densities are unlikely to deliver the high quality of life that centrists promise. Although some reduction in energy consumption might be expected from compaction, they argue that a large centralised city can often result in greater traffic congestion with fuel efficiency greatly reduced. Another important aspect mentioned is that even if vehicle emissions are reduced, they may be concentrated in the precise areas where they cause most damage and adversely affect most people [14].

1.1.2. Urban sprawl in Europe

Historically, urban dispersion rose from the struggle against the 19th century industrial cities, which were congested, polluted, and foci of crime and disease [15]. After that, the growth of cities has been driven by the growth of population; however, in Europe today there is little or no population growth, while sprawl shows no signs of slowing down. A variety of factors such as the negative environmental (pollution and noise) and social factors (poverty and insecurity) related to city cores, rising living standards, changing living preferences, and a new mobility paradigm are now driving sprawl [2, 16].

Since the mid-1950's, European cities have expanded on average by 78% whereas the population has grown by only 33%; also, more than 90% of the new residential areas are low density areas; inevitably European cities have become much less compact [4]. Figure 1 shows the European areas with higher urbanization rates, where urban land cover has been increasing between four to six times faster than the European average, and the population density in residential areas declining six times faster [17].

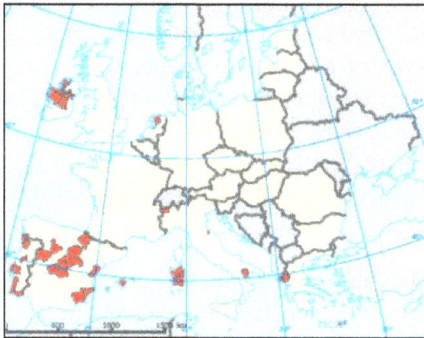

Figure 1. European areas with very rapid urbanization [17].

Clearly for these areas the term sprawl is well fitted. Regions of this type can be found along the Portuguese coastline, in Madrid and its surroundings as well as in some coastal regions in Spain, in the north of the Netherlands, north-western Ireland, Italy and Greece. Sprawl is particularly evident in countries or regions that have benefited from EU regional policies, such as Portugal, Ireland, and Spain.

1.2. Urban air quality

Problems regarding air pollution in urban areas have been known for millennia, but the attitude towards them was ambiguous, since they were even considered a symbol of growth and prosperity, and the attempts to combat them were scattered and ineffective. It was only after the occurrence of a few major air pollution episodes in the 20th century (such as the Meuse Valley (Belgium) accident in December 1930 and the London December 1952 smog episode) that a greater awareness and the consequent development of air pollution policies took place.

1.2.1. Main atmospheric pollutants and sources

Atmospheric pollutants (gaseous and particulate) can be divided in primary pollutants, which are directly emitted to the atmosphere by a natural or anthropogenic emission source, and secondary pollutants, which result from primary pollutants transformation through chemical reactions highly dependent on meteorological conditions and/or solar radiation [18]. Currently, the two air pollutants of most concern for public health are surface particulate matter and tropospheric ozone, therefore receiving special attention in this review and also throughout this chapter.

There is increasing evidence that fine dust particles have deleterious effects on human health, causing premature deaths and reducing quality of life by aggravating respiratory conditions such as asthma [19]. One reason why particulate matter (PM) is of such concern is the absence of any concentration threshold below which there are no health effects. Evidence suggests that fine particulates, with an equivalent aerodynamic diameter less than 2.5 micrometres (PM2.5), do most damage to human health, and that effects depend further on the chemical composition or physical characteristics of the particle [20]. Particulate matter (PM) includes as principal components sulphate, nitrate, organic carbon, elemental carbon, soil dust, and sea salt. The first four components are mostly present as fine particles, and these are of most concern for human health. Sulphate, nitrate, and organic carbon are produced within the atmosphere by oxidation of sulphur dioxide (SO_2), nitrogen oxides (NO_x) and non-methane volatile organic compounds (NMVOC); carbon particles are also emitted directly by combustion. The seasonal variation of PM is complex and location-dependent; in general, PM needs to be viewed as an air quality problem year-round [21].

While ozone (O_3) in the upper atmosphere provides an essential screen against harmful UV radiation, at ground level it is lung irritant causing many of the same health effects as particulate matter, as well as attacking vegetation, forests and buildings. Observed effects on human health are inflammation and morphological, biochemical, and functional changes in the respiratory tract, as well as decreases in host defence functions. Effects on vegetation include visible leaf injury, growth and yield reductions, and altered sensitivity to biotic and abiotic stresses [22]. Ozone is produced in the troposphere by photochemical oxidation of carbon monoxide (CO), methane (CH_4), and NMVOC by the hydroxyl radical (OH) in the presence of reactive nitrogen oxides. The relation between O_3, NO_x and VOC is driven by complex nonlinear photochemistry, with the existence of two regimes with different O_3-NO_x-VOC sensitivity: in the NO_x-sensitive regime (with relatively low NO_x and high VOC), O_3 increases with increasing NOx and changes little in response to increasing VOC; in the NO_x-saturated or VOC-sensitive regime O_3 decreases with increasing NOx and increases with increasing VOC [23]. Also, in the vicinity of large nitrogen monoxide (NO) emissions, ozone is destroyed according to the reaction NO + O_3 = NO_2 + O_2, generally referred as O_3 titration by NO. This situation usually takes place in heavily polluted areas, with ozone consumption taking place immediately downwind of the sources, and becoming elevated as the plume moves further downwind [24]. Ozone pollution is in general mostly a summer problem because of its photochemical nature [23].

1.2.2. Emissions and air quality trends in Europe

Emissions of air pollutants decreased substantially during the period 1990–2009 across Europe (Figure 2). PM emissions fell by 27 % for PM10 and 34 % for PM2.5. Emissions of the precursor gases SO_x and NO_x declined by 80 % and 44 % respectively. Emissions of ammonia (NH_3), have fallen less: only about 14 % between 1990 and 2009. It is estimated that current European policies reduced NO_x emissions from road vehicles by 55 % and from industrial plants by 68 % in the period 1990–2005 [25].

Notwithstanding the emissions decrease in Europe, the analysis of PM10 concentrations since 1999 for a total of 459 European air quality monitoring stations reveals that 83 % of the stations presents a small negative trend of less than 1 $\mu g.m^{-3}$ per year [25]. For ozone there is a discrepancy between the substantial cuts in ozone precursor gas emissions and the stagnation in observed annual average ozone concentrations in Europe [26]. Reasons include increasing inter-continental transport of O_3 and its precursors in the northern hemisphere, climate change/variability, biogenic NMVOC emissions, and fire plumes from forest and other biomass fires [26].

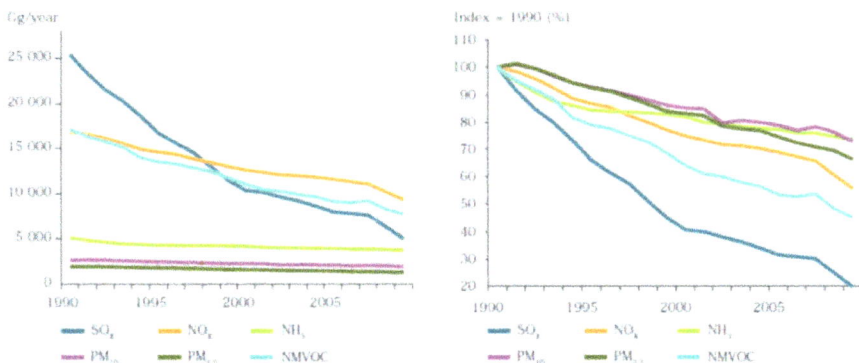

Figure 2. EU emissions of PM and ozone precursor gases 1990-2009 [25].

The target value threshold for ozone of 120 $\mu g.m^{-3}$ (daily maximum of running 8-hour mean values) was exceeded on more than 25 days per year at a large number of stations across Europe in 2009 (Figure 3a). The map shows the proximity of recorded ozone concentrations to the target value. At sites marked with dark orange dots, the 26th highest daily ozone concentration exceeded the 120 $\mu g.m^{-3}$ threshold, implying an exceedance of the threshold and the number of allowed execeedances by the target value [25]. The EU limit and target values for PM were exceeded widely in Europe in 2009, as evidenced in Figure 3b. The annual limit value for PM10 was exceeded most often (dark orange dots) in Poland, Italy, Slovakia, several Balkan states and Turkey. The daily limit value was exceeded (light orange dots) in other cities in those countries, as well as in many other countries in central and western Europe.

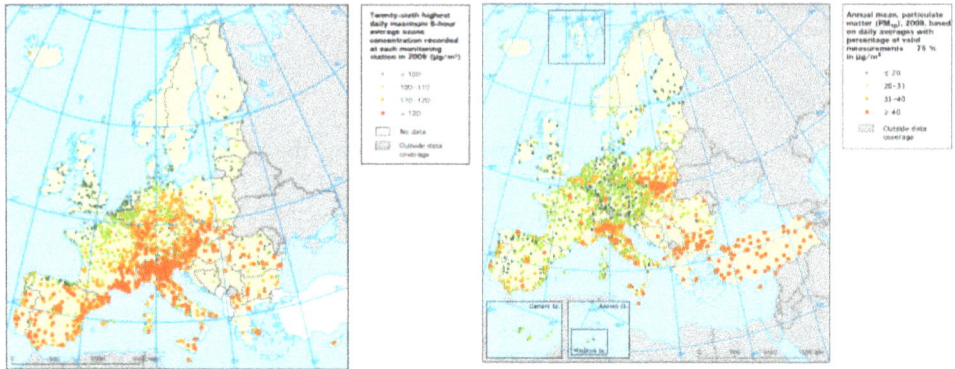

Figure 3. a) Twenty-sixth highest daily maximum 8-hour average ozone concentration; b) Annual mean concentration of PM10 recorded at each monitoring station in 2009 [25].

Across Europe, the population exposure to air pollution exceeds the standards set by the EU (Figure 4). For ozone there has been considerable variation along the period 1997-2009, with 14% to 61% of the urban population exposed to concentrations above the target value. In 2003, a year with extremely high ozone concentrations due to specific meteorological conditions, the exposure was higher. Regarding PM10, in the period 1997-2009, 18 to 50% of the urban population was potentially exposed to ambient air concentrations higher than the EU limit value set for the protection of human health [25].

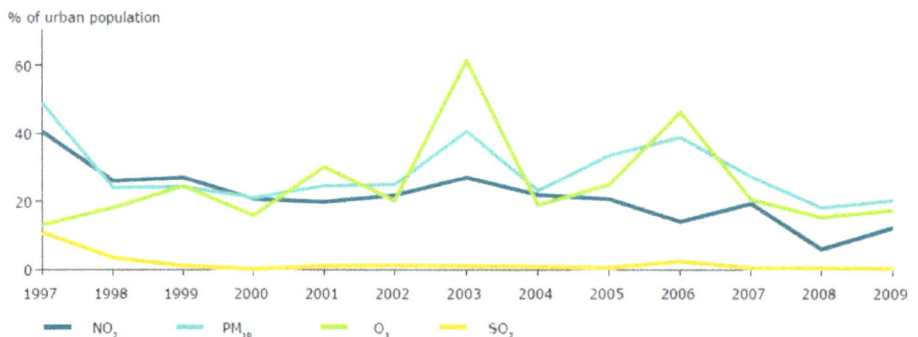

Figure 4. Percentage of the EU urban population potentially exposed to air pollution exceeding acceptable EU air quality standards [25].

1.3. Integrating urban planning and air quality

Since the world's cities are the major consumers of natural resources, the major producers of pollution and waste, and the focus of most other human activities, various governments realised that much of the sustainable debate has an urban focus. Solving the problems of the city would be a major contribution to solving the most pressing global environmental problems, since it is in cities that we find the greatest concentration of population and economic activity, and it is in cities that the crucial long term and often irreversible decisions on infra-structure investments (related to energy supply and waste treatment) are made. After the Brundtland Commission report [27] the notion that the natural environment should become a political priority, and the pursuit of sustainable development received a remarkable attention. In many countries there have been profound changes in policies and in political and popular attitudes, as the commitment to the sustainable development idea has increased. The question now is which urban form or structure will be likely to deliver more environmental benefits or will be less harmful to the human health and the environment. The most important work conducted in the field in the last two decades is reviewed next.

1.3.1. Data analysis studies

Much of the technical arguments for compact cities have revolved around the allegedly lower levels of travel, and hence lower levels of fuel consumption and emissions, associated with high urban densities. [28, 29] have related fuel consumption per capita to population density for a large number of cities around the world, and found a consistent pattern with higher densities associated with lower fuel consumption. [30] compared a group of world cities over the period 1980 to 1990 regarding its land use and transport characteristics. The study demonstrated the importance of urban density in explaining annual per capita auto use, with annual kilometres travelled per capita strongly inversely correlated with urban density. Similar conclusions emerged from [11], which found a clear inverse correlation between total distances travelled per week and population density. People living at the lowest densities were found to travel twice as far by car each week in comparison to those living at the highest densities.

The studies by [28, 29] have been criticised for focusing on the single variable of density, when other factors are likely to be important in explaining travel behaviour. [31] argues that household income and fuel price are important determinants of such behaviour, making it difficult to clearly identify the link between density and fuel consumption.

While several additional studies [32-35] have related travel behaviour, traffic, energy consumption and emissions with land use patterns, only few were found relating land use with air quality, i.e., with atmospheric pollutant concentrations. [36] examined the relationship between the degree of sprawl and ozone levels for 52 metropolitan areas in the United States. While there was evidence regarding the association between lower population densities and higher vehicle miles of travel, only moderate evidence was found relating sprawl and increased ozone levels. [37] analysed the impact of changes in land area and population on per capita exposure to motor vehicle emissions, concluding that infill development has the potential to reduce motor vehicle emissions yet increasing per capita inhalation of those emissions, while sprawl has the potential to increase vehicle emissions but reduce their inhalation. [38] explored the implications of sprawl for air quality through the integration of data on land use attributes and air quality trends recorded in 45 of the 50 largest US metropolitan regions. The results of this study indicate that urban form is significantly associated with both ozone precursor emissions and ozone exceedances. Overall, the most sprawling cities experienced over 60% more high ozone days than the most compact cities.

Most of the above work relied on empirical studies to provide descriptive comparisons of current cities and to find evidence that certain types of urban forms are correlated with desirable levels of energy consumption and emissions. These approaches integrate the land use and transport aspects of urban form, but lack the extra step that translates energy efficiency into indicators of air quality, via pollutant concentrations.

1.3.2. Numerical modeling studies

As just mentioned, several empirical and modelling studies integrate land use and transport issues and its relation with urban structure, however, few were found that explore the connection to air quality and human exposure. Conclusions from most of the studies done so far have been harmed by the lack of knowledge about the complex path between an initial action for the reduction of atmospheric emissions and the final benefit in terms of air quality and human exposure [39]. Health effects of air pollution are the result of a chain of events, going from the release of pollutants leading to an ambient atmospheric concentration, over the personal exposure, uptake, and resulting internal dose to the subsequent health effect. It is important to make a distinction between concentration and exposure; concentration is a physical characteristic of the environment at a certain place and time, whereas exposure describes the interaction between the environment and a living subject, referring to an individual's contact with a pollutant concentration.

Emissions reduction conducts to changes in atmospheric pollutant concentrations, but those changes will have different spatial and temporal magnitudes and signs, due to differences in emissions, weather patterns and population exposed to pollution according to the time of the day, day of the week or month of the year, and also according to the population age structure (children, adults and elderly suffer different effects due to their different respiratory frequencies). Exposure is the key factor in assessing the risk of adverse health effects, since high pollutant concentrations do not harm people if they are not present, while even low levels may become relevant when people are present [19].

Recent advances in computer technology have allowed the integration of land-use and traffic models with air quality models; these modelling tools assume a particular importance to the subject under study, since they allow the integration of the most important variables that have to be analysed. One of the earliest investigations in this field was carried out for Melbourne in [39]. The authors developed a framework for linking urban form and air quality, integrating land use, transport and air quality models, and the results of the study shown that any of the several strategies designed to deliberately channel and concentrate additional population and industry into specific zones, when supported by simultaneous investments in transport infrastructure, will deliver environmental and efficiency benefits that consistently outperform those associated with the "business-as-usual" approach.

Through the application of dispersion and photochemical models, [40] concluded that compact cities with mixed land uses promote a better air quality when compared with dispersed cities with land use segregation. A subsequent study, conducted by the same team [41], investigated the influence of urban structure on human health, estimating the human exposure to atmospheric pollutants. Results reveal that the compact city presents more people exposed to higher pollution levels due to the existent high population densities. [42] investigated the potential effects of extensive changes in urban land cover, in the New York City (NYC) metropolitan region, on surface meteorology and ozone concentrations. Results from the study suggest that extensive urban growth in the NYC metropolitan area has the potential to increase afternoon temperatures by more than 0.6°C leading to increases in episode-average ozone levels by about 1–5 ppb, and episode-maximum 8 h ozone levels by more than 6 ppb. [43, 44] investigated the effects of urban sprawl on road traffic, air quality and population exposure for the German Ruhr area. The sprawl scenario produced a temperature increase of about half a degree over significant portions of the domain, including beyond the area where the land use changes were implemented. The combination of increased temperature and emissions yielded ozone concentration pattern changes, from -1.5 to +4.5 µg.m^{-3}.

2. Case study presentation

In the report "Urban sprawl in Europe" [2], Porto urban area is identified as one of the top ten European cities where sprawl is growing faster. In the last decades, the Porto area has experienced an accelerated process of land occupation, with the urban area increasing at much faster rates than the population. Also, according to the air quality reports for Portugal's Northern region, the assessment of pollutant concentrations measured in the air quality monitoring network shows that Porto metropolitan region presents a poor air quality, with ozone thresholds and PM10 limit values exceeded [45]. It seems therefore that the Porto region is an interesting and challenging case to be studied in the framework of the topic urban structure and air quality.

The region selected for the analysis is showed in Figure 5 and includes 21 municipalities, with a total area of almost 240 000 hectares. The Porto municipality constitutes the study region's centre around which a first metropolitan ring is formed by the municipalities of Matosinhos, Maia, Gondomar and Vila Nova de Gaia; the municipalities of P. Varzim, V.N. Famalicão, Lousada, Felgueiras, Penafiel, M. Canavezes, C. Paiva and S.J. Madeira can be considered part of a peripheral ring, while the remaining intermediate municipalities constitute a second metropolitan ring.

2.1. Patterns of urban growth and change

This section explores the path of the recent urban expansion in the Porto area. For that purpose the process of urban growth in this area is analysed in detail, with the use of two digital Corine Land over (CLC) maps – CLC90 and CLC2000.

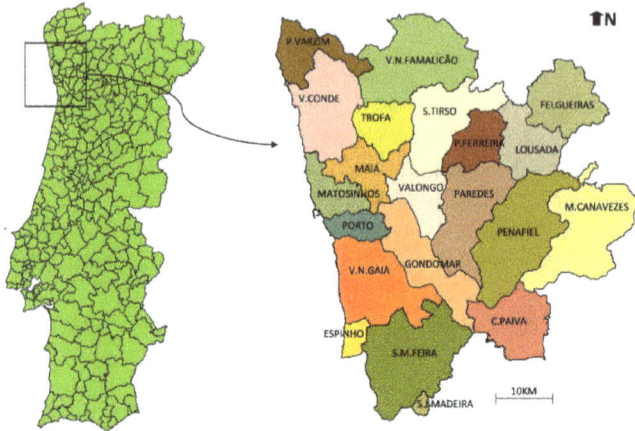

Figure 5. Study region, including 21 municipalities.

The CORINE (COordination of INformation on the Environment) programme of the European Commission includes a land cover project – CLC - [46] intended to provide consistent localized geographical information on the land cover of the Member States of the European Community. CLC is a standardised land cover inventory derived from satellite imagery for 24 countries, with 250 m resolution. For Portugal, CLC 1990 (CLC90) was produced with satellite images from 1985 to 1987, depending on the region, while CLC2000 concerns the year 2000. The two datasets are here analysed for the study region, in order to produce a thorough characterization of the land use evolution in the period between 1987 and 2000. Figure 9 presents the study region land cover maps for 1987 and 2000, resulting from the processing of CLC90 and CLC2000 data, respectively. To obtain a clearer picture of the land cover, the 44 CLC classes were grouped in 5 large categories: 1) artificial surfaces; 2) agricultural areas; 3) forests and shrub areas; 4) other non-artificial surfaces (areas of little or no vegetation, and inland and coastland wetlands); 5) water bodies.

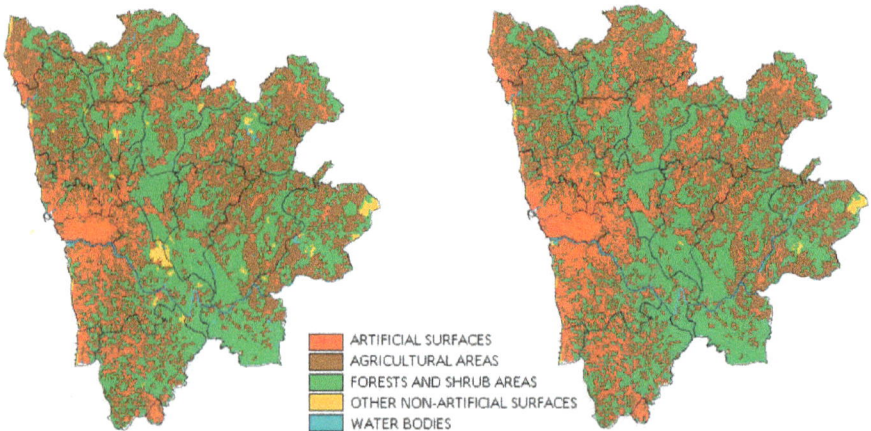

ARTIFICIAL SURFACES
AGRICULTURAL AREAS
FORESTS AND SHRUB AREAS
OTHER NON-ARTIFICIAL SURFACES
WATER BODIES

Figure 6. Study region land cover maps for 1987 and 2000.

The land cover maps reveal the expansion of artificial areas throughout the study region, mainly occupying land previously dedicated to agriculture, due to its proximity to the already existent urban areas. In order to have a clearer picture of the magnitude and nature of this growth, Table 1 presents the numbers behind the maps, including the total area for each of the four large land use categories and corresponding share (%) for each dataset, as well as the magnitude of the change between 1987 and 2000. Furthermore, artificial surfaces area is analysed with more detail by looking at its composition: continuous urban fabric; discontinuous urban fabric; industrial or commercial units; other artificial surfaces. From 1987 to 2000, built-up land uses increased 41.5%, around 13 000 new hectares have become artificial during this period, with urbanized land rising from 13% to 18% of the total area of the region. The analysis by municipality shows that the largest artificial surface increases are particularly observed outside the urban centre confirming the previous assertions about the existence of urban sprawl processes in the region. Municipalities in the first metropolitan ring around Porto reveal the largest absolute increases of artificial surfaces. Municipalities outside the first metropolitan ring, with very low shares of urbanised areas in 1987 presented the highest growth rates between 1987 and 2000. As expected, Porto municipality presents the highest percentage of artificial land uses, with 91.5% of the total area in 2000 (83% in 1987). As urbanization advanced, many non-urban hectares disappeared: agriculture land loss represents more than half of the entire non-urban losses (12820 ha); forest and shrub areas come next with 26%.

Land uses	CLC90 (1987 data)		CLC2000		Change	
	hectares	%	hectares	%	hectares	%
Artificial surfaces	30908.2	12.9	43727.9	18.3	+ 12819.7	+41.5
Continuous urban fabric	3369.0	10.9	4059.2	9.3	+690.2	+20.5
Discontinuous urban fabric	23583.0	76.3	32895.0	75.2	+9312.0	+39.5
Industrial or commercial units	2719.9	8.8	4973.1	11.4	+2253.2	+82.8
Other artificial surfaces	1236.3	4.0	1800.7	4.1	+564.3	+45.6
Agricultural areas	101350.1	42.3	93766.2	39.1	-7584.0	-7.5
Forests and shrub areas	101598.7	42.4	98319.4	41.0	-3270.3	-3.2
Other non-artificial surfaces	5750.4	2.4	3784.9	1.6	-1965.4	-34.2
TOTAL AREA	239598.4	100	239598.4	100	-	-

Table 1. Study region land cover data for 1987 and 2000.

A more detailed analysis of the new artificial uses between 1987 and 2000 reveals little changes in the urbanization trends. The discontinuous or low density urban fabric ranks first for both years, summing around 75% of the total artificial area. While in 1987 continuous urban fabric was the second land use category, with 11% of the total artificial area, in 2000 the industrial and commercial units took over the second place. This land use category showed the highest growth rate between 1987 and 2000 (83%), followed by other artificial surfaces (46%). The discontinuous urban fabric is the first in terms of area growth, representing 73% of the new artificial areas. The land use category compact or continuous urban fabric showed the lowest growth.

Evidence therefore suggests that Porto region is undergoing a process of urban sprawl; to further confirm it, it is important to look at the relation between the artificial areas growth and the population growth in the same period. Making use of the population data and of the residential area, obtained through the sum of continuous and discontinuous urban fabric, the residential density (number of residents per residential square kilometre) was calculated for 1987 and for 2000 (Figure 7), for a limited group of municipalities with available data for 1987 and 2000 simultaneously.

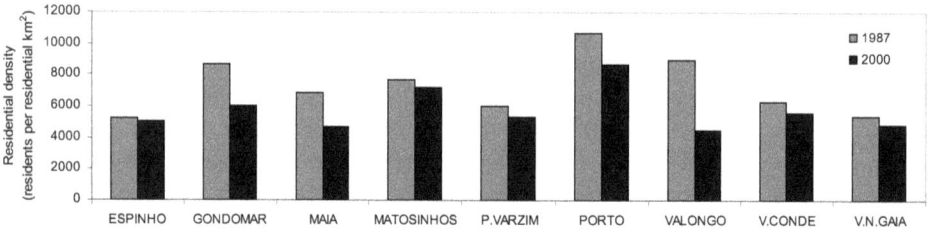

Figure 7. Residential density calculated for 1987 and 2000 for a group of municipalities in the study region.

A trend towards lower residential densities is observed, revealing that the population growth has lost importance as an explanatory factor of the urbanization process, while the generalization of dispersed urban patterns has risen. An important sprawl process in the region is the proliferation of new industrial and commercial areas. Extensive industrial areas and mega commercial structures punctuate the Porto region, with the traditional tendency of locating commercial uses within the urban fabric rapidly fading. There is no longer a real mixture of uses; instead, commercial activities are now segregated and concentrated in large portions of land orientated to commercial and leisure activities.

2.2. Mobility and attractiveness

In metropolitan areas, the need for daily-travel or commuting is a reality steaming from the progressive distancing between residential areas and work and study areas. Hence, in a study whose aim is to link urban structure with emissions and air quality, it is essential to look not only at the number of residents per municipality but also at the population flow between municipalities. It was therefore necessary to characterize the commuting characteristics of the region and the relative attractiveness/ repulsiveness of each municipality in the study area. For that purpose, a study from the National Statistics Institute [47] for the year 2001 was the main source of data. The study demonstrates the existence of important commuting movements in the Porto Metropolitan Area, through the analysis of the main interaction axis and the accounting of workers and student's flows between municipalities. Of great significance are the interactions between Porto, the centre of the region, and the municipalities of the first metropolitan ring; these interactions are strongly unbalanced in favour of Porto [47]. The mentioned study compiled the rates relating the number of individuals entering/ exiting a given municipality with the number of individuals residing in the municipality. The described data was processed and attraction and repulsion rates re-calculated for the municipalities in the case study region. It was assumed that the study region acts as a tight zone, and the possible interactions between it and the surrounding areas are not considered. As an example, Figure 8 presents the data for Porto municipality, with a net attraction rate of 38.2%.

These attraction and repulsion rates are essential for the definition and construction of the urban development scenarios for the region since, in order to determine the total amount and distribution of atmospheric pollutant emissions in the study region, it is necessary to consider not only the number of inhabitants or residents per municipality but also the flow between municipalities.

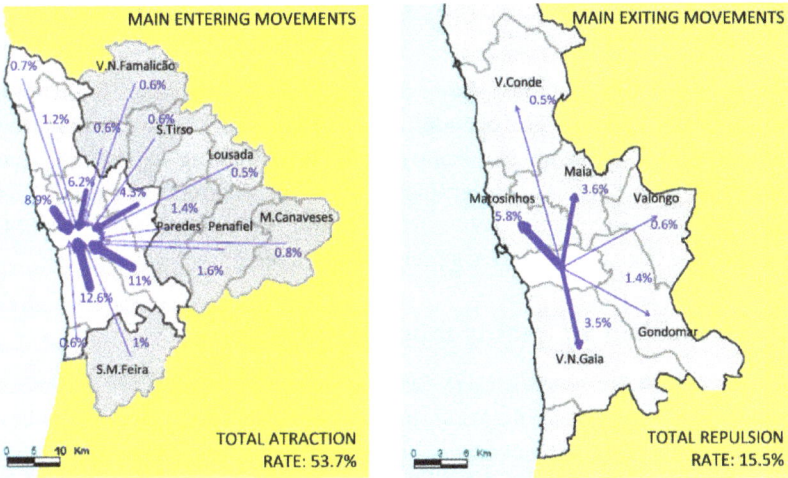

Figure 8. Porto main entering and exiting movements and attraction and repulsion rates for 2001.

2.3. Air quality levels

Portugal's northern region, in accordance to the established in the Air Quality Framework Directive (96/62/EC), was classified [48] in two zones (Interior North and Coastal North) and four agglomerations (Coastal Porto, Braga, Vale do Ave and Vale do Sousa). Since 2005, the air quality monitoring network covers all the zones/agglomerations, with a total of 24 stations in 2006, the large majority of them (15) located in Coastal Porto due to the high number of inhabitants.

Figure 9 shows the air quality monitoring stations for which PM10 daily legal requirements were not fulfilled [49]. High PM10 concentrations are measured in urban and suburban monitoring stations; regarding the daily limit value the number of annual exceedances goes well beyond the allowed 35. As a result of these exceedances, and accordingly to the determined in the Air Quality Framework Directive, the Northern Region of Portugal is currently under the obligation of developing and implement Plans and Programs for the Improvement of the Air Quality [49].

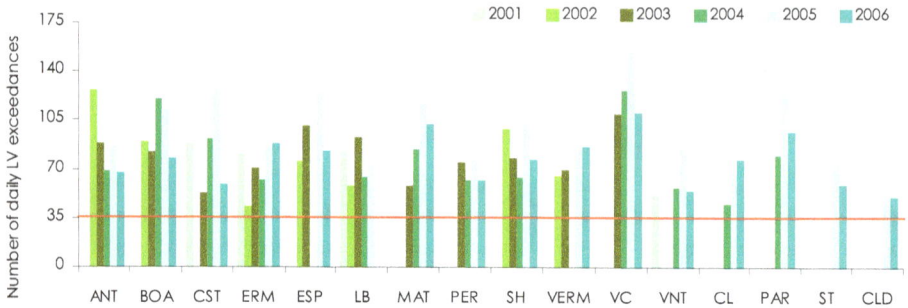

Figure 9. Monitoring stations not fulfilling PM10 legal requirements for daily LV + MT in 2001-2006 in the study area (the red line indicates the allowed number of daily exceedances) (data from [49]).

The analysis of ozone measured data shows that concentration values are higher outside the urban centre of the region, i.e. outside Porto municipality. Nevertheless the ozone information threshold is exceeded in the majority of the monitoring stations, and often along a high number of hours per year. Concerning the seasonal occurrence of exceedances, ozone limit values are generally higher between April and September, while for PM10 high concentrations have been found both in summer and winter.

3. Setup of the urban air quality modelling system

This section describes the meteorological (MM5) and chemical (CAMx) numerical models, used in the atmospheric simulations for the Porto study region. Both models are freely available, and have been extensively used and validated worldwide, being subject of constant improvement and update. These facts, together with the good performance of the models obtained for different regions, including the present study region, justify their selection. Moreover, these models are ready to be applied in long-term simulations with acceptable computing times.

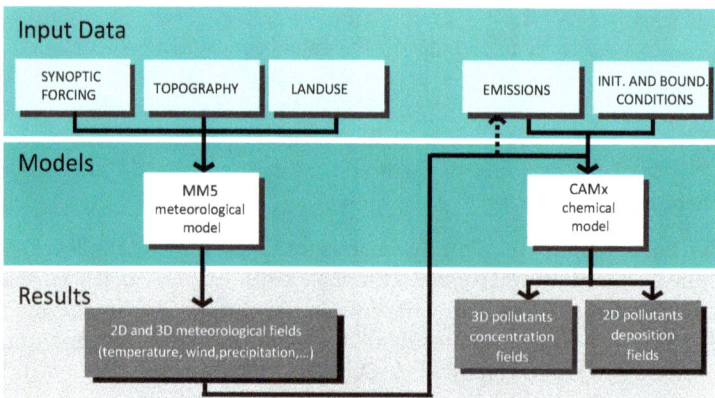

Figure 10. Simplified scheme of the MM5-CAMx modelling system.

Figure 10 presents a simplified scheme of the MM5-CAMx modelling system applied to the simulation of the atmospheric flow and air quality in the study region.

3.1. MM5 meteorological model

The PSU/NCAR mesoscale model was developed at the Pennsylvania State University and the National Centre for Atmospheric Research (NCAR). The model is supported by several pre- and post-processing programs, which are referred to collectively as the MM5 modelling system [50]. The MM5 modelling system software is freely provided and supported by NCAR, therefore it is widely used internationally [51, 42]. The MM5 is a three-dimensional non-hydrostatic prognostic model that simulates mesoscale atmospheric circulations. Important features in the MM5 modelling system include: (i) a multiple-nest capability; (ii) non-hydrostatic dynamics; (iii) a four-dimensional data assimilation capability; (iv) increased number of physics options; and (v) portability to a wide range of computer platforms [52]. The program numerically solves the pressure, mass, momentum, energy and water conservation equations; it presents different parameterization schemes for clouds, planetary boundary layer and diffusion, moisture, radiation, and surface. MM5's nesting capability allows the consideration of several domains in a single simulation or in consecutive simulations; therefore, the first domain can present a more regional dimension with a coarser mesh, while the next domain will cover a smaller area but with a higher resolution.

Since MM5 includes several parameterizations, users can choose among the multiple options of model physics and parameterization schemes; some are based on the scale of the motion, such as the cumulus parameterizations, while others are dependent on users preferences, such as the planetary boundary layer (PBL) schemes [53]. Based on previous MM5 applications for the West Coast of Portugal, namely by [54], the chosen MM5 physical options include: Grell cumulus scheme for the coarser resolution domain and no cumulus parameterization for the smaller grids, RRTM radiation scheme, Reisner-Graupel moisture scheme, MRF BPL scheme for the coarser resolution domain and Gayno-Seaman PBL scheme for the smaller grids. The used land surface model is the five-layer soil model. The initial and boundary conditions are from the National Centre for Environmental Predictions (NCEP) global 1-degree reanalysis data, updated every 6-hours [55].

The MM5 modelling system has two types of land use data with global coverage available from the United States Geological Survey (USGS): 13-category, with a resolution of 1 degree, 30 and 10 minutes; and 24-category, with a resolution of 1 degree, 30, 10, 5 and 2 minutes, and 30 seconds. The USGS 24-category data is referred to 1990, and some of the components are originated from a dataset compiled in the 1970s [52]. This data was compared with the data from Corine Land Cover 2000 [56], and it was possible to conclude that the land use in the study area is weakly represented in the USGS24 original dataset: in CLC2000, Porto and the surrounding municipalities are presented as a large urban area, while in USGS24 the urbanized area is much more restricted and concentrated over Porto. Therefore, the USGS24 default land use data was replaced by CLC2000 data in the present study.

3.2. CAMx air quality model

The Comprehensive Air quality Model with extensions (CAMx) was developed by ENVIRON International Cooperation, from California, United States of America. CAMx [57] is an Eulerian photochemical dispersion model that allows the integrated "one-atmosphere" assessment of gaseous and particulate air pollution over many scales ranging from sub-urban to continental. CAMx simulates the emission, dispersion, chemical reaction, and removal of pollutants in the troposphere by solving the pollutant continuity equation for each chemical species on a system of nested three-dimensional grids. The Eulerian continuity equation describes the time dependency of the average species concentration within each grid cell volume as a sum of all of the physical and chemical processes operating on that volume [58]. The nested grid capability of CAMx allows cost-effective application to large regions in which regional transport occurs, yet at the same time providing fine resolution to address small-scale impacts in selected areas [58]. The CAMx chemical mechanisms are based on Carbon Bond version 4 and SAPRC99.

CAMx requires input files that configure each simulation, define the chemical mechanism, and describe the photochemical conditions, surface characteristics, initial/boundary conditions, emission rates, and various meteorological fields over the entire modelling domain. Preparing this information requires several pre-processing steps to translate "raw" emissions, meteorological, air quality and other data into the final input files for CAMx. Some changes have been performed over the last years in order to implement MM5-CAMx system for Portugal [59].

The MM5-CAMx pre-processor generates CAMx meteorological input files from the MM5 output files, including land use, altitude/pressure, wind, temperature, moisture, clouds/rain and vertical diffusivity. The vertical structure in CAMx will be defined from the MM5 sigma layers, and therefore will vary in space, also vertical layer structures can vary from one grid nest to another. Topographic and land use information is also provided by the MM5 model through the MM5-CAMx pre-processor.

In this study initial concentrations and hourly boundary conditions were created from output concentration files from the LMDz-INCA chemistry-climate global circulation model [60] for gaseous species, and from the global model GOCART [61] for aerosols.

Finally, pre-processors are also used to calculate the hourly variation of emissions from point and area sources, respectively. The processing of the atmospheric emissions is described in the following section.

3.3. Atmospheric emissions processing

Emission inventories are crucial ingredients to successfully simulate atmospheric pollutants concentrations, although including substantial uncertainties related to the spatial and temporal allocation of emissions, as well as the chemical speciation [62, 63]. Besides the degree of completeness of the inventory and the quality of the emission factors, the accuracy of the inventory's temporal and spatial patterns is of major importance for successful air quality modelling. The Portuguese National Inventory Report (NIR) [64] compiles total annual quantities of atmospheric emissions, which are assigned by municipality and SNAP (Selected Nomenclature for sources of Air Pollution) category. For air quality modelling purposes it is therefore necessary to further spatially disaggregate emissions to the model's grid cell resolution level. In previous air quality studies for Portugal [62, 65] the NIR was disaggregated at the sub-municipality level using data given by Census 2001, concerning population and fuel consumption [66].

In the present study a new methodology is designed and implemented using spatial surrogates to disaggregate national emission totals onto a spatially resolved emission inventory, which can be used as input for any air quality model domain over Portugal. A spatial surrogate is a value greater than zero and less than or equal to one that specifies the fraction of the emissions of a particular country, in this case Portugal, which should be allocated to a particular grid cell of the air quality model domain of interest [67]. Typically, some type of geographic characteristic is used to weight the attributes into grid cells in a manner more specific than a simple uniform distribution. In this study, based on the methodology described in [68], CLC2000 land use data in combination with national statistics (for population, industry and agriculture employment) are applied as spatial surrogate variables for disaggregating non-point emission sources over Portugal. The surrogate value is calculated as the ratio of the attribute value in the intersection of the country and the grid cell to the total value of the attribute in the country.

The methodology developed and applied is now described. First, point source emissions were allocated on the air quality domain of interest. Next, non-point emissions, for each SNAP category, were spatially distributed using specific quantitative spatial surrogate data, based on statistics from the National Statistics Institute (INE), and other source specific activity data, and on CLC2000 data for Portugal. The emissions considered in the present study concern the following atmospheric pollutants: NO_x, NMVOC, CO, NH_3, PM10 and PM2.5. The first step consisted in disaggregating population according to land use. Population density data are available in Portugal at the sub-municipality level, or commune. The size of communes in Portugal is very heterogeneous, ranging from 4 ha to 42500 ha; hence this level of spatial resolution is insufficient for air quality modelling purposes. Moreover, a certain commune may contain, for instance, parts of dense urban nucleus, agricultural land with some sparse population, and natural vegetation areas with very little or no population. CLC2000 gives useful geo-referenced information for disaggregation, since its geographic database provides information that is spatially much more detailed than the commune limits. Different population densities were attributed to different land cover categories, following the methodology described in [69]. The methodology was then applied to the Portuguese inland territory, with population and employment statistics given by CENSUS 2001 [70] being disaggregated over the CLC2000 and emissions disaggregated with population density using GIS. This procedure is illustrated in Figure 11 for NO_x emissions from non-industrial combustion (SNAP2).

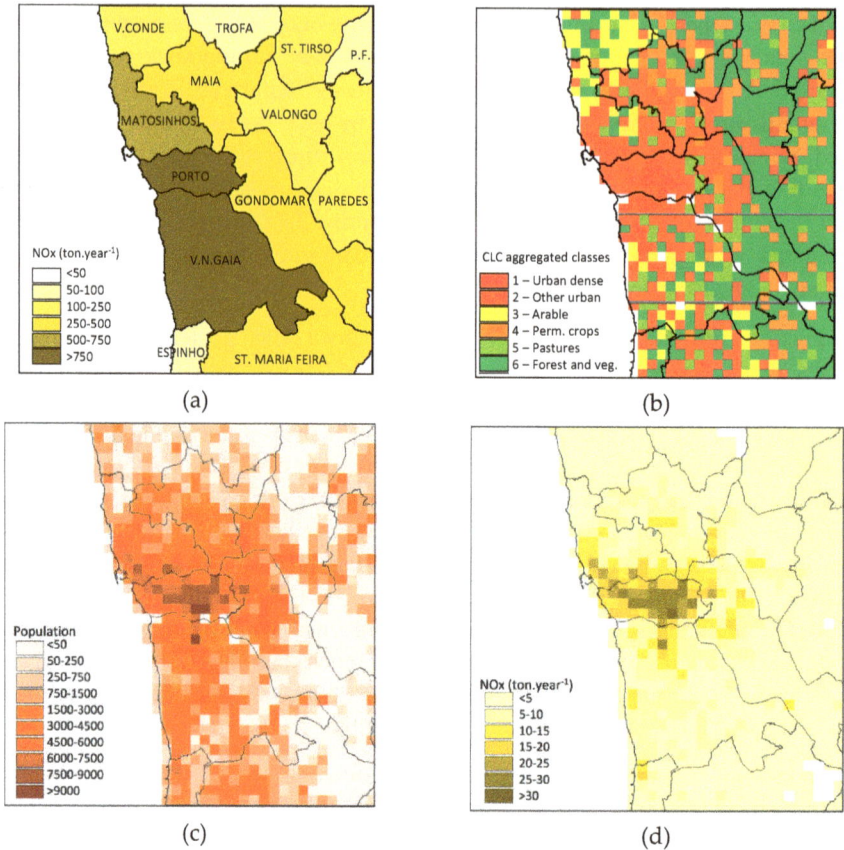

Figure 11. Spatial allocation of NOx emissions from SNAP2 for domain 3: a) Input data: emissions at municipality level; b) CLC aggregated classes; c) calculated population for each grid cell of the domain; d) gridded emissions at 1 km resolution.

The same procedure was followed for disaggregating emissions for the remaining SNAP categories, except for SNAP7, since the NIR distinguishes road transport emissions in two sub-categories: motorway emissions and non-motorway emissions. Non-motorway emissions were spatially distributed using the population disaggregated over the CLC2000 data as described above. Motorway emissions were disaggregated over the national motorway network, again using GIS.

For the biogenic emissions a bottom-up approach was used. The methodology for Portugal is described in [71], and requires the knowledge of the temperature, solar radiation and forest area density. For the CAMx simulations, biogenic emissions are given as isoprene and monotherpenes.

As the NIR provides annual emission totals, time-varying profiles were developed describing variations in monthly (12-element), daily (2-element, weekday and weekend) and hourly (24-element) anthropogenic emissions, transforming time-averaged man-made emissions into hourly fluxes. The information to construct representative and meaningful temporal profiles was taken from National official statistics (energy, industrial production, transport, etc).

3.4. Case study domain definition

For the meteorological simulation, the MM5 capability of doing multiple nesting is used, and the model is applied for four domains, using the two-way nesting technique. Figure 12 shows the model domain setup and the location of the meteorological stations to be used in the validation process: domain 1 (D1) at 27 km resolution covering the Iberian Peninsula and France; D2 at 9 km resolution over Portugal; D3 at 3 km resolution over NW Portugal; and D4 with 1 km resolution over Great Porto Area.

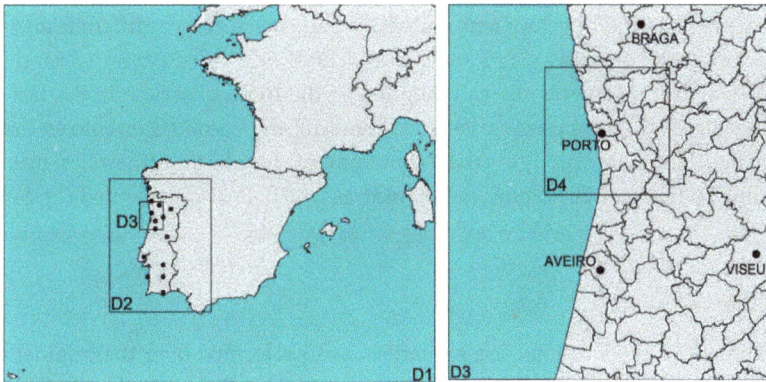

Figure 12. Simulation model domains.

Table 2 summarizes the corresponding grid configurations. Considering previous research studies performed for NW Portugal [72], 25 unequally spaced vertical levels are used in order to optimize the simulation through the increase of vertical resolution near the surface.

Domain	No. of cells in x-direction	No. of cells in y-direction	Z levels	Resolution (km)
D1	91	77		27
D2	63	81	25	9
D3	45	51		3
D4	51	51		1

Table 2. MM5 domains configuration.

Regarding the air quality simulations, CAMx is applied for three domains, slightly smaller than the corresponding MM5 domains, using its two-way nesting capability: domain 1 (D1) at 9 km resolution covering Portugal; D2 at 3 km resolution over NW Portugal; and D3 with

1 km resolution over Great Porto Area. Table 3 summarizes the corresponding grid configurations. Considering previous research studies performed for Portugal [59], 17 unequally spaced vertical ⊛levels are used.

Domain	No. of cells in x-direction	No. of cells in y-direction	Z levels	Resolution (km)
D1	40	70		9
D2	35	41	17	3
D3	38	38		1

Table 3. CAMx domains configuration.

4. Urban development scenarios

This section presents the development and characterization of two different and opposite urban development scenarios - SPRAWL and COMPACT - in terms of land use and population. The first represents the continuation of the trend observed in the last decades, and can be described as a business-as-usual scenario; the second symbolizes the rupture with the current situation through urban containment. In addition, the reference situation corresponding to the year 2000, now on referred as BASE, is also presented for comparison purposes.

4.1. Land use

The development of the two land use scenarios, was performed over the original CLC2000 land use map, through the alteration of land use type parcels, using the ArcGis software. The SPRAWL scenario corresponds to the business-as-usual scenario, representing the continuation of the last decades trend, with urban areas continuing to expand at much faster rates than population, and urban development spreading throughout the study area, by filling up existing gaps and expanding the boundaries of existing urban areas. All the new residential areas (or urban fabric) take place in the form of discontinuous urban fabric. This urban sprawl scenario results in the smearing out of the region's inhabitants over a large area, thus effectively simulating the sprawl-related growth process. The urban development process in the period 1987-2000 was analysed for each municipality separately and replicated for SPRAWL; the original CLC2000 land use map was changed through the creation of new artificial surface areas, which replaced natural and semi-natural areas. The combined SPRAWL land use from each municipality resulted in a new land use map for the study region presented in Figure 13, side-by-side with the BASE map (CLC2000). The built-up area (artificial surfaces) was increased from 18% to 25% of the total area; a number that can be considered realistic given current trends and the fact that in 1987 the share was 13%. The artificial areas expansion took over agricultural and forested landscapes located in the proximity of already existent urban areas.

ARTIFICIAL SURFACES
AGRICULTURAL AREAS
FORESTS AND SHRUB AREAS
OTHER NON-ARTIFICIAL SURFACES
WATER BODIES

(a)   (b)

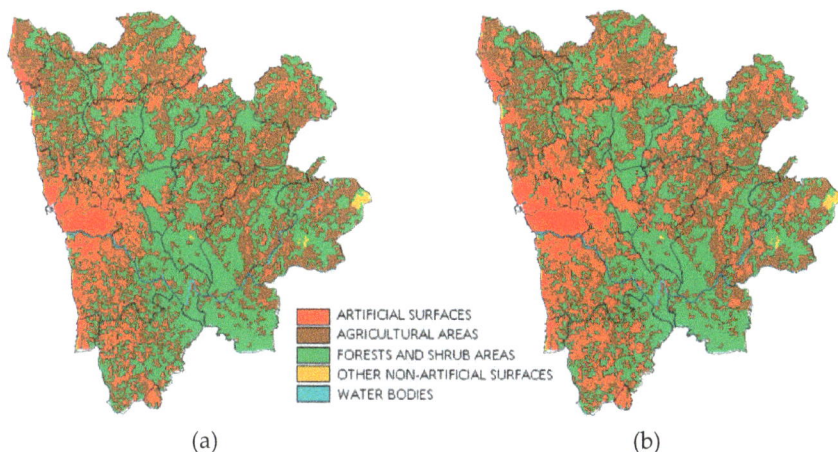

Figure 13. Study region land cover maps for a) BASE and b) SPRAWL scenario.

The land cover maps reveal the expansion of artificial areas not only in the urban centre of the region (Porto, Matosinhos, Gondomar and Vila Nova de Gaia), but also throughout the entire study region. Table 4 presents the comparison between the BASE and the SPRAWL scenario in terms of the total area for each of the 4 large land use categories, and sub-categories, and corresponding share (%), as well as the magnitude of the change.

Land uses	BASE		SPRAWL		Change	
	hectares	%	hectares	%	hectares	%
Artificial surfaces	43727.9	18.3	60139.2	25.1	+ 16411.3	+37.5
Continuous urban fabric	4059.2	9.3	4059.2	6.7	0	0
Discontinuous urban fabric	32895.0	75.2	44647.7	74.2	+11752.7	+35.7
Industrial or commercial units	4973.1	11.4	9571.7	15.9	+4598.6	+92.5
Other artificial surfaces	1800.7	4.1	1860.6	3.1	0	0
Agricultural areas	93766.2	39.1	83201.4	34.7	-10564.8	-11.3
Forests and shrub areas	98319.4	41.0	92472.9	38.6	-5846.5	-5.9
Other non-artificial surfaces	3784.9	1.6	3784.9	1.6	0	0

Table 4. Study region land cover data for the BASE and SPRAWL scenario.

In comparison with BASE, in the SPRAWL scenario built-up land uses increase 37.5%. Agricultural areas present the largest decrease, representing now less than 35% of the total area of the region; forest and shrub areas continue to be the dominant land use in the region, with a share around 39%. Regarding the composition of artificial surfaces, the continuous urban fabric loses importance, with no additional areas of this type being created, representing now less than 7% of the artificial surfaces. Discontinuous urban fabric presents the largest increase, almost 12 000 hectares; industrial and commercial units continue the growth trend verified between 1987 and 2000, with the highest relative growth, almost doubling its presence in the study area.

In COMPACT the totality of urban growth is accommodated within already existent urban areas, i.e., no additional artificial surfaces are created. The only land-use changes implemented in this scenario concern limited changes from discontinuous to continuous urban fabric (around 40 hectares). Therefore, no spatial representation of the COMPACT scenario is presented here, since it coincides with the BASE maps.

4.2. Population

The population of the study region has been increasing; however, this increase has not been uniform along the region, with municipalities growing at different rates and even decreasing in Porto municipality. From 1991 to 2006 the study region population increased from 1.86 million people in 1991 to 2.07 million in 2006 (11.3% growth); the rate of growth however has decreased from around +1% per year in 1991-2001, to 0.2% per year, in 2001-2006. In the 25-years period under analysis, in Porto municipality population presented a decrease of 27%; an important feature of this decrease is that its rate has been accelerating: in the period 1981-1991 the rate was around -0.8%, in 1991-2001 the rate increased to -1.3%, and in 2001-2006 around -1.8% [56].

Considering the previous population evolution, both scenarios are developed for a population of 2.2 million people, corresponding to an increase of 220'000 inhabitants (13% increase) in relation to the base year 2000, in what can be considered a 20-year period. This population increase is differently distributed through the municipalities, according to the land use scenario. Since the SPRAWL scenario corresponds to the perpetuation of the past 20 years trend, the population will change accordingly in each of the municipalities, presenting the same growth rates as observed between 1991 and 2001. In the COMPACT scenario however, the trend is interrupted; Porto municipality attracts new residents, and its population is increased. The remaining cities will continue to attract people, but at a smaller rate than the verified in the last years (and therefore also in SPRAWL). Figure 14 presents the population observed in 1991 and 2000, and considered in SPRAWL and COMPACT. In COMPACT all the municipalities present a growth in their population, but at a smaller rate than the verified for SPRAWL. The exception is Porto, with more inhabitants than those in 2000, but still less than those registered in 1991.

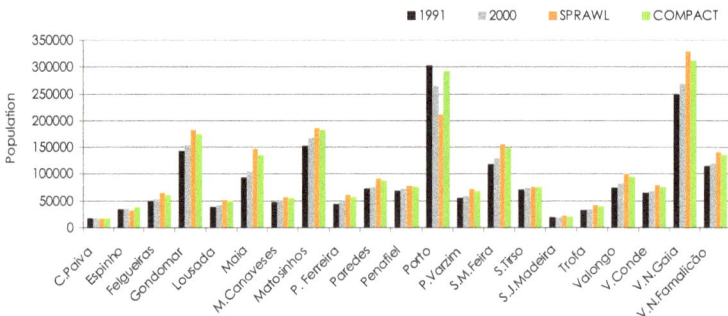

Figure 14. Population for the SPRAWL and COMPACT scenarios and its comparison with the population in 1991 and 2000.

The population in each municipality is distributed over the land use data for BASE, COMPACT and SPRAWL, according a disaggregation methodology described in §3.3. SPRAWL presents the lowest population density (maximum values are below 9000 inhab.km^{-2}), while BASE and COMPACT show a similar situation, but higher densities are found in the later with maximum values of 11 000 inhab.km^{-2} in comparison with 10 000 inhab.km^{-2} in BASE. This data is fundamental for the further determination of the population affected by air pollutants concentrations in each of the studied scenarios.

4.3. Emissions

As a result of the population growth and the land use changes established for each urban development scenario, new emission totals have to be calculated, as well as their spatial distribution. Atmospheric pollutants emissions for the BASE situation were the basis for estimating the scenarios emissions. New emissions were recalculated for each scenario considering the new population in each municipality, and also land use changes. Emission rates per inhabitant per municipality were kept equal to the BASE rate, as well as emission rates per land use type per municipality.

Land use differences are particularly important for three emission categories - mobile, agriculture and biogenic sources -, therefore these will be given particular attention in the next sections.

4.3.1. Road transport emissions

Since road transport emissions are highly dependent not only on population distribution but mainly on the mobility of the population, ideally a traffic model should be applied to simulate the effect of urban sprawl on traffic volumes and their spatial distribution. These modelling techniques fall out of the scope of the present work and therefore are not used. Here, to calculate transport emissions resulting from land use changes, a methodology is developed taking into account the population growth, the urban area expansion and the mobility attractiveness/repulsion rates between municipalities. These three factors influence emissions and are considered as follows:

i. The growth of the population causes an increase in the number of trips. For each municipality it was assumed that the emissions are proportional to the number of trips, which in turn is proportional to the number of residents.
ii. The growth of the urban area causes an increase in the mean distance from home to employments and leisure destinations. The residents in new urbanized areas find themselves more distant from locations where most employments are concentrated, while the residents in already existent urban areas will find possible employment and leisure destinations in the newly built areas in the periphery. For each municipality it was assumed that the emissions are proportional to the mean travel distance, which in turn is proportional to the urban area's radius. For example, in SPRAWL Maia's urban area increases by a factor of 1.4; therefore the mean travelled distance increased by a factor of $1.4^{1/2}=1.185$; in COMPACT the factor is 1 since no urban growth was verified.

iii. An additional factor related to attraction/repulsion rates between municipalities has to be considered since traffic emissions are not only dependent on the population and urban area, but also on the mobility of people between municipalities. The attraction/repulsion rates calculated for BASE, presented in §2.2 are maintained and used for both scenarios.

The distribution of emissions between municipalities is very different for both scenarios, as illustrated in Figure 15, which presents CO yearly emission totals for non-motorways road transport emissions for each municipality and for the entire study area. Resulting emissions are higher for SPRAWL, which are 19% higher than the BASE emissions, while COMPACT emissions are only 4% higher. The largest differences between scenarios are found for Porto (25% lower than the BASE emissions for SPRAWL, and 30% higher for COMPACT), Matosinhos (+38% for SPRAWL, +8% for COMPACT), Vila Nova de Gaia (+20% for SPRAWL, -2% for COMPACT) and Maia (+56% for SPRAWL, +9% for COMPACT).

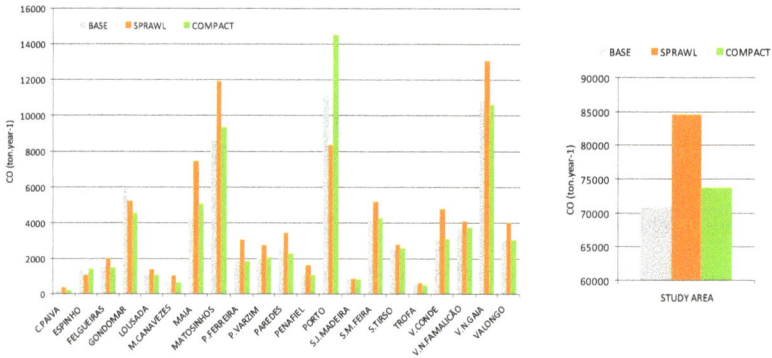

Figure 15. Study region SNAP7 (non-motorways road transport) CO emissions for BASE, COMPACT and SPRAWL, for each municipality and for the entire study area.

Regarding the spatial distribution of emissions, Figure 16 presents SNAP7 non-motorway CO grid emissions at 1 km resolution for SPRAWL and COMPACT. For both scenarios, emissions are concentrated in the Porto, Matosinhos, Maia, NW Gondomar and Vila Nova de Gaia municipalities; however COMPACT presents a greater concentration of emissions, as a result of the urban containment, and therefore higher emission rates.

4.3.2. Agriculture emissions

New emissions for the agriculture category were recalculated considering the new agricultural area in each scenario, with emission rates per agricultural area per municipality kept equal to the BASE rates. Since the COMPACT scenario presents no changes in agricultural area in relation to the BASE, emission totals, as well as their spatial distribution are the same. As a result of the transformation of agricultural areas into artificial land use, agriculture emissions were reduced by almost 10% in SPRAWL.

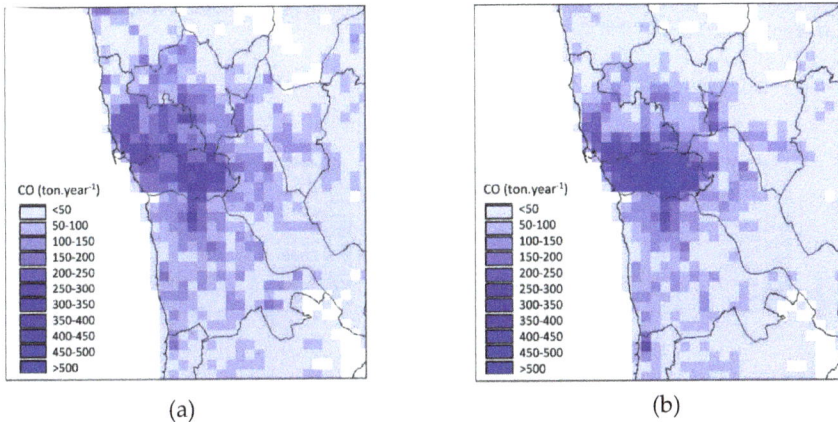

Figure 16. SNAP7 (non motorway road transport) CO grid emissions at 1 km resolution for a) SPRAWL and b) COMPACT.

4.3.3. Biogenic emissions

Biogenic emissions were calculated for the forested areas according to the methodology previously described in §3.3. Differences in relation to BASE result from the conversion of forested areas to artificial areas, and also from temperature changes induced by land use changes; these only take place in the SPRAWL scenario, since in COMPACT, the forest land use are not changed in relation to BASE. Therefore, as a result of land use changes biogenic SPRAWL emissions are lower when compared to BASE (and COMPACT): 20% lower for monotherpene and 16% lower for isoprene.

4.3.4. Total emissions

The above presented methodology results on different emission totals for both scenarios. Figure 17 shows emission totals for the study region for SPRAWL and COMPACT as well as for BASE.

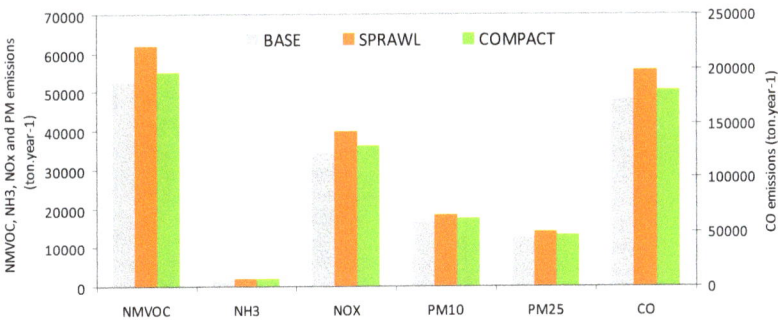

Figure 17. Study region total NMVOC, NH₃, NOₓ, PM and CO emissions for BASE, SPRAWL and COMPACT.

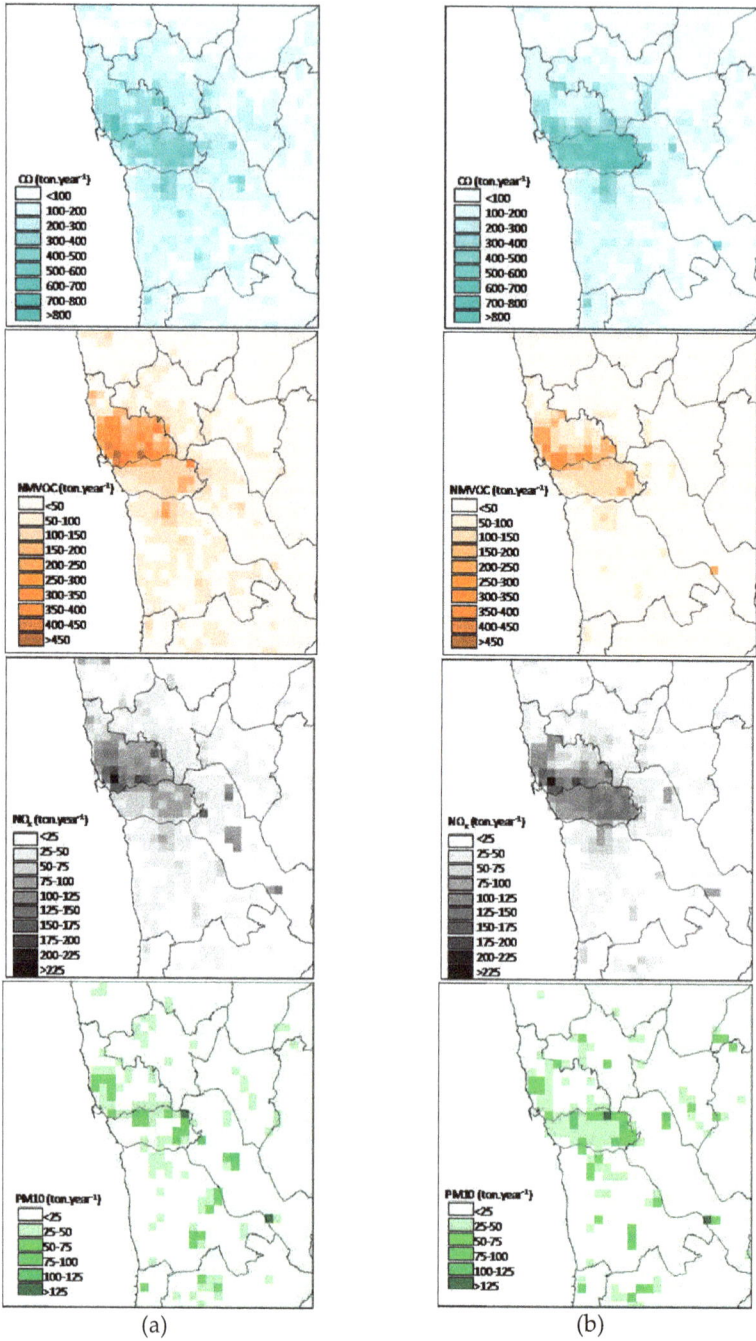

Figure 18. Spatial allocation of CO, NMVOC, NOx and PM10 total emissions at 1 km resolution for a) SPRAWL and b) COMPACT

Lower emissions are obtained for BASE and higher for SPRAWL; SPRAWL emissions are around 9% to 17% higher than BASE emissions (for NH$_3$ and NMVOC, respectively), while COMPACT emissions are 4% to 6% higher (for NH3 and NMVOC, respectively).

Figure 18 shows the spatial distribution of CO, NMVOC, NO$_x$ and PM10 gridded emission totals for the 1 km resolution domain for SPRAWL and COMPACT. COMPACT emissions are more concentrated over Porto municipality and present higher emission rates per grid cell; SPRAWL presents more scattered emissions throughout the simulation domain, and therefore lower emission rates. Emissions of NMVOC constitute an exception, because they are highly related with the port activity in Matosinhos, and therefore present higher values for this municipality in both scenarios.

5. Atmospheric modelling results

Aiming to provide a thorough analysis of the air quality impacts of different urban land use scenarios, the atmospheric simulation of BASE and scenarios is performed for a one-year period, covering a wide range of air pollution conditions. The meteorological year of 2006 was chosen for simulations since it is considered an "average" year, as opposed to others such as 2003 and 2005, which were abnormally dry and/or warm [73, 74]. Meteorological differences between the two scenarios, and between each of the scenarios and BASE, will steam solely from land use changes since the meteorology is the same. The air quality simulations were performed with meteorological inputs given by the respective MM5 annual simulation and emissions described in §3.2 and §4.3 for BASE and for the scenarios, respectively.

5.1. Meteorological modelling

5.1.1. BASE simulations

For BASE the simulation was performed with land use data from 2000 since no data was available at the time for 2006. In order to evaluate the model performance, modelling results were compared with data from Porto/Pedras Rubras meteorological station, located in the municipality of Matosinhos. Figure 19 shows the time-series comparison of surface temperature and wind components for observed and BASE simulated values.

Concerning temperature, simulated values follow the distribution of the observed ones; a general under-estimation of temperature is visible, especially for the higher temperatures registered at the end of May / beginning of June, July and August. Simulated wind components present a smaller variability when compared with observed ones, but also follow the observed trend.

The MM5 skill was also evaluated through the application of the quantitative error analysis introduced by in [75] and widely used in model validation exercises:

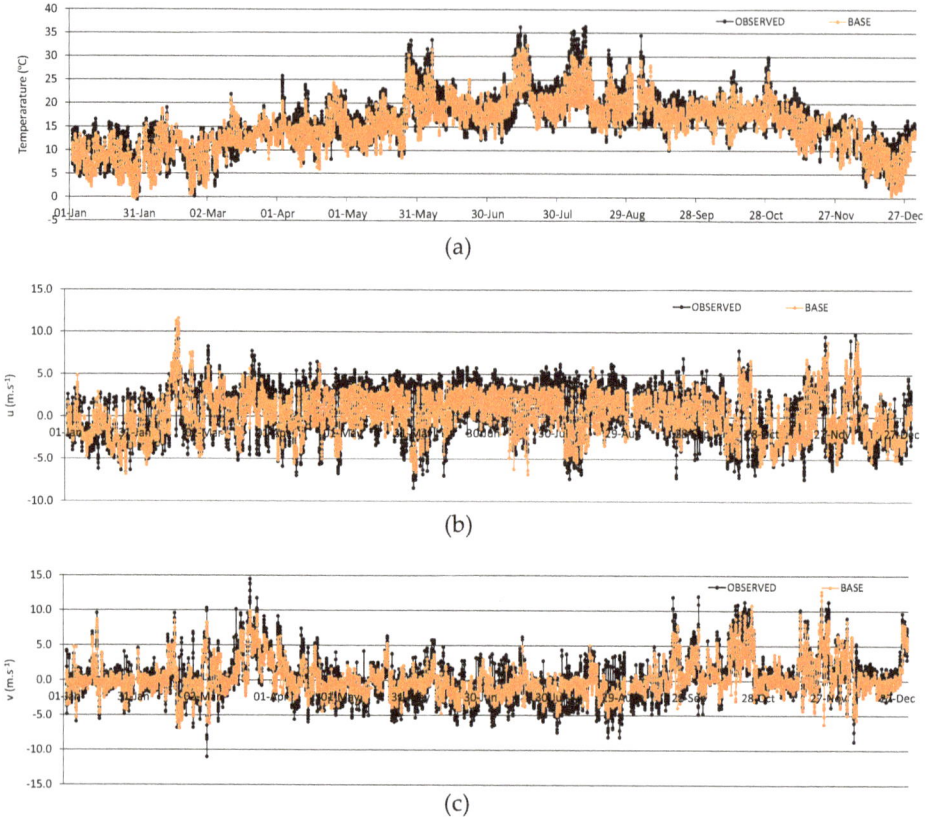

(a)

(b)

(c)

Figure 19. Observed and BASE (1km resolution) time-series comparison of surface a) temperature, b) zonal wind component and c) meridional wind component, at Porto/Pedras Rubras meteorological station.

$$E = \left(\sum_{i=1}^{N} \left(\phi_i - \phi_{iobs} \right)^2 \Big/ N \right)^{1/2} \qquad (1)$$

$$E_{UB} = \left(\sum \left[\left(\phi_i - \phi_0 \right) - \left(\phi_{iobs} - \phi_{0obs} \right) \right]^2 \Big/ N \right)^{1/2} \qquad (2)$$

$$S = \left(\sum_{i=1}^{N} \left(\phi_i - \phi_0 \right)^2 \Big/ N \right)^{1/2} \qquad (3)$$

$$S_{obs} = \left(\sum_{i=1}^{N} \left(\phi_{iobs} - \phi_{0obs} \right)^2 \Big/ N \right)^{1/2} \qquad (4)$$

The parameter E is the root mean square error (rmse), E_{UB} is the rmse after the removal of a certain deviation and S and S_{obs} are the standard deviation of the modelled and observed data. If ϕ_i and ϕ_{iobs} are individual modelled and observed data in the same mesh cell, respectively, ϕ_0 and ϕ_{0obs} the average of ϕ_i and ϕ_{iobs} for some sequence in study, and N the number of observations, then the simulation presents an acceptable behaviour when $S \approx S_{obs}$, $E < S_{obs}$ and $EUB < S_{obs}$. In addition to these parameters the correlation coefficient was also determined for each simulation.

Figure 20 presents the statistical analysis of BASE 1km resolution simulations, for Porto/Pedras Rubras. For temperature the correlation coefficient obtained is 0.9, with S/S_{obs} also near 1, and E/S_{obs} below 0.5. As expected, wind components results are not as good as for temperature, with lower correlation coefficients and higher errors. The meridional wind component is better simulated than the zonal one. Overall, the meteorological simulation reveals a good performance for the three meteorological variables, with statistical parameters presenting a reasonable behaviour.

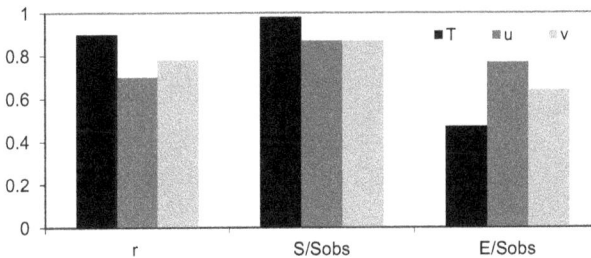

Figure 20. BASE statistical parameters for surface temperature, and zonal and meridional wind components for Porto/Pedras Rubras meteorological station.

5.1.2. Scenario simulations

As for BASE, the SPRAWL and COMPACT meteorological simulations are performed for 2006 meteorological year, using the land use data produced according to the procedure described in §4.1. Since for COMPACT the land use is very similar to that of BASE (the only change concerned the conversion of a few hectares of discontinuous urban fabric to continuous urban fabric), meteorological results from COMPACT only present very small temperature differences in relation to BASE. Therefore, from now on, and for meteorological purposes, no distinction is made between BASE and COMPACT.

Taking into consideration that the most widely recognized meteorological effect of urbanization is the urban heat island effect and because of the recognized influence of urban temperatures on ozone formation, hereafter the meteorological analysis will be focused on surface temperature. SPRAWL meteorological simulations produced a domain-averaged annual temperature increase of approximately 0.4 °C. This is attributed to the increased share of built-up areas in the domain, which convert incoming radiation to sensible heat rather than to latent heat (evaporation), owing to the limited water availability in artificial surfaces characterized by impervious materials. However, in some regions and for certain time-periods differences between scenarios reached significantly higher values than the average.

Figure 21 presents the differences between COMPACT and SPRAWL annual simulations for hourly surface temperature, at Porto/Pedras Rubras meteorological site, with 1 km resolution. Although the land use in Porto/Pedras Rubras was not changed, there were temperature differences as high as 2.5°C between the two simulations. These differences indicate that changes in meteorological parameters are not necessarily confined to the cells where the land use pattern was modified. Also, higher differences are found in the summer months, i.e., from April to September, since higher temperatures are also reached, and therefore meteorological differences are enhanced. While temperature increases would be expected with increasing urbanization, due to the urban heat island effect, temperature decreases are also verified. Local temperature increases in grid cells with modified land use could have lead to higher wind speeds and increased instability which downwind can lead to areas of increased vertical mixing and decreased surface temperatures.

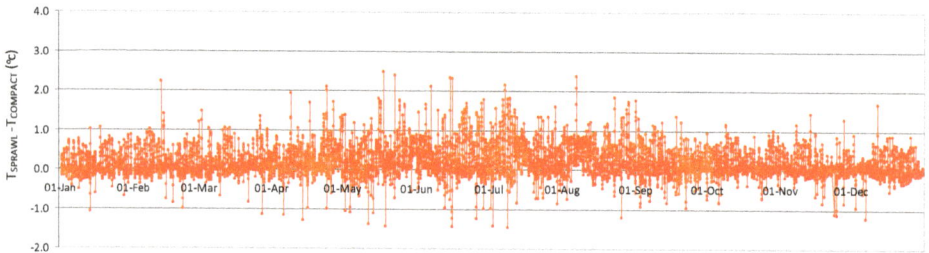

Figure 21. Hourly surface temperature differences between SPRAWL and COMPACT for Porto/Pedras Rubras meteorological site for 1-km resolution.

To illustrate the spatial extent of effects of land use changes in temperature, the average afternoon (12:00 – 18:00) temperature differences for July are shown in Figure 22. For July, average afternoon temperature differences range from about -1.2°C to +1.4°C, with largest increases occurring over Vila do Conde, Maia, Matosinhos, Porto and Gondomar, i.e., municipalities in the first metropolitan ring, which present some of the largest urban expansion. The observed changes are consistent with the substantial increases in urban surfaces across large parts of the model domain, and the spatial pattern of the temperature changes generally matches the area of increased urbanization. This is quite evident for the coastal part of Vila do Conde, NE Matosinhos and SE Vila Nova de Gaia.

The temperature differences obtained as a result of land use changes are consistent with previous research by [42, 44], although these authors conducted research only for episodic air pollution situations. Although not presented, the SPRAWL scenario with its increased urban land cover also had a noticeable effect on surface layer winds across the metropolitan region, generally leading to a slight increase in wind speed.

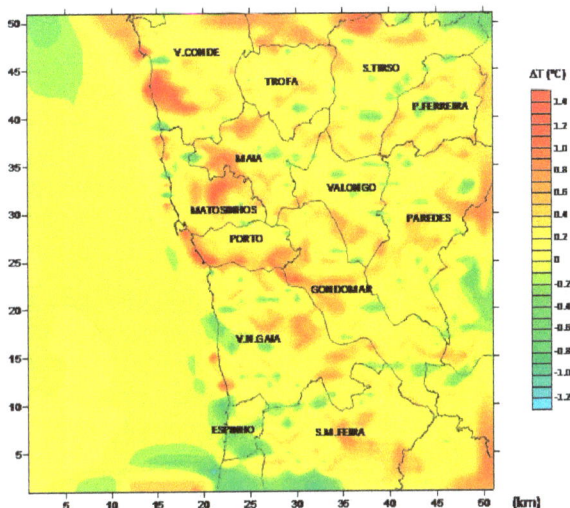

Figure 22. July differences between SPRAWL and COMPACT afternoon (12:00 – 18:00) average surface temperature fields between at 1 km resolution.

5.2. Air quality modelling

5.2.1. BASE simulations

Here the air quality results for the annual simulation of BASE are presented. The air quality model configuration and its application are those described in §3.2. For BASE the simulation used emissions data for 2005 (there are no emission estimates for 2006, since the national inventory is updated with a 2-year periodicity). Data from the Northern Region's air quality monitoring network [76] for 2006 was used for the validation of BASE simulations. [45] recommends a group of statistical parameters for air quality models evaluation; from the proposed group, three parameters were selected for a quantitative error analysis: the correlation coefficient (r), the root mean square error (RMSE) and BIAS:

$$r = \frac{\sum_{i=1}^{n}\left(C_{obsi} - \overline{C}_{obs}\right)\left(C_{modi} - \overline{C}_{mod}\right)}{\sqrt{\sum_{i=1}^{n}\left(C_{obsi} - \overline{C}_{obs}\right)^2 \sum_{i=1}^{n}\left(C_{modi} - \overline{C}_{mod}\right)^2}} \tag{5}$$

$$RMSE = \sqrt{\frac{1}{n}\sum_{i=1}^{n}\left(C_{obsi} - C_{modi}\right)^2} \tag{6}$$

$$BIAS = \frac{1}{n}\sum_{i=1}^{n}(C_{obsi} - C_{modi}) \tag{7}$$

where: n is the total number of sample pairs, C_{obsi} is the observed value at time i and C_{modi} is the respective simulated concentration. These three parameters offer complementary information: the correlation factor (r) translates the linear relation between concentrations, reflecting a better or worst reproduction of physical and chemical atmospheric processes; RMSE and BIAS give an indication of the deviation between observed and simulated concentrations, either in absolute (RMSE) or in systematic terms (BIAS), allowing the inference of the magnitude and trend of the errors, respectively. For both the ideal value is zero.

Table 5 shows the statistical results for ozone and PM10, averaged over the air quality monitoring sites, already mentioned in §2.3 for the 1km resolution simulation. For ozone statistical parameters are given considering the entire year (from January to December) and considering only the summer months (April to September). These statistical parameters are within the range of those obtained with this and other air quality modelling systems [62, 77].

Air quality monitoring station	PM10			O_3		
	r	BIAS ($\mu g.m^{-3}$)	RMSE ($\mu g.m^{-3}$)	r	BIAS ($\mu g.m^{-3}$)	RMSE ($\mu g.m^{-3}$)
Espinho	0.41	-5.2	26.9	n.a	n.a	n.a
Baguim	n.a	n.a	n.a	0.66	-27.2	35.8
V.N.Telha	0.44	1.1	23.6	0.62	-20.1	31.8
Vermoim	0.63	1.1	23.2	0.62	-25.1	34.9
Custoias	n.a	n.a	n.a	0.64	-19.5	32.5
L.Balio	0.65	-16.9	17.9	0.65	-26.6	36.8
Matosinhos	0.58	-10.2	32.7	0.66	-32.2	41.1
Perafita	0.43	1.8	14.4	0.62	-16.4	30.6
Antas	0.56	-18.8	49.1	0.62	-24.7	36.4
S.Hora	0.49	-12.9	34.4	n.a	n.a.	n.a
Boavista	0.44	-6.3	24.7	0.62	-15.9	18.8
Ermesinde	0.61	-10.2	25	0.64	-21.8	36.4
AVERAGE	0.53	-7.7	27.2	0.64	-22.9	33.5

Table 5. CAMx statistical results obtained for O_3 and PM10.

In addition to the statistical analysis of the model performance, another possible and interesting exercise is the comparison of observed and simulated BASE concentrations in terms of the legislated values for O_3 and PM10. In this scope, Figure 23a presents the number of exceedances to the PM10 daily limit value (50 $\mu g.m^{-3}$, not to be exceed more than 35 days along the year, indicated by the red line) observed and BASE simulated; Figure 23b shows the number of annual exceedances to the ozone information threshold (180 $\mu g.m^{-3}$) observed and BASE simulated.

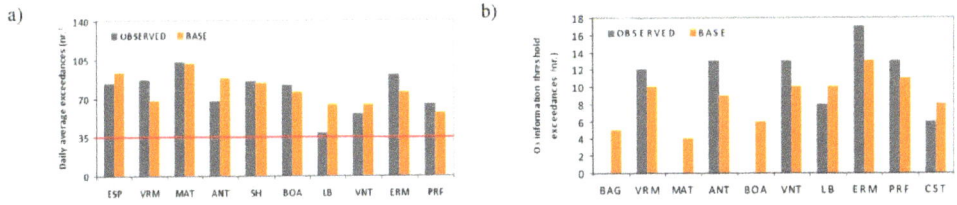

Figure 23. Observed and BASE a) number of exceedances to PM10 daily limit value, and b) number of exceedances to O₃ information threshold

Regarding the number of daily average PM10 exceedances the model, although the higher over-prediction at Antas, and Leça do Balio, and the under-prediction at Vermoim and Ermesinde, correctly identifies that all the air quality monitoring sites are not in compliance with the legislation. Model results point to exceedances to the ozone information threshold in Baguim, Matosinhos and Boavista, while these have not been observed; for the remaining air quality sites, the model presents a good agreement with observations.

5.2.2. Scenario simulations

For SPRAWL and COMPACT, simulations are performed with land use and emissions data produced according to the procedures previously described. Meteorological inputs are given by the respective MM5 annual simulation. Results from the two scenarios are analysed against the BASE simulation and against each other in order to identify the main differences between them.

Figure 24 presents the spatial distribution of PM10 annual average concentrations calculated for BASE, SPRAWL and COMPACT, highlighting the areas for which the legislated annual limit value (40 $\mu g.m^{-3}$) is exceeded. BASE and COMPACT present a larger area of high PM10 annual averages (> 40 $\mu g.m^{-3}$) over Porto municipality and its immediate surroundings, in comparison to SPRAWL. This is because the SPRAWL scenario implies a further decrease in Porto's population, and therefore emissions, and a consequent increase in neighbouring municipalities. The result is a decrease of emissions in Porto and therefore in pollutants concentrations. Nevertheless, considering the entire simulation domain, SPRAWL shows the highest PM10 annual concentrations (> 70 $\mu g.m^{-3}$), and larger areas above the annual limit value in Gondomar and Vila Nova de Gaia. The comparison between COMPACT and BASE suggests that the higher concentrations take place in exactly the same areas, with COMPACT revealing slightly higher concentrations (> 65$\mu g.m^{-3}$). This is due to the population concentration in already urbanized areas, with the consequent increase of emissions.

To better analyse the differences between the scenarios, the spatial distribution of the concentration differences are presented in Figure 25. Air quality monitoring stations are also represented for further analysis. Differences between annual averages from SPRAWL and BASE range from -15 to +24 $\mu g.m^{-3}$, with negative values mainly over Porto, as a result of the

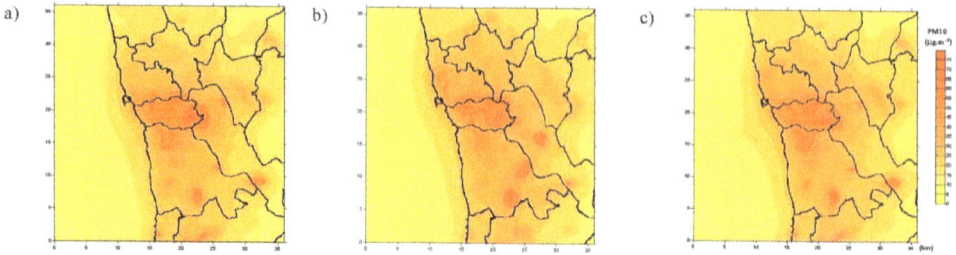

Figure 24. PM10 annual average for a) BASE, b) SPRAWL and c) COMPACT (the orange lines surround the areas for which the legislated annual limit value is exceeded).

decrease in emissions from traffic in this municipality. Higher positive differences are found over certain parts of the municipalities in the first metropolitan ring corresponding to areas of urban expansion. Differences between COMPACT and BASE range from -5 to +8 $\mu g.m^{-3}$, with higher positive differences over Matosinhos, in areas previously urbanized but with a greater population density in COMPACT. However, for the most part of the simulation domain differences are small.

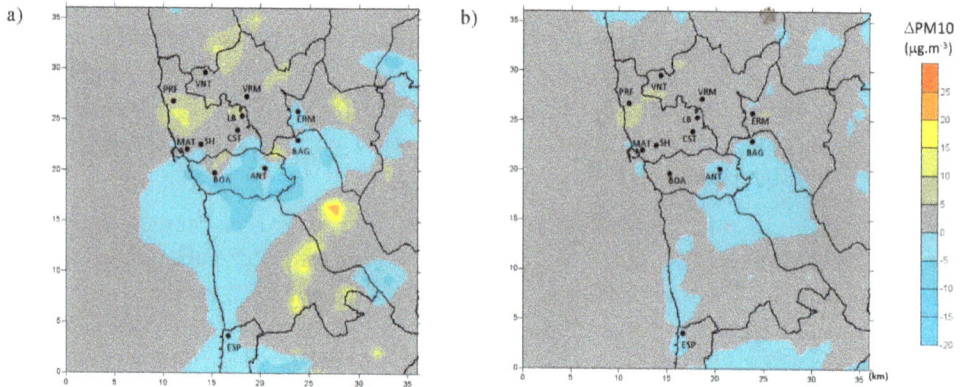

Figure 25. PM10 annual average differences between a) SPRAWL and BASE, and b) COMPACT and BASE.

Figure 26 presents the results for PM10 annual averages for BASE, SPRAWL and COMPACT for each air quality monitoring site located in the simulation domain.

For the majority of the air quality sites, SPRAWL presents the highest annual average of the three simulations. The results for Baguim, located in Gondomar are not representative of the municipality, with areas of increased PM10 concentrations, not captured by the air quality monitoring site. Also, sites which in BASE did not exceed the legislated annual average, such as Boavista and Leça do Balio, now exceed the limit with SPRAWL and COMPACT. Other sites which were already in non-compliance show a deterioration of their situation (such as Matosinhos and Senhora da Hora). In Antas, Baguim, and Ermesinde both scenarios improve the PM10 levels.

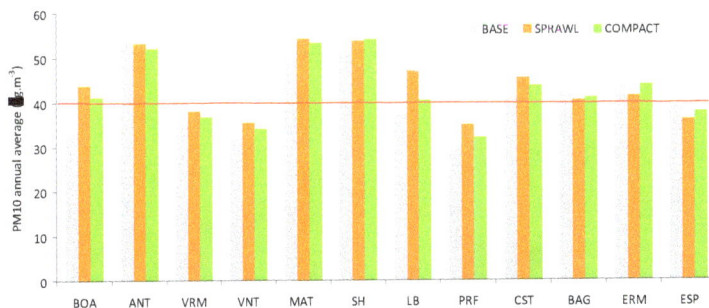

Figure 26. PM10 annual average for BASE, SPRAWL and COMPACT (the red line indicates the legislated annual limit value, 40μg.m⁻³), at the air quality monitoring sites.

Besides the obtained concentrations for each scenario it is also important to assess the number of individuals affected by high PM10 concentrations, since the population distribution across the study area is quite different for BASE, SPRAWL and COMPACT. Therefore, the maps of annual average concentrations (Figure 24) were crossed with population data per grid cell, to calculate the number of individuals affected by PM10 concentrations above the annual limit value. The results in terms of percentage of population (and not absolute since BASE has a lower population) are shown in Figure 27.

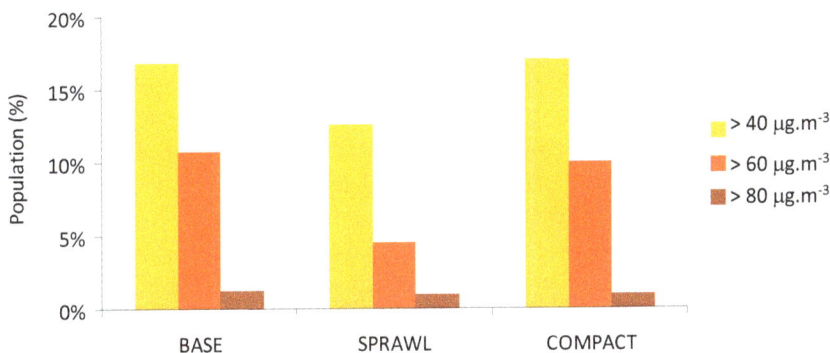

Figure 27. Population affected by PM10 concentrations above the annual limit value in BASE, SPRAWL and COMPACT.

COMPACT presents the greatest share of population affected by PM10 concentrations above 40 μg.m⁻³ (17%, corresponding to 370 000 inhabitants), while SPRAWL has the lowest number (12.5%, around 270 000 inhabitants). For the three considered concentration ranges, SPRAWL has the lowest share of people affected, while BASE and COMPACT show similar concentrations, although generally lower for BASE. Notwithstanding the existence of higher PM10 concentrations in SPRAWL, results indicate that the dispersion of the population along the study region withdraws people from the areas of higher concentrations. In turn, the COMPACT scenario places a greater part of the region's population in areas of highest PM10 levels.

The combination of increased temperatures (for SPRAWL) and different emissions (for both scenarios) produces the ozone concentration pattern changes displayed in Figure 28. The spatial distribution of the ozone summer (April to September) average concentration differences between BASE, SPRAWL and COMPACT are shown. Air quality monitoring stations location is also depicted for further analysis.

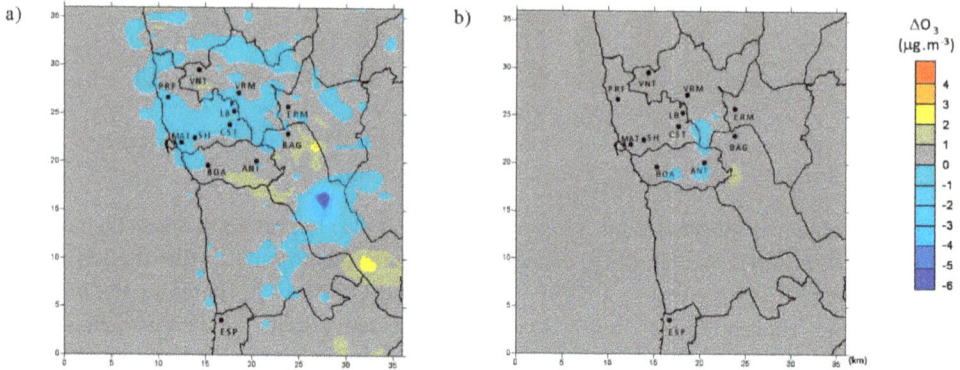

Figure 28. Ozone summer average differences between a) SPRAWL and BASE, and b) COMPACT and BASE.

The immediate analysis of the maps reveals that differences between the scenarios and BASE are much smaller than those obtained for PM10. Differences between SPRAWL and BASE range from -6 to +4 $\mu g.m^{-3}$, with negative values mostly found over Matosinhos, Maia and Gondomar (centre), in areas where the population expanded and emissions increased. In fact, comparing this map with the one for PM10 (Figure 25), negative differences for ozone are found in the areas of positive PM10 differences. Still regarding SPRAWL, ozone increases occur over Porto and part of Gondomar (N and S) in areas downwind the largest emission increase, such as Matosinhos, Maia and the centre of Gondomar municipality, as a result of air pollutants transport and consequent ozone formation. This is consistent with the prevailing NW wind direction in the region. Differences between COMPACT and BASE range from -1.5 to +2 $\mu g.m^{-3}$. Negative differences take place in Porto municipality as an outcome of the population densification in that area and the corresponding emissions increase, which lead to the local consumption of ozone. For both scenarios the largest part of the simulation domain presents very small positive differences, less than 1$\mu g.m^{-3}$, meaning that average concentrations are slightly higher in comparison to BASE.

Under the combined effects of increased urbanization and increased emissions, ozone decreases are not completely unexpected and have been found in previous research works [42, 44] This is probably due to the higher ozone removal by titration caused by higher anthropogenic emissions in an already emissions-dense region. Also, as investigated in [78], the non-linear response of ozone concentrations to changes in precursor emissions was found to increase with tonnage and emission density of the source region; this seems to be the case in the study region. According to the modelling study conducted by [79], the synergy among precursor's emission source categories may sometimes suppress O_3, acting as negative source contributions. These authors concluded that the full potential of each source category in O_3 formation (the pure contribution) is not achieved when emissions from the other source categories are accounted for.

Figure 29 presents the number of exceedances to the hourly ozone information threshold (180 µg.m^{-3}), obtained for BASE, SPRAWL and COMPACT. SPRAWL presents the lowest number of exceedances, except in Espinho where the three simulations produced similar results. COMPACT is the worst scenario, with more exceedances than BASE for Boavista, Vila Nova da Telha, Senhora da Hora and Perafita.

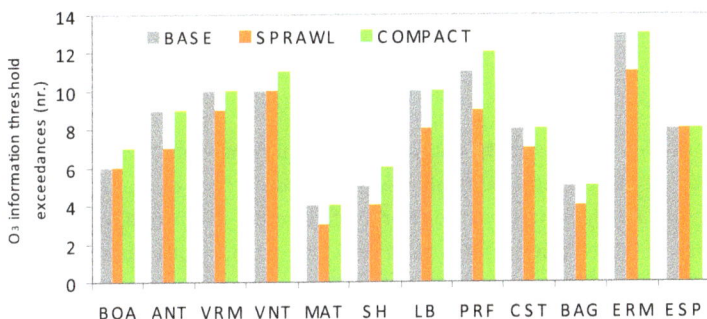

Figure 29. Number of exceedances to the ozone information threshold for BASE, SPRAWL and COMPACT.

The comparison of these results with the concentration patterns presented in Figure 27, reveals that there are no air quality sites in the areas of concentration increases, mainly for SPRAWL. However, if the same analysis is carried out for Gondomar in an area where no monitoring stations exist and for which higher positive differences are observed in the map of Figure 38, results are quite different: SPRAWL yields more exceedances (8) to the ozone information threshold in comparison with BASE (5) and COMPACT (6).

Regarding the number of persons affected by high ozone concentrations, the combination of the annual average concentrations maps with population data per grid cell, allows the determination of the number of individuals affected by ozone summer average concentrations above 70 µg.m^{-3}. This value was chosen because it is the concentration above which differences between the three situations are more substantial. The results are presented in Figure 30.

Once more, differences between scenarios and BASE are smaller than those observed for PM10. COMPACT presents the highest share of inhabitants affected by ozone summer average concentrations above 70 µg.m^{-3} (48.5%, corresponding to roughly 1 million people). However, looking at other concentration ranges the situation is different, since above 75 µg.m^{-3} BASE is the worst situation, with 21% of the population.

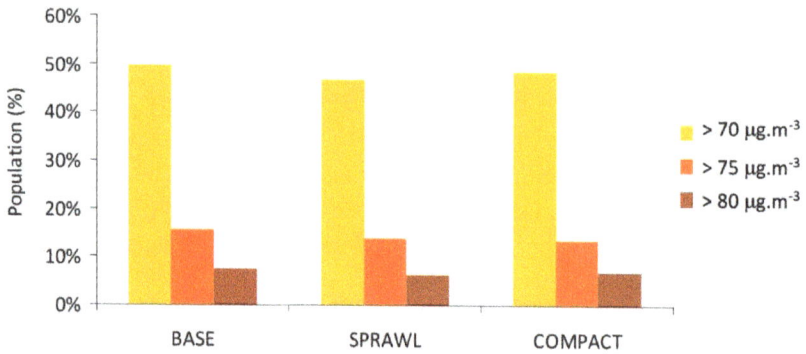

Figure 30. Population affected by ozone summer average concentrations above 70, 75 and 80 µg.m^{-3} in BASE, COMPACT and SPRAWL.

6. Main findings

The main aim of this study was to explore the relationship between the structure of the urban area and its air quality. Several research studies had demonstrated already that compact cities with mixed land uses are energetically more efficient and are responsible for lower emissions of atmospheric pollutants in comparison with sprawling cities. But a fundamental question remained unanswered: do compact cities promote a better air quality when compared to sprawling cities? And, given the ever-growing concentration of population in urban areas, do compact cities promote a healthier atmospheric environment? Given the signs provided by the energy and emissions aspects, the answers may seem obvious and straight forward but, as it was demonstrated along this study, they are not.

To answer these questions a strategy was drawn. The strategy, or approach, relied on the use of advanced atmospheric modelling tools for the evaluation of different urban development scenarios.

Aiming to assure a correct and complete analysis, a step-by-step methodology was defined and applied. First, it was necessary to characterize the current state of knowledge on the subject, including the genesis and growth of the problem, the tools available to tackle it, and gain insight from the studies previously conducted by several researchers on the field.

The selected working area is located in Portugal's Northern region, covering the Porto urban region, which is composed of a regional conglomerate of cities with a total population of over two million. Maps of land use and population parameters and an emission inventory were established for the situation as it is today (BASE). Moreover, two distinct future urban development scenarios - COMPACT and SPRAWL - were created, based on population and land use changes. The population of the study region was increased, to reflect a 20-years period, and differently distributed among municipalities according to each scenario. The land use patterns of the area were modified following a scenario of urban sprawl (SPRAWL) and maintained through the concentration of people in already existent urban areas

(COMPACT). New emissions were estimated for each scenario, taking into account population growth and land use changes.

The modelling system was then applied for SPRAWL and COMPACT, and also BASE, for a full-year simulation. The analysis of the meteorological results revealed that, owing to the land use changes in SPRAWL, the average temperature increased by 0.4°C. However local increases reaching 3°C were also detected; and some were even estimated in areas where land use changes were not implemented.

Regarding air quality, SPRAWL presented the highest PM10 concentrations, with an aggravation of the annual average values especially over areas of urban expansion and increasing emissions. Also, in the sites corresponding to the current monitoring stations, an increase in the number of exceedances to the daily limit value was found. For COMPACT slightly higher PM10 concentrations than BASE were estimated, due to the population increase in already urbanized areas, and consequent increase of emissions in those same areas.

For ozone, while the largest part of the domain had small concentration increases for both scenarios, smaller concentrations are found in areas where the population expanded and emissions increased, as a result of ozone titration by NO in the polluted atmosphere. Instead higher ozone levels are estimated for areas downwind the greatest emission increases, as a result of air pollutants transport and consequent ozone formation. Differences between scenarios and BASE were smaller than those found for PM10.

Finally, the population affected by higher PM10 and O_3 concentrations was determined for each scenario and for BASE. The analysis revealed that although the existence of higher PM10 concentrations in SPRAWL, the increase of the population density in COMPACT places a greater part of the inhabitants in areas of highest PM10 levels. This means that individually each inhabitant is exposed to lower PM10 concentrations in COMPACT, however, looking at the population as a whole, in terms of public health, the situation is inverted and SPRAWL presents a lower number of people affected by the highest concentrations. For ozone, results are not so clear, with BASE and COMPACT sharing the highest number of individuals affected, and SPRAWL clearly presenting the lowest number of total inhabitants affected by higher concentrations.

In conclusion, it seems clear that changes in land use patterns in urban areas lead to changes in meteorology, emissions, air quality, and population exposure. The signal of the change is evident: sprawling urban areas, when compared to contained urban development, are responsible by higher temperatures, higher emissions of pollutants to the atmosphere and higher atmospheric pollutants concentrations. However, compact urban developments imply a higher number of individuals exposed to the higher concentrations.

According to the review of the literature on the present subject, this was the first time a long term study was performed to analyse the impacts of urban growth, and consequent land use changes, on air quality, through the development of alternative urban development scenarios and the application of an air quality modelling system. Also, the methodology can be applied to any city or urban area for which the required data is available. However, the methodology presented here can be improved. Future work shall focus on the use of land use models for the simulation of land use changes, and traffic modelling to simulate the effect of land use changes on traffic volumes and their spatial distribution.

Along the next decade, it is expected that changes in the land use will take place. More likely, as revealed by the current trends, urban sprawl, the destruction of agricultural lands, and forestation and deforestation are expected to alter the landscape. These patterns will, in turn, lead to changes in population, energy consumption, traffic and anthropogenic and biogenic emissions. The results of this work suggest that changing land use patterns should be taken into consideration when using models to evaluate changes in quality levels (in particular ozone and PM10) stemming from various emissions reduction scenarios in urban areas.

Also, it is important to note that, such as technology alone has not been able to tackle the air quality problems, more compact urban development patterns alone will not be sufficient to fully address urban air quality problems. Technological advances in emissions control have proven to be highly effective in reducing emissions over the last decades, and emerging technologies, such as hybrid or electric vehicles and alternative fuels, are expected to continue these reductions. The importance of land use-oriented approaches to air quality management lies in the potential for these strategies to limit the dramatic growth in traffic, which has greatly diluted the benefits of technological improvements so far, and also in addressing the local meteorological drivers of air pollution, such as temperature.

In the years to come, cities will continue to be the main centres of economic activity, innovation and culture. Therefore, managing the urban environment and the quality of life of its inhabitants goes well beyond the concern for the well-being of the urban population, affecting instead the well-being of humanity as a whole.This work presents an achievable approach to urban sustainable development supporting the object8ive of the UN Conference Rio+20.

Author details

Helena Martins*, Ana Miranda and Carlos Borrego
CESAM & Department of Environment and Planning, University of Aveiro, Aveiro, Portugal

Acknowledgement

The authors acknowledge the Foundation for Science and Technology for the financing of the post-doc grant of H. Martins (SFRH/BPD/66874/2009).

---

* Corresponding Author

7. References

[1] UN-United Nations, 2004. World urbanization prospects: The 2003 revision. UN Department of Economic and Social Affairs, New York.

[2] EEA – European Environment Agency, 2006. Urban sprawl in Europe – the ignored challenge. EEA Report No. 10/2006. European Environment Agency, Copenhagen.

[3] Williams, K., Burton, E., Jenks, M. (Eds.), 2000. Achieving Sustainable Urban Form. E & FN Spon, London, 388p.

[4] Kasanko, M., Barredo, J.I., Lavalle, C., McCormick, N., Demicheli, L., Sagris, V. Brezger, A., 2006. Are European cities becoming dispersed? A comparative analysis of fifteen European urban areas. Landscape and Urban Planning 77, 111–130.

[5] Kühn, M., 2003. Greenbelt and Green Heart: separating and integrating landscapes in European city regions. *Landscape and Urban Planning* 64, 19–27.

[6] Schoffman, E., Vale,B., 1996. How compact is sustainable - how sustainable is compact?, in The compact city a sustainable urban form? Jenks, M., Burton, E., Williams, K. (Eds.). Taylor and Francis Group, Oxford, 350p.

[7] Ewing, R., Pendall, R., Chen, D., 2003. Measuring sprawl and its transportation impacts. *Transportation Research Record* 1831, 175–183.

[8] Stone, B., Bullen, J.L., 2006. Urban form and watershed management: how zoning influences residential stormwater volumes. *Environment and Planning B: Planning and Design* 33(1), 21 – 37.

[9] Heimlich, R., Anderson, W., 2001. Development at the urban fringe and beyond: Impacts on agriculture and rural land. Agriculture Economic Report (AER803). United States Department of Agriculture.

[10] Frumkin, H., Frank, L., Jackson, R., 2004. Urban Sprawl and Public Health: Designing, Planning, and Building for Healthy Communities. Island Press,Washington, DC.

[11] ECOTEC, 1993. Reducing Transport Emissions Through Planning. HMSO, London.

[12] Jenks, M., Burton, E., Williams, K. (Eds.), 1996. The compact city a sustainable urban form? Taylor and Francis Group, Oxford, 350 p.

[13] Breheny, M.1992. The contradictions of the compact city: A review, in: Breheny, M. (Ed.), Sustainable Development and Urban Form. Pion, London, pp. 138-159.

[14] Barret, G., 1996. The transport dimension, in: Jenks, M., Burton, E., Williams, K. (Eds.), The compact city a sustainable urban form? E & FN Spon, London, pp.171-180.

[15] Neuman, M., 2005. The compact city fallacy. *Journal of Planning Education and Research* 25, 11 – 26.

[16] Catalán, B., Saurí, D. , Serra, P., 2008. Urban sprawl in the Mediterranean? Patterns of growth and change in the Barcelona Metropolitan Region 1993–2000. Landscape and Urban Planning 85, 174-184.

[17] PBL — Netherlands Environmental Assessment Agency, 2008. Urbanisation dynamics and quality of place in Europe, EURBANIS report 1. Planbureau voor de Leefomgeving (NEAA), Bilthoven, the Netherlands.

[18] Alley, E.R., Stevens, L.B., Cleland, W.L., 1998. Air Quality Control Handbook. McGrawHill.

[19] WHO - World Health Organization, 1999. Monitoring Ambient Air Quality for Health Impact Assessment. World Health Organization Regional Publications, European Series, No.85, Copenhagen.

[20] Burnett, R.T., Brook, J., Dann, T., Delocla, C., Philips, O., Cakmak, S., Vincent, R., Goldberg, M.S., Krewski, D., 2000. Association between particulate- and gas-phase components of urban air pollution and daily mortality in eight Canadian cities. *Inhalation Toxicology* 12 (Suppl. 4), 15–39.

[21] Jacob, D.J., Winner, D.A., 2009. Effect of climate change on air quality. *Atmospheric Environment* 43, 51-63.

[22] Jacobson, M.Z. 2002. Atmospheric Pollution: History, Science and Regulation. Cambridge University Press, Cambridge, 375 p.

[23] Seinfeld J.H., Pandis S.N., 1998. Atmospheric chemistry and physics – From air pollution to climate change. ISBN 0-471-17816-0. John Wiley & Sons, inc. Wiley Interscience.

[24] Derwent, R.G., 1999. Atmospheric Chemistry. In Air Pollution and Health, Holgate, S., Samet, J., Koren, H., Maynard, R. (Ed.) London Academic Press, 51-62.

[25] EEA, 2011. Air quality in Europe – 2011 Report. EEA Technical report 12/2011. EEA, Copenhagen.84pp

[26] EEA, 2010. The European environment state and outlook 2010, European Environment Agency (http://eea.europa.eu/soer) accessed 27 July 2011.

[27] WCED - World Commission on Environment and Development, 1987. Our Common Future, Oxford University Press, Oxford.

[28] Newman, P., Kenworthy, J.R., 1989b. Cities and automobile dependence: An international sourcebook. Gower, England.

[29] Newman, P., 1992. The compact city – an Australian perspective. *Built Environment* 18, 285-300.

[30] Kenworthy, J., Laube, F., 1996. Automobile dependence in cities: an international comparison of urban transport and land use patterns with implications for sustainability. *Environmental Impact Assessment Review* 16, 279-308.

[31] Gomez-Ibanez, 1991. J.A. Gomez-Ibanez, A Global View of Automobile Dependence (book review). *Journal of the American Planning Association* 57, 376–379.

[32] Cervero, R., 2000. Transport and Land Use: Key Issues in Metropolitan Planning and Smart Growth. UCTC Report #436, University of California Transportation Center, Berkeley, CA. Available from http://www.uctc.net/papers/436.pdf

[33] Crane, R., 2000. The influence of urban form on travel: an interpretative review. *Journal of Planning Literature* 15 (1), 3-23.

[34] Irving, P., Moncrieff, I., 2004. Managing the environmental impacts of land transport: integrating environmental analysis with urban planning. Science of the Total Environment 334-335, 47-59.

[35] Young, W., Bowyer, D., 1996. Modelling the environmental impact of changes in urban structure. Computers, Environment and Urban Systems 20 (4/5), 313-326.

[36] Emison, G., 2001. The relationships of sprawl and ozone air quality in United States' metropolitan areas. *Regional Environmental Change* 2, 118-127.

[37] Marshall, J.D., McKoneb, T.E., Deakind, E., Nazaroff, W.W., 2005. Inhalation of motor vehicle emissions: effects of urban population and land area. *Atmospheric Environment* 39, 283–295.

[38] Stone, B., 2008. Urban sprawl and air quality in large US cities. *Journal of Environmental Management* 86 (4), 688-698.

[39] Marquez, L., Smith, N., 1999. A framework for linking urban form and air quality. Environmental Modelling and Software 14 (6), 541-548

[40] Borrego, C., Martins, H., Tchepel, O., Salmim, L., Monteiro, A., Miranda, A.I., 2006. How urban structure can affect city sustainability from an air quality perspective. Environmental Modelling and Software 21, 461-467.

[41] Ferreira, J., Martins, H., Miranda, A.I., Borrego, C., 2005. Population Exposure to Atmospheric Pollutants: the influence of Urban Structure. In 1st International Conference Environmental Exposure and Health, Atlanta, EUA, 5-7 October 2005 - Environmental Exposure and Health. Eds. M.M. Aral, C.A. Brebbia, M.L. Masila and T. Sinks. WITPress, p. 13-22.

[42] Civerolo, K., Hogrefe, C., Lynn, B., Rosenthale, J., Ku, J-Y, Solecki, W., Cox, J., Small, C., Rosenzweig, C., Goldberg, R., Knowlton, K., Kinney, P., 2007. Estimating the effects of increased urbanization on surface meteorology and ozone concentrations in the New York City metropolitan region. Atmospheric Environment 41, 1803-1818.

[43] De Ridder, K., Lefebre, F., Adriaensen, S., Arnold, U., Beckroege, W., Bronner, C., Damsgaard, O., Dostal, I., Dufek, J., Hirsch, J., IntPanis, L., Kotek, Z., Ramadier, T., Thierry, A., Vermoote, S., Wania, A., Weber, C., 2008a. Simulating the impact of urban sprawl on air quality and population exposure in the German Ruhr area. Part I: Reproducing the base state. Atmospheric Environment 42 (30), 7059-7069.

[44] De Ridder, K., Lefebre, F., Adriaensen, S., Arnold, U., Beckroege, W., Bronner, C., Damsgaard, O., Dostal, I., Dufek, J., Hirsch, J., IntPanis, L., Kotek, Z., Ramadier, T., Thierry, A., Vermoote, S., Wania, A., Weber, C., 2008b. Simulating the impact of urban sprawl on air quality and population exposure in the German Ruhr area. Part II: Development and evaluation of an urban growth scenario. Atmospheric Environment 42 (30), 7070-7077

[45] Borrego, C., Miranda, A.I, Sousa, S., Carvalho, A., Sá, E., Martins, H., Valente, J., Varum C., Jorge, S., 2008a. Planos e Programas para a Melhoria da Qualidade do Ar na Região Norte - Uma visão para o período 2001-2006. Departamento de Ambiente e Ordenamento da Universidade de Aveiro, Portugal.

[46] EEA– European Environment Agency, 2000. CORINE land cover technical guide – Addendum 2000. M. Bossard, J. Feranec and J. Otahel Technical report No 40. European Environment Agency, Copenhagen.

[47] INE – Instituto Nacional de Estatística, 2003. Movimentos pendulares e organização do território metropolitano – Área Metropolitana de Lisboa e Área Metropolitana do Porto: 1991/2001. Instituto Nacional de Estatística, Lisboa.

[48] IA - Instituto do Ambiente , 2001. Delimitação de Zonas e Aglomerações para Avaliação da Qualidade do Ar em Portugal, Instituto do Ambiente, Lisboa.

[49] Borrego, C., Monteiro, A., Ferreira, J., Miranda, A.I., Costa, A.M., Carvalho, A.C. , Lopes, M., 2008b. Procedures for estimation of modelling uncertainty in air quality assessment. Environment International 34 (5), 613-620. DOI:10.1016/j.envint. 2007.12.005.

[50] Dudhia, J., 1993. A nonhydrostatic version of the Penn State/NCAR mesoscale model: Validation tests and simulation of an Atlantic cyclone and clod front. Monthly Weather Review 121, 1493-1513.

[51] Vautard, R., Beekmann, M., Bessagnet, B., Blond, N., Hodzic, A., Honoré, C., Malherbe, L., Menut, L., Rouil, L., Roux, J., 2004. The use of MM5 for operational ozone/NOx/aerosols prediction in Europe: strengths and weaknesses of MM5. Paper 5th WRF /14th MM5, Users' Workshop NCAR, June 22-25, 2004.

[52] Dudhia, J., Gill, D., Manning, K., Wang, W., Bruyere, C., 2005. PSU/NCAR Mesoscale Modelling System Tutorial Class Notes and Users' Guide (MM5 Modelling System Version 3) (updated for MM5 Modelling System Version 3.7 - Released January 2005).

[53] Mao, Q., Gautney, L.L., Cook, T.M., Jacobs, M.E., Smith, S.N., Kelsoe, J.J., 2006. Numerical experiments on MM5-CMAQ sensitivity to various PBL schemes,. Atmospheric Environment 40, 3092–3110.

[54] Aquilina, N., Dudek, A.V., Carvalho, A., Borrego, C., Nordeng, T.E., 2005. MM5 high resolution simulations over Lisbon. Geophysical Research Abstracts. SRef-ID: 1607-792/gra/EGU05-a-08685. European Geosciences Union. 7:08685.

[55] University Corporation for Atmospheric Research: http://dss.ucar.edu/datasets/ds083.2/data

[56] Martins, H., 2009. Exploring the links between urban structure and air quality. PhD Thesis. University of Aveiro. (available at: http://biblioteca.sinbad.ua.pt/Teses/2010000010)

[57] Morris, R.E., Yarwood, G., Emery, C., Koo., B., 2004. Development and Application of the CAMx Regional One-Atmosphere Model to Treat Ozone, Particulate Matter, Visibility, Air Toxics and Mercury. Presented at 97th Annual Conference and Exhibition of the A&WMA, June 2004, Indianapolis

[58] ENVIRON, 2008. CAMx v4.5.1 User's Guide. ENVIRON International Corporation, Novato, California, EUA. June 2008.

[59] Ferreira, J., Carvalho, A., Carvalho, A.C., Monteiro, A., Martins, H., Miranda, A.I., Borrego, C., 2003. Chemical Mechanisms in two photochemical modelling systems: a comparison procedure. In Int. Tech. Meeting of NATO-CCMS on "Air Pollution Modelling and its Application", 26th, Istanbul, Turkey, 26-30 May 2003 - Air Pollution Modelling and its Application XVI, Eds Carlos Borrego and Selahattin Incecik, Kluwer Academic/ Plenum Publishers, New York, p. 87-96.

[60] Hauglustaine, D. A., Hourdin, F., Walters, S., Jourdain, L., Filiberti, M., Larmarque, J., Holland, E., 2004. Interactive chemistry in the Laboratoire de Météorologie Dynamique general circulation model : description and background tropospheric chemistry evaluation. Journal of Geophysical Research 109, D04314, doi:10.1029/2003JD003957

[61] Ginoux, P., Chin, M ., Tegen, I., Prospero, J. M., Holben, B., Dubovik, O., Lin., S. J., 2001. Sources and distributions of dust aerosols simulated with the GOCART model. Journal of Geophysical Research 106, 20255-20274

[62] Monteiro, A., Miranda, A.I., Borrego, C., Vautard, R., 2007. Air quality assessment for Portugal. Science of the Total Environment 373, 22-31.

[63] Webster, M., Nam, J., Kimura, Y., Jeffries, H., Vizuete, W., Allen, D.T., 2007. The effect of variability in industrial emissions on ozone formation in Houston, Texas. *Atmospheric Environment* 41, 9580-9593.

[64] Agência Portuguesa do Ambiente – Inventário 2005
http://www.apambiente.pt/politicasambiente/Ar/InventarioNacional/Documents/Aloca cao_Espacial_Emissoes_2005.xls

[65] Ferreira J., Martins, H., Monteiro, A., Miranda, A. I., Borrego, C., 2006. Air quality modelling application to evaluate effects of PM air concentrations on urban population exposure. Epidemiology. 17, 6, S252-S253.

[66] Monteiro, A., Borrego, C., Tchepel, O., Santos, P., Miranda, A.I., 2001. Inventário de Emissões Atmosféricas – base de dados POLAR2. Aplicação à modelação atmosférica. In: Actas da 7ª Conferência Nacional sobre a Qualidade do Ambiente, 18-20 Abril, Universidade de Aveiro, Aveiro, Portugal, 954-958.

[67] Eyth, A.M., Habisak, K., 2003. The MIMS spatial allocator: a tool for generating emission surrogates without a geographic information system, in: 12th International Emission Inventory Conference – Emission Inventories – Applying New Technologies San Diego, April 29–May 1, 2003. Available from:<http://www. epa.gov/ttn/chief/conference/ei12/modeling/eyth.pdf>.

[68] Maes, J., Vliegen, J., Vel, K.V., Janssen, S., Deutsch, F., Ridder, K., Mensink, C., 2009. Spatial surrogates for the disaggregation of CORINAIR emission inventories. Atmospheric Environment 43, 1246–1254

[69] Gallego, J., Peedell, S., 2001. Using CORINE Land Cover to Map Population Density. Towards Agri-Environmental Indicators. Topic Report 6/2001. EEA, Copenhagen. Available at: <http://reports.eea.europa.eu/topic_report_2001_06/en/Topic_6_2001.pdf>, pp. 92–103.

[70] Instituto Nacional de Estatística: http://www.ine.pt/

[71] Tchepel, O., 1997. Application of Geographical Information Systems to Mesoscale Atmospheric Pollution Modelling. MsC Thesis. University of Aveiro, Portugal.

[72] Carvalho, A.C., Carvalho, A., Gelpi, I., Barreiro, M., Borrego, C., Miranda, A.I., Perez-Munuzuri, V., 2006. Influence of topography and land use on pollutants dispersion in the Atlantic coast of Iberian Peninsula. *Atmospheric Environment* 40, 3969-3982.

[73] Trigo, R.M., Pereira, J.M.C., Pereira, M.G., Mota, B., Calado, T.J., DaCamara, C., Santo, F.E., 2006. Atmospheric conditions associated with the exceptional fire season of 2003 in Portugal. International Journal of Climatology 26, 1741-1757

[74] Viegas, D.X., Abrantes, T., Palheiro, P., Santo, F.E., Viegas, M.T., Silva, J., Pessanha, L., 2006. Fire weather during the 2003, 2004 and 2005 fire seasons in Portugal. In V International Conference on Forest Fire Research. Ed D.X. Viegas, Figueira-da-Foz, 2006. Proceedings in CD.

[75] Keyser, D., Anthes, R.A., 1977. The applicability of a mixed-layer model of the planetary boundary layer to real-data forecasting. Monthly Weather Review 105, 1351-1371.

[76] Base de dados on-line sobre a qualidade do ar : http://www.qualar.org

[77] Vautard, R., Builtjes, P., Thunis, P., Cuvelier, K., Bedogni, M., Bessagnet, B., Honoré, C., Moussiopoulos, N., Pirovano G., Schaap, M., Stern, R., Tarrason, L., Van Loon, M., 2007. Evaluation and intercomparison of ozone and PM10 simulations by several chemistry-transport models over 4 European cities within the City-Delta project, Atmospheric Environment 41, 173-188.

[78] Cohan, D.S., Hakami, A., Hu, Y., Russell, A.G., 2005. Nonlinear response of ozone to emissions: source apportionment and sensitivity analysis. Environmental Science and Technology 39, 6739-6748.

[79] Tao, T, Larson, S. M., Williams, A., Caughey, M., Wuebbles, D. J., 2005. Area, mobile, and point source contributions to ground level ozone: a summer simulation across the continental USA. *Atmospheric Environment* 39, 1869-1877.

Chapter 4

Pollutants and Greenhouse Gases Emissions Produced by Tourism Life Cycle: Possible Solutions to Reduce Emissions and to Introduce Adaptation Measures

Francisco A. Serrano-Bernardo, Luigi Bruzzi,
Enrique H. Toscano and José L. Rosúa-Campos

Additional information is available at the end of the chapter

1. Introduction

Tourism and Travel (T & T) is a vital contributor to the global economy and considered particularly important for developing countries. It is regarded as an effective way of redistributing wealth and, if conducted according to sustainability directions, may promote cultural heritage conservation and contribute to nature preservation. Tourism industry has experienced a significant development in the last 50 years and, presently, represents around 260 million jobs worldwide, 100 million of whom work directly in the tourism industry and the rest in induced activities. Moreover, tourism accounts for about 9% (direct & induced) of the global GDP [1,2] (more than the automotive industry, 8.5% and slightly less than the banking sector, 11%). The economic and cultural importance of tourism is now widely recognized.

However, negative impacts from tourism may take place, for instance, when the level of visitor use is greater than the capability of the environment to cope with this use, operating beyond the acceptable limits of change or regeneration capacity of a given territory, e.g., by the sheer effect of the number of visitors [3]. A good example is constituted by the Mediterranean coast, where in a narrow strip (50-100 km) about 130 millions of residential habitants are incremented seasonally by about 100 million of tourists. In marine areas tourist activities such as diving or cruising, may cause damage of fragile ecosystems such as coral reefs, which are also affected by CO_2-emissions due to the change in the pH-value of seawater (coral bleaching).

Waste handling and disposal, increment of noise (related mainly to transportation to and at destinations), increased use of water resources [4], loss of biodiversity and wild life habitats by tourism leisure activities, represent part of the stresses put on visited areas, beside the pressure on local resources like energy, food, and other raw materials that might be locally already in short supply.

One of the most negative impacts of tourism is on climate through so-called **Greenhouse Gases (GHG)** emissions, in particular CO_2 [3,5]. In fact, it is now widely recognized that climate change is a global issue and one of the most serious threats to society, the economy and the environment, being by now for decades a constant issue of concern [6]. The Inter-Governmental Panel on Climate Change (IPCC) has reported that warming of the global climate system is unequivocal and that it is likely that anthropogenic GHG production (mainly from energy conversion) have caused most of the observed global temperature rise since the middle of the 19th Century. Hence, ambitious emissions reduction targets for developed countries and an effective framework that addresses the needs of developing countries has been already adopted (e.g., the objectives 20/20/20 in the European Energy Program for Recovery).

Relating these two important issues, it is now recognized that T & T constitutes also a vector of climate change since, according to current estimations, tourism accounts for approximately five per cent of global carbon dioxide emissions, establishing in this way the synergy between T & T and climate, which – on the other side - may define the length and quality of tourism seasons, affect tourism operations, hence attracting or deterring visitors depending on climate conditions. It can be, then, asserted that tourism is a highly climate sensitive economical sector, being of paramount importance the assessment of the possible influence of tourism on climate change through emissions and on environment in general through its implementation.

The principal environmental impact of GHG emissions is climate change but many secondary effects which affect, for instance, coastal areas, have been identified. These are sea surface temperature and sea level rise, changes in temperature and precipitation, as well as biodiversity loss mainly in the marine environment. These changes threaten the quality of destinations, which is at the core of the tourism product. It therefore makes sense for stakeholders in tourism and tourism mobility, not only environmentally but also from the point of view of business, to act more sustainably.

In the tourism sector, energy consumption at destinations and the related GHG emissions strongly depends, e.g., on the infrastructure of the accommodation, particularly installations for heating, cooling and hot water [7]. On the other hand, by definition tourism is impossible without transportation. At destinations the impact of the GHG emissions can be challenged by improving new concepts and/or changing existing infrastructure.

On the other side, for tourism mobility, the type of transport and the distance to be covered determine the amount of energy consumed and, consequently, the emissions generated. Transport to and on destinations represents a high percentage of energy consumption (currently about 30%), and a large fraction of it is represented by travels for tourism. If we consider that almost all transport vehicles are fuelled by liquid fuels, such as diesel oil, kerosene and gasoline, it appears clear that travels are certainly responsible of large quantities of greenhouse gases emitted into the atmosphere. Hence, for transportation the number of journeys, the distribution over transport modes, total passenger kilometrage travelled, the efficiency of transport means, etc., are the most important parameters to be taken into consideration in assessing emissions.

In this chapter the focus is concentrated on the significant contribution to the emissions of pollutants and greenhouse gases both in the destination and in travels needed to reach the destination and around the destination itself. The approach followed in this analysis is based on the *Life Cycle Assessment* [8], including the effects produced in reaching the destination, the staying in the destination for a certain number of days and travel back to the starting point. The quantification of emissions is performed for different distances between the starting place and the destination, for different period of staying in the destination and for different means of transportation (car, bus, train, ship and airplane). In addition, *Tourism Indicators* [9] are introduced to establish the sustainability level of tourism. Finally, founded on studies and research on the effects produced by pollutants and GHG emissions into the atmosphere, changes that T & T should undergo to improve its sustainability are proposed.

2. Tourism market

Motivations for tourism are multiple; they include: travel, leisure, business, cultural, educational and/or religious purposes. Religion and culture have been key stimulants for many tourist destinations; religious travel has been popular for decades and has allowed for scores of people to take pilgrimages. For example, many Roman Catholics visit the Vatican City annually, while Hindus trek to the Ganges and other spiritual spots across India. Jerusalem and Israel are also popular spots for Christian pilgrims, as well as the Mecca for the Muslims. Due to these reasons, the countries benefit from tourist arrivals and, in many cases, also neighboring countries or cities. Different nations have various histories and unique cultures and traditions that accompany them. The cultural distinctiveness and 'unusual' traditions attract curious or interested travelers to certain places. In addition, if one visits a country for other reasons, the cultural aspects contribute toward a unique experience. The various art forms (song, dance, sculpture and artwork, drama, opera, etc.) and festivities have great influence on visitors' experiences. Often overlapping with other types of tourism, food tourism plays an important part in the industry. Visitors generally prefer sampling local cuisine and many set off on trips to experience food made by locals. As a result, local restaurants and food stands thrive off this. Locals benefit from employment opportunities either directly (servers, cooks and managers for example) or indirectly (e.g., agriculturists and aquaculturists) and the economy of surrounding communities, are boosted.

The economic contribution of tourism has two elements: direct and indirect. The direct contribution is solely concerned with the immediate effect of expenditure made by visitors.

This, for example, accounts for expenditures in hotels, restaurants, souvenir shops, transport services and attractions entrance fees. Indirect contributions are often underestimated: they include, for example, expenditures on fuels for transport and power generation, utility bills for hotels and guest houses (to maintain the electricity and water supply), purchase or rental of equipment for various activities (such as diving, hiking and beach sports), among others.

Tourism in its various forms is currently recognized as the world's largest single industry with a direct worldwide contribution to GDP of about 6% [10]. A world forecasts for the near future estimates a number of international arrivals by the year 2020 of about 1.6 billion. *Figure 1* shows the contribution to the global tourism activities of different geographical areas together with the forecasted increase up to 2020.

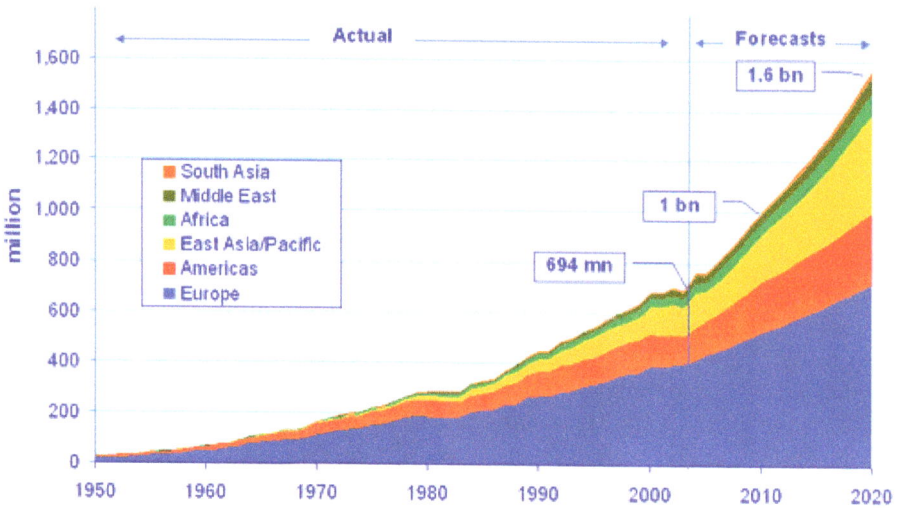

Figure 1. Evolution of world tourism from 1950, in billions of arrivals. Source: [11]

Tourism is one of the strongest economic sectors in the member states of the European Union (EU), where it involves around 2 million businesses (mostly small and medium-sized enterprises) generating up to 12% of the GDP (directly plus indirectly), 6% of employment (directly) and 30% of external trade. All of these figures are expected to further increase as tourism demand is expected to experience a substantial growth. An analysis of changes in tourism in the EU over the past 20 years shows that the numbers of bed-places and overnight stays have increased by almost 64%.

An important sector of tourism is the coastal [12], based on a unique resource combination of the appealing of landscape and sea environment: sun, water, beaches, outstanding scenic views, rich biological diversity (birds, whales, corals, etc.), sea food and good transportation infrastructure. In the middle of the 20th century coastal tourism in Europe turned into mass tourism and became affordable for an increasing portion of the population. Today, more than 60% of the European tourists favor the seashore for vacation. Coastal tourism sector in

Europe is getting increasingly competitive, with tourists expecting increasing quality for lower price [13]. Nowadays tourists expect more than sun, sea and sand, demanding a wide variety of associated leisure activities and experiences, including sports, cuisine, culture and natural attractions. Tourism is becoming more and more important for the economy of the communities at the destinations; it is also a strong employment generator with a total of almost 20 million jobs (direct and indirect employment, in Europe).

Mass tourism is the most common aspect of the industry. It is an assembly of standardized low cost tourism packages appealing to tourism masses traveling to popular geographical areas. It involves tourists on pilgrimages or visiting places of religious interest, tourists visiting beaches or coastal areas, visitors to popular nightlife and casino areas, tourists to popularized landmarks or structural wonders and tourists seeking shopping and leisure in internationally hyped locations.

The growth in world tourism is related to three main factors: increased personal incomes and leisure time; improvements in transportation systems and greater public awareness of other areas of the world due to improved communications. Many destinations have a wealth of assets to give them a distinctive appeal: combinations of activities (leisure activities, sports, cultural and natural heritage, cuisine, etc.); at the same time, local people are increasingly anxious to preserve their own identity, their environment and their natural, historic and cultural heritage, from the impact of unrestrained tourism. In this context, it has been acknowledged that the global tourism industry is a "massive consumer of energy and resources" and, since it is expected to continue to grow significantly in the future, the question of its sustainability has been recognized.

3. Tourism sustainability

3.1. Need for tourism to be sustainable

In the last two decades there has been growing recognition of the importance for tourism to be sustainable. Tourism is one of the oldest industries, it has become integrated into everyday life for many countries and, as discussed in the previous paragraph, is undeniably a major contributor to economic and social development. However, increasing tourist pressure and overexploitation of natural resources endangers the existence of this industry in many countries. In fact, one of the most diffused types, mass tourism, often leads to severe degradation of natural landscapes (e.g., through construction of massive infrastructure), pollution of coastal zones and reduction in water supply. Ecological development with respect to tourism is known as sustainable tourism and it encompasses the development of an industry in such a manner that can sustain itself while improving the quality of life for all concerned stakeholders, such as indigenous populations. Hence, sustainable tourism entails the search for a more productive and harmonious relationship between the visitor, the host community and the residents [14]. In addition, tourism to be sustainable must remain competitive and attract first time as well as reiterate visitors. In this context, recognize, accept and implement limits on tourism development is one way to counteract the potential overuse and exploitation of destinations natural resources and cultural heritage.

Sustainable tourism (ST) can be, hence, defined as an industry which attempts to make a low impact on the environment and local culture, while aiming to generate income, employment, and the conservation of the local environment. ST should be both, ecologically and culturally sensitive, producing minimum impact on the eco-system and culture of the host community. According to The World Tourism Organization, sustainable tourism is a sort of tourism that leads to the management of all resources in such a way that economic, social and aesthetic needs can be fulfilled while maintaining cultural integrity, essential ecological processes, biological diversity and life support systems. The United Nations Environment Programme (UNEP) refers to the environmental, economic, and socio-cultural aspects of tourism development, and recommends a suitable balance between these three dimensions to guarantee its long-term sustainability [15]. The United Nations World Tourism Organization (WTO) defines ST as an activity that meets the needs of present tourists and host regions while protecting and enhancing opportunities for the future. The objective of ST is, hence, to retain the economic and social advantages of tourism development while reducing or mitigating any undesirable impacts on the natural, historic, cultural or social environment. This can be achieved by balancing the needs of tourists with those of the destination. Summarizing, ST is a tourism that is economically, socio-culturally and environmentally sustainable, with impacts are neither permanent nor irreversible.

3.2. Difficulties and opportunities in making tourism more sustainable

An increasing number of tourists are aware of the environmental impacts that tourism may cause particularly in Europe. They expect a high environmental quality in their destination, usually prefer eco-labeled accommodation services, look for certified products in the travel catalogues and "green" destinations. The direct local impacts of tourism on people and environment at destinations are strongly affected by concentration in space and time (seasonality). There are different quality characteristics requested by tourists, such as clean beaches and water, cleanness in the resorts and in the surrounding areas, reduce urbanization of rural areas, nature protection in the destination, low noise pollution from traffic or discothèques, reduce traffic and good public transport in the destination, possibility of reaching the destination easily by bus or train, environmentally-friendly accommodation, etc. Construction of hotels, recreation and other facilities often leads to increased pressure on sewage facilities, in particular because many destinations have several times more inhabitants in the high season than in the low season. Waste water treatment facilities are often not built to cope with the dramatic rise in volume of waste water during the peak [13]. In some locations, conventional tourism has been accused of failing to integrate its structures with the natural features and indigenous architecture of the destination. One of the most difficult challenges tourism is facing is the ability to combine sound economic development with the protection of natural resources. There will be an increasing need to analyze the trade-offs between native cultural integrity and the benefits of employment, and the need to understand the impact of rapid climatic changes on prime vacations sites, such as coast lines. Nevertheless, looking at the whole picture it can recognize that tourism can help sustainability. In fact, tourism can facilitate the restoration, conservation and protection of physical environments; it can provide the incentives and the income necessary to restore and rejuvenate historic buildings and to create and maintain national parks. Hence, tourism can be a force for the development of better infrastructure such as improvements to roads, water supply and treatment and waste management systems which can improve environmental quality, facilitating the development of attractions through restoration and protection of natural and built heritage.

3.3. Mobility and sustainability of tourism

The knowledge and proper management of all adverse impacts are extremely important to make tourism sustainable. Generally, they are factors contributing to create environmental pressure exerted locally but not only. For instance, it can be surely asserted that that fuel and electricity consumption in tourism are usually very high, but – as will be discussed below - the travel to reach the destination is the most important contributor to GHG emissions [14]. The approach to sustainability of tourism has been so far concentrated mainly on destination, ignoring that tourism, increasingly oriented toward destinations far away (many thousands km). Therefore fuel consumption and GHG emissions due to mobility have become the most important factors for sustainability assessment.

4. Tourism life cycle

The evaluation or assessment of the "life cycle" of a product or a service in general can be defined as the technique to assess the environmental impact associated with all the phases of the product manufacturing or service provision. The assessment includes all the stages needed to manufacture the product (or delivering the service): extraction of the raw materials, processing, manufacturing of the product itself (or service delivery), distribution, use, maintenances and repair, and – most important – perform its safe disposal or recycling at the end of life. To accomplish the evaluation, the compilation of an inventory of relevant energy and material inputs and environmental releases in each phase has to be performed. Then, the impacts associated with the inputs and releases have to be evaluated and, finally, an interpretation of the results has to be performed with the goal of allowing the decision makers and stakeholders in general to adopt an informed decision [16]. This important process is sometimes also called "Eco balance" and it is also described using the illustrative expression "from the cradle-to-grave".

The assessment process has been internationally standardized (ISO 14040 and 14044) by including four main phases: goal and scope definition, inventory analysis, assessment of the impacts and, finally, interpretation of the results of the previous phases, as schematically illustrated in *Figures* 2 and 3.

In order to assess tourism activities in terms of environmental effects, the possibility to adopt the LCA process is analyzed. According to the scheme in *Figure* 2, goal definition and scope could be interpreted as: the goal is the assessment of the environmental impacts of a tourism activity (e.g., a vacation in a destination), whereas the scope is the ideal space in which the touristic activities (travel, permanence, etc.) are performed.

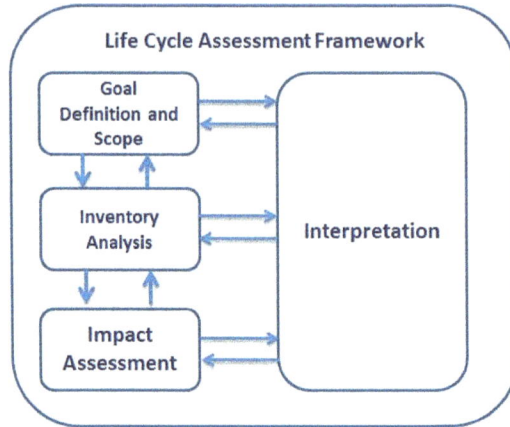

Figure 2. Schematic view of LCA analysis. Source: [16]

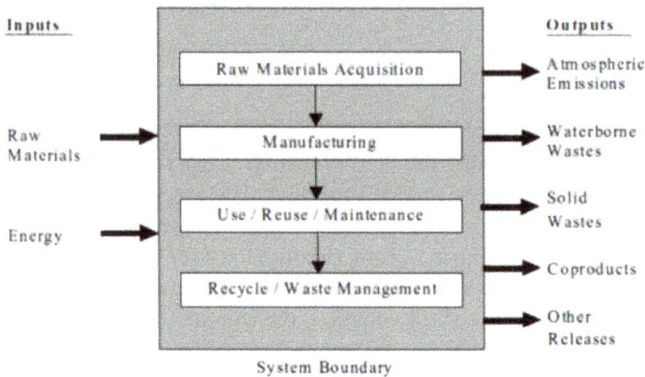

Figure 3. Inventory energy and material. Source: [16]

Before continuing into what would be a life cycle analysis adequate for tourism, it is worthwhile to remember that the concept of tourism life cycle has been already defined in some previous papers. According to [8], tourism life cycle is a six stage model to be performed for the assessment of a destination. It includes the exploration, involvement of the local people, the development of a tourism resort in a given country, the consolidation through the integration of the resort into the local economy, the stagnation (e.g., competition from other resorts, saturation, etc.), and, finally, either the declination or the rejuvenation of the tourism site. This approach, being essentially correct, takes into account only the life cycle of a destination (in this sense, it should be called, tourism area life cycle), without considering the whole cycle of the most classical tourism activity, the seasonal vacation (reaching the destination, staying and traveling back).

The Butler model is strictly connected to the concept of carrying capacity (CC) of a site; it represents the maximum burden that can be accepted by a territory without producing a crisis of the local ecosystem. WTO has provided a specific definition of tourism CC [17]. [18] defined Tourism Carrying Capacity (TCC) as "...the level of human activity an area can accommodate without the area deteriorating, the resident community being adversely affected or the quality of visitors experience declining". Looking at this definition it appears clear the connection with the Butler model: the stagnation of the area where the tourism is taking place starts when the TCC is reached and the further development is not possible without intervention addressed to a rejuvenation process.

The model proposed in the present paper does not completely fit into the concept contained in the LCA, but has as its ultimate goal the purpose to quantify one of the most important factor, consumption of fossil fuel and related CO_2 and other emissions, mostly produced during travels to reach the destination. The approach implemented in the present paper is an attempt to adopt a methodology able to include all the aspects of a very common type of tourism which consist in a vacation in a destination that can be reached by a traveling round trip.

To make the model operative, identification of the different phases in which pollutant emissions are taking place have to be identified. This type of analysis can be done for different destination and for given data characterizing the period spent in the destination. One valuable approach is the comparison of different ways to spend vacations, trying to select the solution which minimizes the CO_2 emissions. To make the comparison meaningful a functional unit has to be introduced, e.g. the amount of CO_2 emitted for one day spent in vacation (per person in a given place and for a given distance).

On the basis of what has been discussed hitherto, the concept of a vacation life cycle can be schematically defined by the following parameters:

- Number of visitors arriving at a destination and from where.
- Distance from the tourist home to the destination.
- Diffe5rent scenarios for reaching the final destination (airplane, car, train, bus, ship, etc.).
- Differ5ent types of accommodation ranging from camping places, rented homes, mobile homes, mountain huts to five star hotels, if present.
- Number of days staying in the destination.
- Different types of activities that can be performed at the destination, such as diving, sailing, cruises, mobile homes or visiting zoos, cinemas, indoor pools, museums or theatres.
- Mobility in the destination (private car, public transportation, sightseeing tours, etc.).
- Other issues related to the holidays (such as the practice of "zero km" food for catering).

To assess the impacts of all these activities the classical LCA requests the assessment of all the impacts related to the inputs and outputs of the mentioned items. Impacts are calculated by dividing them into different categories: greenhouse effect (potential global warming), stratospheric ozone depletion, acidification, eutrophication, summer smog, natural resource depletion, aquatic toxicity, etc. The phases of classification, characterization, and interpretation of the results and identification of significant issues are very complicated and it is beyond the scope of this chapter (being its specific purpose the assessment of the GHG emissions) to discuss all the impacts from tourism activities. On the other hand, it is worthwhile to notice that tourism and vacations are often spent in remote destinations and, therefore, large amount of fossil fuels for transportation are requested, hence, to assess, e.g., the contribution of transport on the total impact produced by GHG, the proposed model appears adequate and, hence, considered a valuable approach, able to compare different types of vacation by quantifying the parameters above.

To this goal, the different phases of tourisms related activities generating emissions have to be identified. In the present chapter inventory analysis is limited to the amount of fuel supplied to transportation means, the related CO_2 emissions, neglecting other environmental impacts due to, e.g., maintenance of the carriers.

The evaluation of potential impacts associated with the inputs of tourism life cycle and their related impacts constitutes important information needed for the analysis of the observed and predicted future climate change. In evaluating impact sources it appears clear the paramount influence of the type of transportation means selected. For instance, aviation emissions have a greater climate impact that the same emissions at ground level due to the fact that, at altitude, they can activate a series of chemical and physical processes that can have enhanced consequences on the climate change and have to be taken into account through multiplying factors. By the same token, for automobiles not only their number but also their age and technology, play a paramount role in evaluating emissions.

The outcome of this form of analysis appears useful for a better global managing of tourism and, in general, for making sustainable choices towards a reduction of the atmospheric pollution, limiting CO_2 and other greenhouse gases emissions.

5. Tourism sustainability indicators

In the context of the present chapter it has to be mentioned that indicators are commonly used by organizations to evaluate their success or the success of a particular activity in which they are engaged. Because of their integrative and forward-looking features, Sustainability Indicators (SI) are suitable to measure and evaluate human activities, and more and more businesses, willing to align their activities with the principles of sustainable development, are adopting SI as a powerful tool in addressing a satisfactory development in relationship to the environment.

Some useful indicators to express the level of sustainability of tourism are: carrying capacity, ecological footprint and carbon footprint. The carrying capacity has been already introduced in the previous paragraph.

5.1. Tourism ecological footprint

The Ecological Footprint is a measure of the 'load' imposed by a given population on nature. It represents the land area necessary to sustain a tourism activity in terms of resource depletion and waste discharge by that population [19]. It represents the total area necessary to satisfy all the needs of the population involved in terms of food, water, land, etc. It also includes the large amount of the area necessary to neutralize the effects all the GHGs, e.g., by photosynthesis of plants. This area, (representing the carbon footprint in terms of area, instead in terms of kg of CO_2) results usually very large: the global average amounts at about 50% of the ecological footprint. In the case of long distance tourism the fraction is even more. It is worth to put into evidence that EF can be calculated both for the population of a community and for a single individual. To measure the level of sustainability of tourism it is worth to introduce another indicator able to express the role of GHG emission: the carbon footprint.

5.2. The carbon footprint

The carbon footprint (CF) is usually measured in kilograms of carbon dioxide equivalent; in the case of tourism the unit for this indicator could be kilograms of carbon dioxide equivalent/ person [20]. Tourism, as any other human activity, has either direct or indirect effects on the carbon footprint. The primary CF is the direct measure of carbon dioxide emissions from burning of fossil fuels (such as energy consumption at the destination and transportation). The secondary CF is a measure of the indirect carbon emissions from the entire life cycle of commonly used products (related to their manufacture and eventual disposal/breakdown).

In assessing the CF it should be kept in mind that electricity is an essential part of the tourism industry. Hotels and accommodation for guests must be fully equipped for their comfort; this includes proper lighting, water heaters, basic electronics, elevators, pool pumps, etc. Restaurants are generally run on electric stoves and in some cases, electric dishwashers (both currently considered negative for sustainability). For these reasons and many more, high amounts of electricity is needed hence a readily available, efficient supply is crucial. For electricity generation the CO_2 specific production for different fuels is shown in *Table 1*.

Fuel	f_c (Carbon Fraction in fuel)	Heat of Combustion [kcal/kg]	P [kgCO$_2$/kWh] (Thermal)	P$_s$ [kgCO$_2$/kWh] (Electrical)
Natural gas	0.75	11900	0.20	0.5
Petrol	0.87	10000	0.27	0.67
Coal	0.85	8500	0.31	0.77

Table 1. Specific CO_2 emissions for different fuels.

It is, hence, evident that the specific emissions will depend on the relative amount of the different fuel used for the electricity production (energy mix). In *Table 2*, the specific production of CO_2 per kilowatt-hour for some European countries is reported.

The analysis of the carbon footprint of tourism worldwide shows that the greenhouse emissions are due to: transport (particularly air and motor vehicle) 82%, accommodation 4.5%; other activities 8.6% retail 3.4% [21]. The transportation of visitors to the destination plays an important role in contributing to the carbon footprint. However it should not be forgotten that transport is a promoter for the rest of the industry: if the number of trips declines significantly then all businesses will be affected.

Country	Emissions [kgCO₂/kWh]			
	Coal	Petrol	Natural Gas	TOTAL
Italy	0.118	0.130	0.178	0.426
Austria	0.097	0.019	0.076	0.192
Germany	0.382	0.011	0.046	0.439
Spain	0.223	0.057	0.081	0.361
France	0.036	0.007	0.015	0.057
Sweden	0.005	0.009	0.004	0.018

Table 2. Specific emissions of CO_2 for some EU countries.

6. Role of transportation in tourism

In modern societies mobility plays a fundamental and increasing role in shaping our daily life: the way people interact, work, play, manufacture, and get access to services, leisure amenities and goods, is inextricably linked with transport. Mobility lies at the heart of tourism and, noticeably, there are synergies between transportation and tourism [22], with technological developments and lower prices for the mobility promoting tourism and, conversely, tourism encouraging the expansion of new transportation possibilities. Furthermore, transportation is the link between home, destination, accommodation, attraction, and all other stages of a tourist journey. Its efficiency, comfort and safety determine to a large extent the quality of the tourist's experience and in many cases its cost comprises the largest portion of a tourist's total expenses. Tourism represents a strong sector for the demand of transportation and prospective studies foresee a further increase and, consequently, in the request for mobility to reach the termini and even at the destination itself. The trend to select far destinations has made the travel phase the prevalent part of the total economic and environmental cost.

On the other side, in evaluating emissions - as will be discussed in detail in the next paragraph - there is no doubt that tourism is an important contributor to the emission in general and of GHG in particular. Indeed, data from the WTO Climate Report shows that total CO_2 from tourist activities amounts to 4.9% of total world emissions, with mobility playing a relevant role. A recent study made by the Direction des Études Économiques et de l'Évaluation Environnementale (D4E), revealed that mobility of French tourists gives a contribution to the GHG emissions of about 6% of the total amount (2006) [23]. Moreover, this study concluded, analysing the duration of the permanence in the destination in the last 40 years, that the average duration has changed from 20 to 12 days and that the portion of tourists spending vacation abroad changed from 12 to 19%, implying more frequent travelling and increased use of the airplane to reach abroad destinations.

Transportation is accomplished by different means, such as car, train, bus, ship, or aircraft. According to a study by [24], out of the total car transport, 20%-30% are used for tourism mobility. Similarly, 20%-40% of rail travel serves tourism purposes, whereas 60%-90% of air travelling passenger accounts for tourism mobility. At global level, tourism mobility causes around 75% of total CO_2 emissions out of all emissions from touristic activities, with aviation representing the bulk of it (40%). According to a research performed by [25] GHG emissions from international aviation grew by 87% between 1990 and 2004 (73% increase for 1990-2003), while total GHG emissions decreased by 5.5% between 1990 and 2003. Air traffic is furthermore expected to double in the next 15 years and is anticipated to counteract the reduction of CO_2 emissions achieved in other sectors.

Actually, the development of the air travel industry, especially low-cost airlines, has made affordable and thereby increased the utilization of this type of mobility, making travelling accessible to a growing number of the world's population. The air-travel industry has substantially reduced travel time and travel costs as compared to other transport modes. The most popular air-travel models are the Low Cost Carriers (LCCs) focusing on sea, sand and sun tourism, short stay city trips and cultural destinations [26]. To make flights to a destination cheaper, it is important flying non-stop. The contribution to emission given by the flights used for tourism is the highest (if expressed in terms of kg CO_2/person and km), even if it is not the most selected way to reach destinations for vacations: In France (2006) only 6% of tourists have selected the airplane to reach the destination, usually located very far away. Nevertheless, automobiles are still the most common way used for tourism travels (75%).

Energy consumption for transportation depends on two factors: the type of transport used and the distance to be covered. Due to the overwhelming use of fossil fuels, mobility generates GHG emissions which can cause climate change and engender impacts that harm the environment and is believed to be a primary cause of climate change. Hence, it can be concluded that transport represents an important phase of tourism but is, on the other hand, responsible of a outstanding amount of emissions. Nevertheless, when looking at tourism mobility we have to keep in mind that transport for tourism only accounts for a fraction of all transport. A large portion of general transport serves for moving freight and non-tourist passengers [27].

7. Emissions produced in tourism

Tourism activities, besides the necessary mobility to reach the destination briefly discussed in the previous paragraph, generate emissions also in other phases of its development such as of residing at a destination as a result of the use of energy for heating or air conditioning, illumination, and other services (cooking, cleaning, office, etc.). All these aspects have to be considered to assess the energy consumption and related emissions during the tourism life cycle. In *Figure 4*, the different phases of tourism and its associated fuel and energy consumption are schematically depicted, substantiating that a touristic activity generates emissions in all the phases of its development. The figure evidences that energy consumption is partly due to travel (to the destination and back home) and partly to the activities performed in the destination. It has been determined that, if the distance from the departure location and the destination are relatively short (less than 1000 km) the preferred way for transport is the private car whereas, for long distances, air transportation is preferred. In the last years the relatively low costs of flights for long distance have encouraged journeys to destinations far away.

As already stated, energy consumption and its associated emissions in tourism depends on the type of services offered, the type of accommodation and the energy management approach. For instance, hot water supply, heating and air conditioning, account for a large part of hotels total energy consumption. As will be discussed later, appliances and utilities represent an area where large savings can be made through efficiency improvements. Accommodation providers should have a particularly strong interest in reducing energy consumption in order to save costs and ensure the sustainable future of the destination.

Although there are different ways to provide energy to tourism activities, large amounts of CO_2 are produced mainly due to the fact that energy is largely converted by burning fossil fuels. In *Figure 5*, the global GHG emissions per economic sector and particularly those due to the mobility, discussed in what follows, are shown.

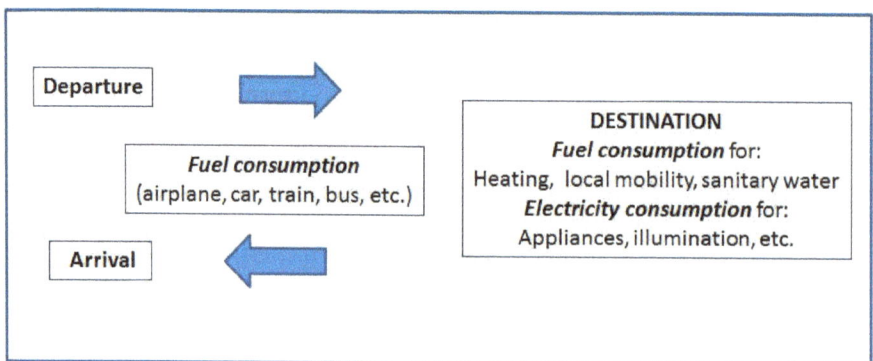

Figure 4. Schematic representation of fuel and energy consumption in the different tourism phases.

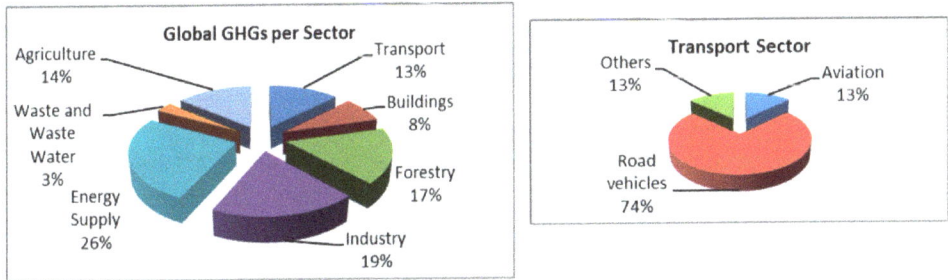

Figure 5. Global and sectorial emissions. Source: [28]

In *Table 3*, the amount of CO_2 emissions (in millions of metric tons) and the share of tourism in the different phases (in percentage of the total), is presented.

	CO₂ - EMISSIONS	
	MT	**SHARE OF TOURISM [%]**
Air transport	515	40
Car	420	32
Other transport	45	3
Accommodation	274	21
Other activities	48	4
Total tourism	1302	100
Total world	26400	-
Share of tourism in total world (%)	4.9	

Table 3. Emissions of the different sectors, according to the WTO. Source: [29]

7.1. Emissions due to mobility

Transportation means emit large quantities of carbon dioxide (CO_2), carbon monoxide (CO), hydrocarbons (HC), nitrogen oxides (NO_x), particulate matter (PM), and very dangerous substances such as benzene, formaldehyde, acetaldehyde, 1,3-butadiene, and lead (where leaded gasoline is still in use). Each of these pollutants, along with secondary by-products (such as ozone), can cause adverse effects on health and the environment. Recognizing the danger the atmospheric pollutants can generate, many developed countries have issued strict emissions controls especially for particulate matter produced by road dust, tire wear, brake wear etc. In recent times much attention has been devoted to non-combustion substances and the so-called Particulate Matter (PM), both of which appear to be dangerous for the human health.

Transportation is a typical system belonging to the so called mobile sources: pollutants emitted are spread out along the pathway followed by the source. For part of the emissions this might represent an advantage since they are diffused in the environment and pollutants undergo to a dilution process (e.g., particulate matter). This is not the case for GHG due to the fact that their effect belongs to the category of global impacts.

7.1.1. Emissions from air transport

Air traffic in the world is growing, and will likely continue to grow. A large part of the expansion of the number of flights is due to tourism, especially to far destinations (more than 1000 km). The majority (60 to 90%, depending on different studies) of air travelling passengers are ascribed to tourism mobility. The total contribution of aircraft emissions to total anthropogenic carbon dioxide (CO_2) emissions was considered to be about 2 percent in the IPCC 4th Assessment Report [28].

Although the contribution of aviation operations to total global CO_2 emissions is relatively small, forecasted traffic growth (4.7% per year) raises questions on the future contribution of aviation activity to emissions and, hence, to climate change, and on the most effective way to address CO_2 releases from the sector.

The effect of emissions from aircraft at high altitudes (especially nitrogen oxides (NO_x) and water vapor) is of particular concern. CO_2 and H_2O are the main combustion products but also products such methane, nitrous oxide and other gases have an important effect on the climate change. The fuel consumption and emissions will be dependent on the fuel type, aircraft type, engine type, engine load and flying altitude.

Emissions from aircraft originate from fuel burned in aircraft engines with two types of fuels used. Gasoline is used in small piston aircraft engine only. Most aircraft run on kerosene, and the bulk of fuel used for aviation is this type of fuel. In an effort to improve efficiency, part of the energy contained in the hot discharged gas is used to drive the turbine that in turn drives the compressor.

GHG emissions of the airplane strongly depend on the type of aircraft and on the distance covered. In analyzing the emissions due to air transport, it is usual to distinguish between the different phases of a flight. The cycle is named Landing/Take-Off (LTO); it includes phases located below 1000 meter (taxi/idle, take -off and landing). The phases of a flight cycle are shown schematically in *Figure 6*.

In short travels the contribution of LTO to fuel consumption and of CO_2 emissions is very high. This is the reason why flights covering long distance become more convenient in terms of amount of CO_2 emitted per km. In fact, an analysis conducted for many types of aircraft show the indicative data for LTO cycle gathered in *Table 4*.

In order to understand better the effect of LTO on the fuel total consumption in covering the distance and the related CO_2 emissions, a modern airplane, traveling a distance of 2000 km with a number of passengers of 200 people is taken as reference. For such a plane, a specific emission of 5 kg/km passenger and a LTO emission of 2000 kg can be assumed. If the travel is performed non-stop 2000 km total emissions will be 10000 (cruise flight) + 2000 for LTO, this means that LTO weight 20%. Obviously, if the number of km is increased to 4000, the percentage of LTO will become only 10%. It appears clear that for short distances the choice of airplane is not favored in comparison with train, bus and car both from economical neither ecological point of view. Obviously, for a long haul a non-stop trip is strongly recommended.

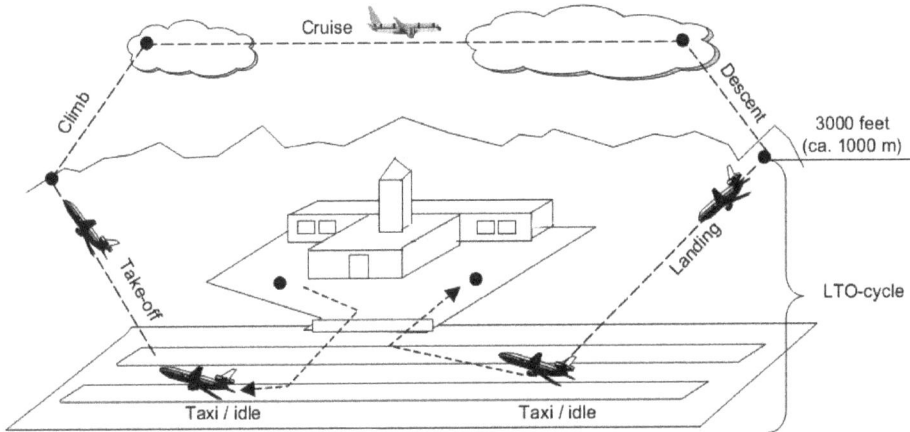

Figure 6. Standard flying cycle. Source: [30]

POLLUTANT	(kg/LTO)
CO_2	3000 – 10000
CH_4	0.1 - 4
N_2O	0.1 – 0.3
NOx	5 - 15
CO	10 - 50
NMVOC	10 - 50
SO_2	1 - 3
Fuel	1000 – 3000

Source: [30]

Table 4. Indicative pollutants and related fuel consumption ranges for different aircraft

Concerning the fuel consumption, passenger airplanes in the year 1998 averaged 4.8 l/100 km per passenger (1.4 MJ/passenger-km). In this context it has to be mentioned that, on average, 20% of seats are left unoccupied. Jet aircraft efficiencies are improving: between 1960 and 2000 there was a 55% overall fuel efficiency gain. Companies using Airbus state a fuel rate consumption of their A380 at less than 3 l/100 km per passenger.

7.1.2. Car transportation

In particular, in developed countries people rely heavily and increasingly on private mobility and the vehicles are expected to become safer but, disappointingly, also more luxurious and powerful. In addition, automobiles and other means of transportation are driven and used progressively more frequently. This individual and collective attitude often does not take into account the resulting consequences: increased traffic congestion land occupation for parking lots in urban areas, increased fuel consumption, greater emissions of air pollutants and greater exposure of people to hazardous contaminations that might cause serious health problems.

As awareness concerning the potential health effects of air pollutants has grown, many countries have implemented more stringent emissions controls and made steady progress in reducing the emissions from cars, buses, airplanes in a perspective of improving air quality and limiting GHG emissions. However, the rapid growth of the world's transportation fleet due to population and economic growth, the expansion of metropolitan areas, and the increasing dependence on motor vehicles because of changes in land use, has resulted in an increase in the fraction of the population living and working in close proximity to busy highways and roads, counteracting to some extent the expected benefits of pollution control regulations and technologies. Pollution produced by cars, buses and ships in tourism activities are giving a great contribution to total GHG emissions. According to a study by [24] out of total car transport, 20%-30% are used for tourism mobility.

Pollutants from vehicle releases are related to vehicle type (e.g., light- or heavy-duty vehicles) and age, operating and maintenance conditions, exhaust treatment, type and quality of fuel, wear of parts (e.g., tires and brakes), and engine lubricants used. Concerns about the health effects of motor-vehicle combustion emissions have led to the introduction of regulations and innovative pollution control approaches throughout the world that have resulted in a considerable reduction of exhaust emissions, particularly in developed countries. These reductions have been achieved through a comprehensive strategy that typically involves emissions standards, leading to the introduction of cleaner fuels and accurate vehicle inspection programs.

The European Union has introduced stricter limits on pollutant emissions from light road vehicles, particularly for emissions of nitrogen oxides and particulates. In order to limit pollution caused by road vehicles, specific regulations have been introduced for emissions from motor vehicles. The European Regulation No 715/2007 deals with vehicles with a mass not exceeding 2610 kg. It includes both positive-ignition engines (petrol, natural gas) and compressed ignition (diesel engines). In order to limit as much as possible the negative impact of road vehicles on the environment and health, the regulation covers a wide range of pollutant emissions: carbon monoxide (CO), non-methane hydrocarbons and total hydrocarbons, nitrogen oxides (NO_x) and particulates (PM). It covers tailpipe emissions, evaporative emissions and crankcase emissions. For each category of pollutant and for the different types of vehicle limits are given. In *Table 5* the limits (**Euro 5 Standard**) fixed for light road vehicles are shown. The regulation in force for European cars requires the respect of the limits.

An important figure for all transportation means is the CO_2 emission directly connected to the chemical composition (carbon %) of the fuel and to the efficiency of the engine. If a car needs 5 liter of gasoline to travel 100 km this means that emissions are 5*0.86*(44/12) = 15.8 kg CO_2 per

100 km, or 158 g CO_2/km. Usually CO_2 specific emissions are expressed in g/km, so our car would have emissions not particularly good i.e. 158 g CO_2/km. Designer of modern cars are giving utmost attention to this performance of cars, advertising of new models with specific emissions of less than 100 g CO_2/km. If this target will be reached for a large portion of car park, the global problem of climate change would be strongly reduced. Emission less than 100 g/km are considered acceptable for the environment and above 200 g/km have to be considered too high. Obviously, the lower the CO_2 output, the lower environmental impact. In the following some characteristic of small cars with low specific emissions are shown. In *Table 6*, the advertised CO_2-emissions for some commercial small cars are gathered.

	POLLUTANT	EURO 5 EMISSION LIMIT (mg/km)
EMISSIONS FROM DIESEL VEHICLES	Carbon Monoxide (CO)	500
	Particulate Matter (PM)	5 (80% reduction of emissions in comparison to the Euro 4 standard)
	Oxides of Nitrogen (NO$_x$)	180 (20% reduction of emissions in comparison to the Euro 4 standard)
	Combined emissions of Hydrocarbons and nitrogen oxides (HC+NO$_x$)	230
EMISSIONS FROM PETROL VEHICLES OR THOSE RUNNING ON NATURAL GAS OR LPG	Carbon Monoxide (CO)	1000
	Non-Methane Hydrocarbons (NMHC)	68
	Total Hydrocarbons (THC)	100
	Oxides of Nitrogen (NO$_x$)	60 (25% reduction of emissions in comparison to the Euro 4 standard
	Particulates (solely for lean burn direct-injection petrol vehicles)	5

Table 5. Limits for diesel and petrol vehicles. Source: [31]

MODEL CAR (STANDARD)	emissions (CO_2) range (g/km)
Italian small car	109-113
French small car	87-153
Japanese small car	95-125

Table 6. Specific emissions of modern small cars according to the advertising

7.1.3. Other transportation means

Among the carriers used for tourism also cruise ships, able to transport several thousands of people, have to be included.

Apart from aviation, the worldwide booming cruise ship industry has also come under increased criticisms. Cruise ships that can carry up to 5000 tourists are not only notorious for creating tremendous amounts of waste and sewage but also belong to the biggest contributors to greenhouse gas emissions within the travel and tourism industry. A single cruise ship can generate emissions equivalent to more than 12400 cars. The ship smokestacks release toxic emissions that lead to acid rain, global climate change, and damaging health effects to communities situated near ports. Despite the fact that ocean cruise liners are more energy efficient than other forms of commercial transportation, marine engines operate on extremely dirty fuels, known as 'bunker oil'. To compound the problem, engines on these ocean-going ships are currently not required to meet the same strict air pollution controls, as cars and trucks are required to do.

Referring to the fuel consumption per single passenger and unit of distance covered (*Table 7*), it is found out that the specific consumption and the related CO_2 emissions are greater than the emissions of an airplane. Rough estimates indicate for cruise liner emissions of about 0.27 kg of CO_2 per passenger and kilometer, as compared to 0.16 kg for a long-haul flight. The cruise industry is the fastest growing sector of the travel industry. In 2003, 9.3 million passengers took a cruise. These figures indicate that if not enough attention is paid to carbon emissions, due to the increased popularity of this type of vacation, the contribution will become not negligible without appropriate improvements in the design of cruise liners.

Concerning railway transportation, it has been estimated that 20%-40% of rail travels serve tourism purposes. Taking into account that most of the rail traces are electrified, rail transportation seems good for the environment. Nevertheless, taking into account the whole cycle, the source of electricity has to be considered, then, if fossil fuels are used, the emissions at the basis has to be included.

The impact of air transportation on climate is exacerbated by the fact that the emissions happen largely during cruise phase and, hence, mainly in the higher layers of atmosphere. Here the impact is due not only to CO_2 but also to other emissions, such as water vapour and nitrogen oxides. The increase of the effect on climate is usually given through a coefficient called radiative forcing, defined as the change in the net irradiance in the different layers of the atmosphere. The Intergovernmental Panel of Climate Change estimates that the warming effect of aircraft emissions is about 1.9 times that of carbon dioxide alone, due to the other gases produced by planes.

The contribution of tourism to global climate change through GHG emissions from the transportation of millions of tourists was first discussed in the middle of 90's. Subsequently, a direct interest by the IPCC has started, devoting attention to tourism in the some regional such as Africa, Australia and New Zealand, Europe and small island states. Later on, a tourism-focused climate change assessment was commissioned by some international organizations to evaluate the relative regional vulnerability of tourism destinations, discussing the state of adaptation within the sector and providing the first quantitative estimate of the contribution of the global tourism sector to climate change, aiming to set out options for decoupling future growth in the tourism sector from GHG emissions. Although recent events such as seismic incidents, hurricanes and tornadoes, the Asian tsunami, and even terrorism attacks, suggest a relatively high adaptive capacity of the sector, whether the touristic sector will be able to cope successfully with future climate regimes and the broader environmental impacts, remains relatively unknown.

MEANS OF TRANSPORT	KILOMETRES PER LITRE [kpl]	EMISSIONS [g CO_2/km passenger]
Car - the most efficient	18-23	130-100
Car - average models	9–16	260-145
Car large models, SUVs, etc.	3-9	500-250
Rail - normal suburban	18-52	130-145
Rail - high speed, few stops	14-28	165-180
Bus - well used service	28-50	80-145
Airplane - (below 500 miles)	4–8*	460-330
Airplane - (long journeys)	8–12*	330- 210

* including radiative forcing index at 1.9

Table 7. Summarizes the present situation concerning mobility. Source: [32]

7.2. Accommodation

The emissions due to the consumption of energy in the destination can be expressed in terms of heat and electricity consumption in the period of staying (number of days). The electricity consumption in the destination can be safely assumed equivalent to the typical consumption of a user at home, which amounts 3 kWh per person and per day. This figure changes with the type of hotel or resort and depends also from the existing degree of energy saving of the accommodation. Heating consumption can also be estimated taking into account meteorological conditions and the thermal isolation provisions isolation of the building. Sound figures for modern building range between 70 to 100 kWh per m² per year. Data in kg of CO_2 can be obtained by conversion factors.

Pollutants produced in tourism destinations have a limited importance if we consider only the impact produced in the accommodation structures. Pollution produced by electricity is not a local problem since its effect takes place directly on the site where the power stations are located; the small amounts of pollutants produced locally can be controlled by adequate systems based on high efficiency and right behavior. Moreover, a complete analysis of local air pollution should take into account the contribution given by cars of tourists circulating in the destination determining an overburden of air pollution, noise and traffic jams.

8. Sensitivity of tourism to climate change

From what we have discussed in the previous paragraphs, the synergy between tourism and environment results evident, particularly due to the interrelation between energy consumption for tourism in all its phases and the emissions produced in the process of energy conversion, believed to be the cause for climate change. In fact, the previous paragraphs demonstrated that tourism activities produce a significant amount of greenhouse gases, contributing thereby to global warming which, in turns, may affect the local climate. Moreover, it is now widely recognized that, among the different causes of greenhouse gases emissions due to tourism activities, travels to long distance destinations (which are increasingly requested in the current tourism market) generate most greenhouse gas emissions and are, thus, supposed to contribute strongly to climate change.

Many research studies consider climate as an essential resource for tourism, and especially for beach, nature and winter sport tourism, and the phenomenon of global warming already severely affects the sector in an increasing number of destinations. It is thereby recognized that the impacts of global warming pose a serious threat to tourism, which constitutes one of the world's largest and fastest growing economic sector [1], according to the World Travel and Tourism Council (WTTC) [33]. As already stated, the relationship between climate change and tourism is two-fold. Not only is tourism affected by a changing climate, at the same time it contributes to climate change by the consumption of fossil fuels and the resulting emissions. Hence, additional efforts are underway to develop environmental policies for the tourism sector that can offer adaptation and, where possible, mitigation.

The predicted modifications caused by the climate change in the tourism destinations due to global warming, are anticipated to be predominantly strong for coastal areas, whose environmental conditions appear particularly sensitive. It has been estimated that about 25% of the CO_2 emitted from all anthropogenic sources currently enters the ocean, where it reacts with water to produce carbonic acid. Carbonic acid dissociates to form bicarbonate ions and protons (see *Figure 7A*).

The protons react with carbonate ions to produce more bicarbonate ions, reducing thereby the availability of carbonate to biological systems (e.g., corals). This decalcification phenomenon might affect both skeletal growth and density, with consequences on the extension of the coral reefs and their mechanical endurance (less resistance to storms and erosion). This phenomenon is evident in many coastal zones, particularly in the Australian coralline barrier, manifesting itself through the so-called coral bleaching (*Figure 7B*).

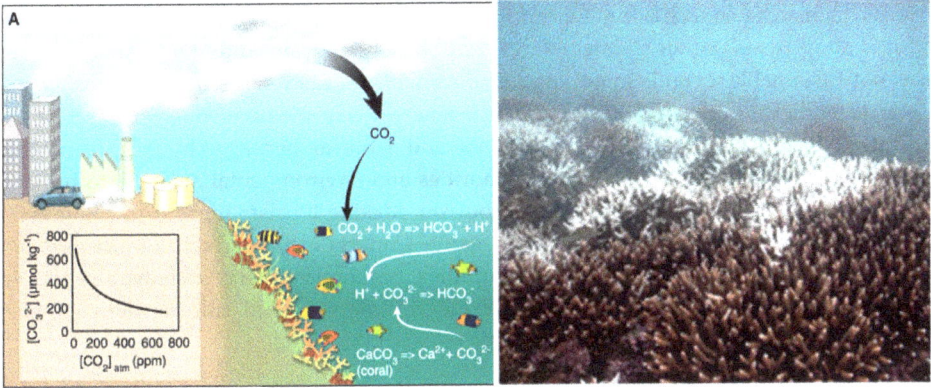

Figure 7. Ocean acidification process (A) and the resulting coral bleaching (B). Source: [34, 35]

Since sea-coast tourism remains one of the dominating market segments, giving a high contribution to the economy of many developing countries, the vulnerability of coastal destinations becomes of paramount importance [1]. In addition to the particular example discussed, it is expected that local effects of global warming, such as the increase of local extreme events (storms, coastal erosion, sea level rise, flooding, water shortages and water contamination), can put in danger beach destinations. As already mentioned, the enhanced vulnerability is often accompanied by a low adaptive capacity, which is particularly true for coastal destinations of developing countries. The seasonality of coastal tourism is an additional facet to be taken into account in the panoply of problems created by the climate change. Generally, coastal areas tourism is concentrated in few months, coinciding e.g., with low water availability, high consumption of fuels, electricity, etc. In some expected conditions global warming could also play a positive role; this could be the case for Mediterranean destinations where the season could be enlarged and the winter period might be more appealing to tourists, providing opportunities to reduce seasonality and expand the tourism product. In addition to the absolute amount of change, the rate at which change occurs is critical to whether organisms and the ecosystem in general will be able to adapt or accommodate to the new conditions.

On the basis of the few examples discussed, it is evident that the interaction between tourism and climate is very complex and has only recently been established as the subject of scientific studies and recognized as the cause of growing contribution to climate change and, hence, the main reason for regional vulnerabilities. In this sense, a recently declaration of UNWTO-UNEP-WMO stated that *"climate change must be considered the greatest challenge to the sustainability of tourism in the twenty-first century"* [29]. Although the interaction between tourism and climate change has been studied to some extent in the last 20 years, there are only few recommendations in specific issues and a real strategy of approach is not yet available.

9. Making tourism more sustainable

Tourists' increasing concerns for environmental issues have also stimulated operators in the sector to adopt sustainability strategies. It is now widely recognized that tourism is in many cases not sustainable, even if some sorts of sustainable tourism and ecotourism are making efforts to enhance and promote local development while simultaneously protecting the natural environment, maintaining traditional and cultural heritage. In fact, many tour operators cooperate with local tourism authorities and environmental agencies to promote ecotourism and other forms of sustainable tourism. Making tourism more sustainable is not just about controlling and managing the negative impacts of the industry; tourism is in a very special position to benefit local communities, economically and socially, and to raise awareness and support for conservation of the environment and cultural heritage, even providing in some cases the basis for scientific research.

Tourism was once viewed as an independent activity, having no impact on environmental resources but in reality, this seldom occurs. Therefore, it is urgent that civic movements concerned with environmental and climate change issues, monitor and respond to these type of activities, since T & T is, for many countries one of the most important industries, not only because of its size and foreseeable growth but also due to the fact that it is considered a driver of globalization and trade liberalization. Nevertheless, in this context, the argument of tourism as a poverty alleviation strategy is doubtful in view of the increasing foreign take-overs of tourism businesses as a result of globalization and liberalization.

In general, the main requirements for improving sustainability in tourism are: to limit resource depletion and degradation including loss of biological diversity, loss of habitat and resources, loss of water resources; fisheries; forests and timber; energy resources; mineral resources. Moreover, improvement of sustainability could be pursued by reducing pollution and wastes production. The process of enhancing sustainability also includes actions addressed at improving the quality of life of host communities, at preserving intergenerational and intra-generational equity and ensuring the cultural integrity and social cohesion of communities, giving at the same time the opportunity to provide a high quality experience for visitors. Other interventions to improve sustainability deal with the promotion of the economic growth connected to tourism activities (hotels, restaurants, beach facilities, entertainment initiatives, etc.).

The measures proposed to reduce environmental impacts at destinations include: avoiding exhaustion and degradation of water resources; deterioration/loss of habitats (i.e. sand dunes), deterioration of terrestrial ecosystems; abandonment of agricultural land, urbanization with loss of urban landscape character, landscape deterioration, soil erosion, desertification, depletion/ significant decrease of fish stock, loss of historic settlements, depletion of low-commercial-value sectors, replacement of pre-existing architecture, concentration of vehicles in the urbanized areas, high level of noise pollution during the day and at night, degradation and fragmentation of natural spaces, loss of open spaces, oversized public services and infrastructures; increased production of waste; deterioration of the shoreline marine environment, bad relations between local population and tourists, depletion of pre-existing economic activities, high human density in the areas generally used by the tourists.

It is evident from what reported in the previous paragraphs that, evaluating the sustainability of tourisms, a major problem is represented by the quantification of tourism GHG emissions. One of the recommendations, suggested in some analysis of the sector, is the possibility to apply the indicator "carbon footprint" (discussed in paragraph 4) to tourism activities to make more comprehensible the role of tourism on GHG emissions. This parameter could be useful to improve the behavior of tourists and tourism operators, guiding them toward "greener forms" of tourism and mobility, such as "slow tourism travel" and different types of ecotourism. Among the different solution to reduce GHG emissions there are some oriented toward a specific goal, the so called carbon neutrality for tourism, proposed by the administration of some famous tourism destinations: *carbon neutral tourism implies the offsetting of a destination's carbon footprint by means of processes balancing carbon emissions, such as planting trees or investing in new, renewable, energy sources.*

10. How to reduce emissions and the environmental impact of tourism

Aiming to reduce air pollution and GHG emission, the tourism industry is usually divided into different sectors: accommodation, catering services, recreation and entertainment, transportation and travel services, etc. In all these sectors actions to reduce the carbon footprint are possible. However, it is widely recognized that two these phases are the main responsible for emissions of pollutants: energy consumption and related emissions at the destination and fuel consumption and related emissions during traveling [36].

10.1. Reduced emissions and environmental impacts at the destination

To diminish emissions at the destination, reduction can be achieved by simple interventions that can be very valuable also from an economic point of view, reducing costs. To this goal, better use of electricity, water and handling of waste can greatly contribute in terms of sustainability and economy, as well as reducing emissions. Some examples for electricity saving are to turn off power of lights and equipment when not in use; install energy efficient fluorescent bulbs; use natural ventilation and fans where possible and when using air conditioning, set it to between 24°C and 28°C in summer. Appreciable amount of heat flow can be reduced by controlling the temperature in the inner spaces, and by an efficient thermal insulation of the wall, doors and windows. An important contribution to the reduction of air pollution and GHG emission can be obtained by limiting the private transport in the destination both for tourist's mobility and freight. An additional important measure is to eat food produced in the destination itself, what it is called "zero km". Other recommendations are concerned with the use of public transport and car-pooling, use of low consumption cars such as hybrid or electric vehicle, encourage cycling and walking where possible, use phone/video conferencing to reduce travel requirements. A further measure to reduce emissions is to change the fuel used for energy conversion from fossil fuels to the adoption of renewable sources such as biomass, eolic and photovoltaic systems. When staying in a hotel turn the lights and air-conditioning off when leaving the room, ask for room towels not to be washed every day which increasingly proposed in many hotels. Main factors taken into account for low consumption energy are the supply devices used for electric lamps, motor-driven appliances and electronic devices as well as heating systems.

To these goals, a new kind of eco-tourism is developing with specific requirements in terms of reduced energy consumption dictated by an extremely high contact with nature in remote destinations. Accommodation is made by the so-called eco-lodges, typical structures designed to have the least possible impact on the natural environment in which it is situated. Since there is no connection with the electricity grid, the eco-lodges are equipped with renewable and non-renewable energy sources and technologies for off-grid facilities. Energy consumed in eco-lodges is very low if compared with the specific consumption of hotels (25 kWh per guest and night in hotel vs. 0.5 kWh per guest and night in eco-lodge).

10.2. Reduce emissions due to mobility

The most advanced program to reduce GHG emissions in tourism have been done in the aviation sector. The *International Air Transport Association* (IATA*)* has advanced a range of very ambitious goals [37], including an average annual aviation fuel efficiency improvement of 1.5%, carbon-neutral growth from 2020 and the reduction of emissions from aviation by 50% by 2050 (compared with 2005 levels).

A reduction of flights would limit the profitability and growth of the tourism sector. However less drastic measures are possible, such as avoiding stops between the starting point and the destination. The question thus arises if it is possible to reduce the fuel consumption of airplanes with technological innovations. The gains reached and expected by technological innovation are represented in *Figure 8* [41] showing a reduction in fuel consumption of about 70% in the period 1960 – 2010. Further improvements are expected in the coming years but with a decreasing steepness of the slope of the curve.

Comparing the fuel consumption of a modern airplane (Airbus A380) with that of an efficient car offers interesting conclusions about the technological improvements in airplane design. The Airbus A380 is a four-engine airliner manufactured by the European corporation Airbus and the largest passenger airliner in the world. It provides seating for 525 people in a typical three-class configuration or up to 853 people in all-economy class configurations. Airbus A380, known under the nickname Superjumbo, is the first aircraft to surpass the 3 liter per 100 seat-km barrier. Taking into account a typical occupancy rate of 70% this translates into 4 liter per 100 km per passenger, about the same as a small car with an average load of 1.25 passengers.

A recent study made in France has analyzed the different ways to reduce fuel consumption [39] arriving to the conclusion that a reduction of 50% can be achieved in the year 2020. Measures that should be adopted to reach this goal are:

- Use of composite materials and ameliorate the aerodynamic design (5 to 15% efficiency improvement)

- Better motors or turbines, such open rotor turbines (15 to 25% efficiency improvement)
- Green taxi of planes (using electrical motors), optimize the traffic management and navigation system (10 to 25% efficiency improvement)

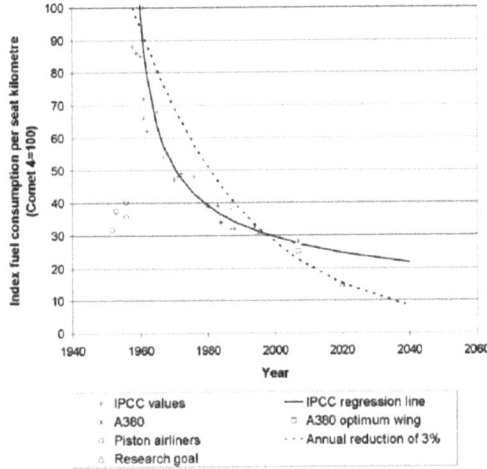

Figure 8. Long haul aircraft fuel efficiency gains since 1950 as an index (100) of De Havilland DH106 Comet 4. Source: based on Peeters et al. (2005) [38]

In Europe the air transportation cost is due for one third to the kerosene, but if the price of oil continues to increase at the present pace, the contribution of fuel could become 50%. A typical aircraft A320 consumes about 15000 litres for a travel of 2000 km. As already mentioned, a considerable saving of energy for the airplane sectors could be obtained by the use of systems operated with electricity instead of using the engines of the aircraft for displacing the airplanes on the taxiway. Further saving can be expected from the reduction of the weight of the aircraft by employing lighter materials and aerodynamic shapes.

Other possibilities to reduce energy consumption from tourism mobility are related to a marginal portion of the tourism market: these include the idea of "slow tourism" or other forms of responsible tourism. In some cases, the use of bus or a train rather than private cars or domestic flights can be advantageous. To fuel cars, airplanes and buses liquid biofuels such as bioethanol and biodiesel can be used.

In tourism activities a further contribution can be obtained with the so called carbon offsets, a process able to reduce a corresponding amount of carbon in the atmosphere by planting trees. If the destination is not far away the contribution to the total emissions can be limited, the production of CO_2 being done in the destination. If the distance is far away, the major part of the emissions are due to transportation, particularly if the travel is by flying.

A comparison of the emissions from different transportation means can be performed by introducing the relationship existing between the specific consumption (X) in km per liter of fuel and passenger and the specific amount of CO_2 in kg per km and passenger (Y); the calculation of the CO_2 emissions from different means of transportation can be performed as follows:

$$X * Y = k$$

Where k is constant that can be expressed for 1000 grams of fuel as:

$$k = 1000 * C * D * MW / AW$$

Where C is the fraction of carbon in the fuel, D the density of the fuel, MW is the molecular weight of CO_2 and AW the atomic weight of carbon, respectively.

Representative values for three different fuels: gasoline (cars), kerosene (airplanes) and diesel (cars) are gathered in *Table 8*, and presented together with the calculated specific constant k and the typical range and average CO_2-emissions.

FUEL	DENSITY RATIO [kg/dm³]	C/CH	SPECIFIC CONSUMPTION [km/liter]	[km/liter·PASSENGER]	k [g CO₂/liter fuel]	CO₂ EMISSIONS [g CO₂/km·passenger] TYPICAL RANGE	AVERAGE
gasoline (cars)	0.752	0.86	15 - 25	12 - 20	2371	110 - 183	146
kerosene (airplanes)	0.795	0.86	-	4 - 12	2507	210 - 460	335
diesel fuel (cars)	0.850	0.86	15 - 25	12 - 20	2680	86 - 95	90

Table 8. Typical values of CO_2 emissions for the three different fuels.

The relationship just introduced can be graphically represented in a series of parametric curves (see *Figure 9*), where the points represent the average values for cars and airplanes and are solely indicative of typical conditions. More accurate figures have to be referred to specific conditions, which will depend on the number on passengers, the length of the travel, the stops in between (for air transportation), the percentage of the seats occupied, etc. For instance, if for cars traveling long distance an average of 1.25 passengers per automobile is assumed, for distances below 1000 km, the best choice seems to be diesel car (due to the more efficient diesel engine) followed by gasoline car and finally by air transportation.

Furthermore, the curves indicated in the graphic can be used to assess the energy requirement for different vacation scenario, assuming a hypothetical destination 1000 km away from the starting point and a resident time for a single tourist of 7 days. According to the data gathered in *Table 9* and *Figure 10*, the way to make a vacation more sustainable from the point of view of emissions is to combine a limited CO_2 emission in the phase of staying at the destination and to travel with a high efficient transportation means such a diesel car. The use of airplane is usually the worst choice from an environmental point of view, since it produces more than three times the emissions of a medium size car.

Pollutants and Greenhouse Gases Emissions Produced by Tourism Life Cycle: Possible Solutions to Reduce
Emissions and to Introduce Adaptation Measures

133

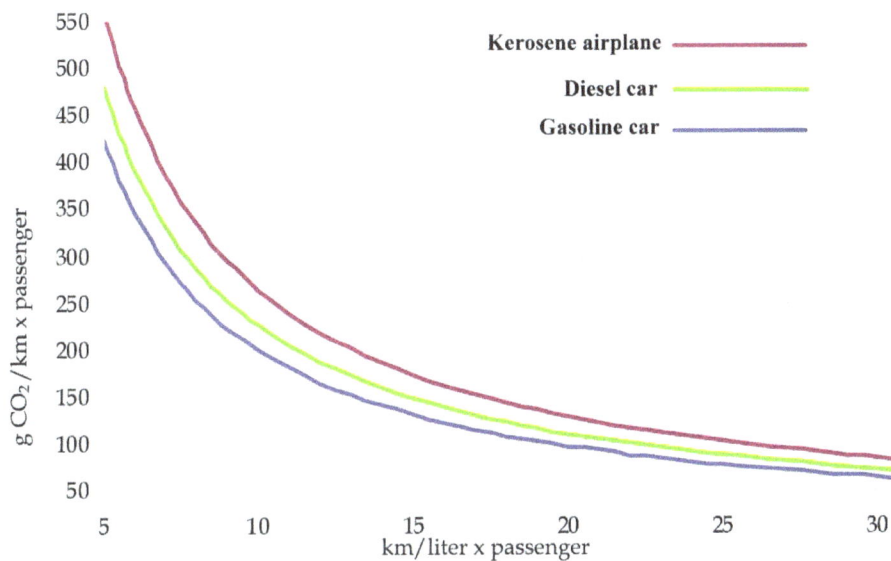

Figure 9. Specific CO_2 emissions for different type of fuel and transportation means.

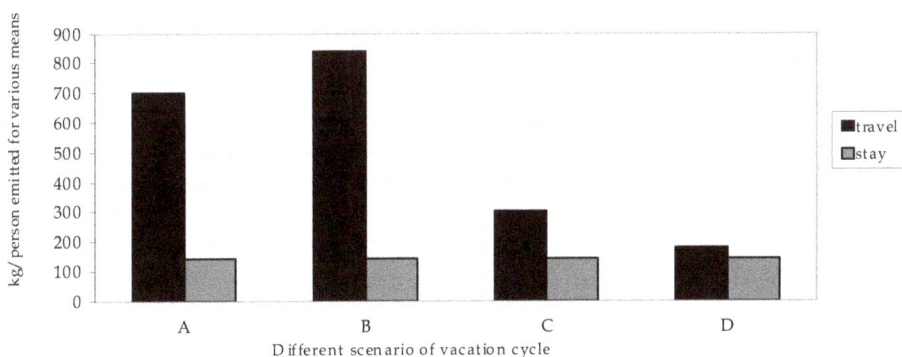

** Energy consumed at hotels. Source: [40]

Table 9. Specific CO_2 emissions different vacation scenarios.

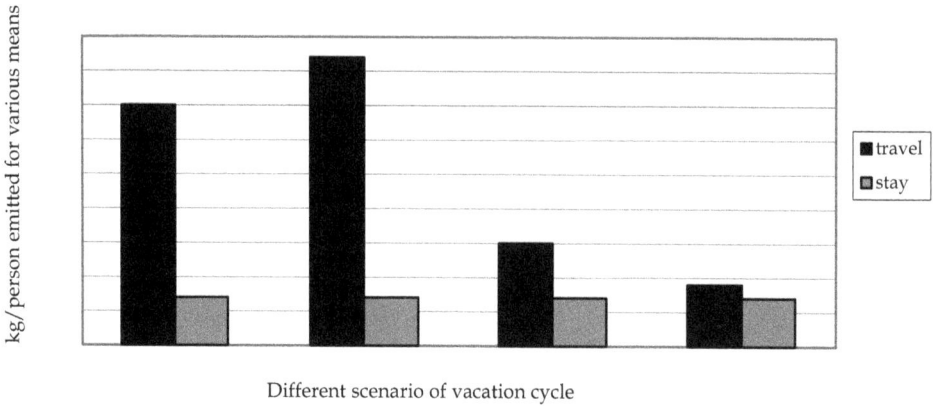

Figure 10. CO_2 emissions from vacation cycle (according to *Table 9*).

It appears that the best way to make sustainable T & T, from the point of view of climate change is to combine a limited CO_2 emission in the phase of staying and to travel with a high efficient transportation means such a diesel car. The use of airplane is usually the worst from an environmental point of view, even if it is more comfortable having a lower duration of the travel, but it produces more than 3 times the production of a medium size car.

11. Conclusions

T & T is a vector of climate change due to the GHG-emissions during the different phases of its development. On the other hand, the resulting climate change can compromise the environmental quality of a tourist destination, since climate conditions co-determines the suitability of locations for a wide range of tourist activities (sun, sea, snow, etc.). Hence, reduction of emissions constitute an essential component of T & T sustainability, particularly in the phase of mobility.

The analysis presented in this chapter shows that the reduction of greenhouse gas emissions from tourism mobility is economically unsustainable. The conclusion that air travel is the main cause of carbon footprint of tourism could bring to a reduction of this kind of transportation but a reduction of flights, would probably limit the profitability and growth of the tourism sector. Such a position would negatively affect the air transportation sector and would also produce a significant negative impact on tourism. However less drastic measures are possible, such as avoiding stops between the starting point and the destination. In addition, in order to reduce the threat of emissions, the aviation sector has already responded with a range of measures able to reduce fuel consumption such as fleet upgrades and changes in environmental practices. The question, thus, arises if it is possible to reduce the fuel consumption of airplanes with technological innovations.

At the destination, the application of different policies and measures to increase the sustainability of the T&T supplies to the consumers solutions that can be easily implemented, with the additional advantage of economic rewards. From a more general point of view, helping local communities to adopt practical strategies to deal with impacts of a changing climate, approach to a holistic, sustainable management through programmes for local development, e.g., protecting children, combating epidemics and promoting healthy eating, and adopting measures that include reduction of water and energy consumption, improvement of waste sorting, recycling and disposal, measures to preserve biodiversity, etc., can significantly improve T & T sustainability.

Author details

Francisco A. Serrano-Bernardo* and José L. Rosúa-Campos
Department of Civil Engineering. University of Granada, Spain

Luigi Bruzzi
Department of Physics. University of Bologna, Italy

Enrique H. Toscano
Joint Research Centre (JRC), European Commission, Karlsruhe, Germany

12. References

[1] Von der Weppen J (2009) Energy consumption in tourism and tourism mobility: An assessment of environmental impacts and prospects for sustainable energy management in coastal areas. MSc Thesis. European Joint Master in Water and Coastal Management. Universidad de Cádiz (Spain). 107 p.

[2] Mercopress. South Atlantic News Agency (2011) Travel and tourism industry account for 9% of world GDP and 260 million jobs. Available: http://en.mercopress.com/2011/09/28/travel-and-tourism-industry-account-for-9-of-world-gdp-and-260-million-jobs. Accessed 2012 Apr 10.

[3] Theobald WF (Ed.) (2005) Global Tourism. Burlington, Massachusetts: Elsevier. 588 p.

[4] Gössling S, Peeters P, Hall M, Ceron JP, Dubois G, Lehmann LV, Scott D (2012) Tourism and water use: Supply, demand, and security. An international review. Tourism Manage. 33: 1-15.

[5] Gössling S, Peeters P, Ceron JP, Dubois G, Patterson T, Richardson RB (2005) The eco-efficiency of tourism. Ecol. Econ. 54: 417-434.

[6] Iordache MC, Cebuc I (2009) Analysis of the impact of climate change on tourism in some European countries. Analele Stiinłifice Ale Universității "Alexandru Ioan Cuza" Din IASI Tomul LVI Stiinńe Economice. Ramnicu Valcea. pp. 270-286.

---

* Corresponding Author

[7] Xin Y, Lu S, Zhu N, Wu W (2012) Energy consumption quota of four and five star luxury hotel buildings in Hainan province, China. Energ. Buildings 45: 250-256.

[8] Butler RW (1980) The concept of a tourist area cycle of evolution: implications for management of resources. Can. Geogr. 24(1): 5-12.

[9] Choi HC, Sirakaya E (2006) Sustainability indicators for managing community tourism. Tourism Manage. 27: 1274-1289.

[10] Aramberri J (2009) The future of tourism and globalization: Some critical remarks. Futures 41: 367-376.

[11] World Tourism Organization (UNWTO) (2012) Tourism 2020 Vision. Available: http://www.unwto.org/facts/menu.html. Accessed 2012 Apr 10.

[12] Bruzzi L, Verità S, Von der Weppen J (2008) Sustainable coastal tourism – the role of energy efficiency and renewable resources. Proceedings International Congress "Sustainable tourism as a factor of local development" 7th-9th November 2008. Monza (Italy).

[13] Bruzzi L, Boragno V, Serrano-Bernardo FA, Verità S, Rosúa-Campos JL (2011) Environmental management policy in a coastal tourism municipality: the case study of Cervia (Italy). Local Environ. 16(2): 93-113.

[14] Rentziou A, Gkritza K, Souleyrette RR (2012) VMT, energy consumption, and GHG emissions forecasting for passenger transportation. Transport. Res. A-Pol. 46: 487-500.

[15] The Global Development Research Center (GDRC) (2012) The sustainable tourism gateway. Available: http://www.gdrc.org/uem/eco-tour/st-whatis.html. Accessed 2012 Apr 10.

[16] Scientific Applications International Corporation (SAIC) (2006) Life Cycle Assessment: Principles and Practice. Available: http://www.cemus.uu.se/dokument/msd/natres/lca%20by%20epa_ch1-2.pdf. Accessed 2012 Apr 10.

[17] O'Reilly AM (1986) Tourism carrying capacity: Concept and issues. Tourism Manage. 7(4): 254-258.

[18] Middleton VTC, Hawkins R (1998) Sustainable Tourism: A Marketing Perspective. Oxford: Butterworth-Heinemann. 224 p.

[19] Gössling S, Hansson CB, Hörstmeier O, Saggel S (2002) Ecological footprint analysis as a tool to assess tourism sustainability. Ecol. Econo. 43: 199-211.

[20] The New York Times. Green: A blog about energy and the environment (2008) Available: http://green.blogs.nytimes.com/2008/06/05/the-carbon-footprint-from-tourism/ Accessed 2012 Apr 10.

[21] United Nations Environment Programme (UNEP) (2012) Tourism and Hospitality. Available: http://www.unep.org/climateneutral/Topics/TourismandHospitality/tabid/151/Default.aspx Accessed 2012 Apr 10.

[22] Dubois G, Peeters P, Ceron JP, Gössling S (2011) The future tourism mobility of the world population: Emission growth versus climate policy. Transport. Res. A-Pol. 45: 1031-1042.

[23] Direction des Études Économiques et de l'Évaluation Environnementale (D4E) (2008) Les déplacements des touristes français ont représenté 6% des émissions de GES en 2006. Available:
http://www.actu-environnementcom/ae/news/etude_D4E_deplacements_touristiques_france_GES_4646.php4. Accessed 2012 Apr 10.

[24] Peeters P (Ed.) (2006) Tourism and Climate Change Mitigation: Methods, greenhouse gas reductions and policies. NHTV Academic Studies No. 6. The Netherlands: Stichting NHTV Breda. 207 p.

[25] Wit RCN, Boon BH, van Velzen A, Cames M, Deuber O, Lee DS (2005) Giving wings to emission trading. Inclusion of aviation under the European Emission Trading System (ETS): Design and impacts. Report for the European Commission, DG Environment. No. ENV.C.2/ETU/2004/0074r

[26] Grossman D (2009) Low-cost airlines face new competition-each other. USA Today. Available: http://www.usatoday.com/travel/columnist/grossman/2009-08-25-low-cost-carriers_N.htm. Accessed 2012 Apr 10

[27] Ministère de l'écologie, du développement durable, des transports et du logement (2008) Gaz à effet de serre : le poids du tourisme. Available:
http://voyages.liberation.fr/rechauffement-climatique/gaz-effet-de-serre-le-poids-du-tourisme. Accessed 2012 Apr 10.

[28] IPCC (2007) Climate Change 2007: Mitigation. Contribution of Working Group III to the Fourth Assessment Report of the Intergovernmental Panel on Climate Change Metz B, Davidson OR, Bosch PR, Dave R, Meyer LA, editors, Cambridge, United Kingdom and New York, NY, USA: Cambridge University Press. 851 p.

[29] World Tourism Organization (UNWTO) and United Nations Environment Programme (UNEP) (2008) Climate change and Tourism: Responding to Global Challenges. Madrid: WTO/UNEP. 256 p.

[30] Rypdal K (2000) Aircraft emissions. In: Penman J, Kruger D, Galbally I, Hiraishi T, Nyenzi B, Emmanul S, Buendia L, Hoppaus R, Martinsen T, Meijer J, Miwa K, Tanabe K, editors. Good Practice Guidance and Uncertainty Management in National Greenhouse Gas Inventories. Background paper, Japan: Institute for Global Environmental Strategies. pp. 93-102.

[31] European Union (2007) Regulation (EC) No 715/2007 of the European Parliament and of the Council of 20 June 2007 on type approval of motor vehicles with respect to emissions from light passenger and commercial vehicles (Euro 5 and Euro 6) and on access to vehicle repair and maintenance information (Text with EEA relevance)

[32] White Energy Services LLP (2012) How does air travel. Available:
http://cobee.gofreeserve.com/white-energy-services-LLP-updated/airtravel.php Accessed 2012 Apr 10.

[33] World Travel & Tourism Council (WTTC) (2012) Travel & Tourism 2011. Available: http://www.wttc.org/site_media/uploads/downloads/traveltourism2011.pdf.   Accessed 2012 Apr 10.

[34] Kuffner IB, Andersson AJ, Jokiel PL, Rodgers KS, Mackenzie FT (2008) Decreased abundance of crustose coralline algae due to ocean acidification. Nat. Geosci. 1: 114-117.

[35] Hoegh-Guldberg O, Mumby PJ, Hooten, AJ, Steneck RS, Greenfield P, Gomez E, Harvell CD, Sale PF, Edwards AJ, Caldeira K, Knowlton N, Eakin CM, Iglesias-Prieto R, Muthiga N, Bradbury RH, Dubi A, Hatziolos ME   (2007) Coral Reefs Under Rapid Climate Change and Ocean Acidification. Science 318: 1737-1742

[36] Queensland Government (2012) Reducing greenhouse gas emissions. Available: http://www.derm.qld.gov.au/sustainability/tourism/reducing-ghg-emissions.html. Accessed 2012 Apr 10.

[37] International Air Transport Association (IATA) (2009) A global approach to reducing aviation emissions. First stop: carbon-neutral growth from 2020. Available: http://www.iata.org/SiteCollectionDocuments/Documents/Global_Approach_Reducing _Emissions_251109web.pdf Accessed 2012 Apr 10.

[38] Peeters PM, Middel J, Hoolhorst A (2005) Fuel efficiency of commercial aircraft. An overview of historical and future trends. NLR-CR-2005-669. Amsterdam: Peeters Advies/ National Aerospace Laboratory NLR. 37 p.

[39] Le Figaro (2011) Comment réduire les émissions des avions. Available: http://www.lefigaro.fr/environnement/2011/10/13/01029-20111013ARTFIG00673-comment-reduire-les-emissions-des-avions.php Accessed 2012 Apr 10.

[40] Gössling S (2002) Global environmental consequences of tourism. Global Environ. Chang. 12: 283-302.

Air Pollution Monitoring and Health Effects

Spatial and Temporal Analysis of Surface Ozone in Urban Area: A Multilevel and Structural Equation Model Approach

S. B. Nugroho, A. Fujiwara and J. Zhang

Additional information is available at the end of the chapter

1. Introduction

Photochemical smog, first identified in Los Angeles in the late 1940s, nowadays is a widespread phenomenon in many of the world's population centers (Jenkin & Chemitshaw, 2000). Photochemical smog occurs when primary pollutants (nitrogen oxides - NOx and volatile organic compound – VOC created from burning of fossil fuel and biomass) interact in the presence of sunlight to produce a mixture of hazardous secondary pollutants (Stern, 1973). Major constituent of photochemical smog is surface (ground-level) O_3, which is not emitted directly into the atmosphere but formed as the product of photochemical reactions of its precursors, NOx and VOC (Seindfeld & Pandis, 1998). At the same time, pollutants also interacts each other to form other secondary pollutants as like acidifying substance and also particulates.

Concentration of atmospheric gases involved in forming O_3 and nitrogen oxides (NOx) changes rapidly with wind speed and direction, ambient air temperature, humidity and solar radiation. Chemical reactions of O_3 production and destruction progresses take place at the same time. O_3 concentrations are affected mainly by photochemical reactions, transport and diffusion process. The photochemical reactions are related to meteorological factors such as solar radiation, temperature and concentration of pollutants. In general, O_3 is closely related to the pollutants like NO_2, NO and NOx according to photochemical oxide interaction in local environment (Wang, 2003). The relationship between precursor pollutants and O_3, thus differ from one place to another due to the emission distribution and meteorology (Zhang & Kim, 2002). It is critical to understand the variability of ozone concentration across location and time.

In a spatial and temporal analysis, it is noteworthy to first clarify several technical terms: heterogeneity, variability, variation and variance. Heterogeneity refers to phenomenon that actual concentration measured at monitoring station changes across individual measurement. This study especially deals with the unobserved heterogeneity. It is well known that variance is a statistical term, representing the degree of variation. The variability means the fact that something being likely to vary. In this study, the later three terms are especially an aggregate of measurement` (or monitoring station`) heterogeneity. To quantitatively assess the properties of unobserved heterogeneity at various situations, we focus on various components, which correspond to the degree of variation caused by unobserved heterogeneity within monitoring station and also among locations by using monitoring data. We use regression-based method a multilevel model to capture temporal variations and spatial heterogeneity caused by land-use characteristics surrounding monitoring stations and its impact on surface ozone. A multilevel analysis was applied to analyze (a) daily event when peak concentration of ozone occurred, (b) daily average concentration of ozone and (c) possibility of phenomena of ozone weekend effect in Jakarta city represented by systematically day-to-day variation of event of peak ozone and daily average concentration of ozone.

In tropical regions, high O_3 level may be expected due to high rate of precursor emissions from anthropogenic and biogenic sources coupled with high sunlight intensity. Yet, there is only a limited research about tropical tropospheric O_3 focusing on Asian cities. The lack of systematic monitoring data of O_3 and its precursors is one of the barriers to scientific research for photochemical smog in most of the developing Asian countries (Zhang & Kim, 2002). In the context of urban areas, NO_2, NO and NO_x, which are generally highly associated with primary sources of air pollution, come from both mobile sources (automobiles) and stationary sources (e.g., household sector and industrial sector). An understanding of ozone (O_3) behavior near surface layer is essential for a study of pollution oxidation processes in urban area (Monoura, 1999). Ground level O_3 is formed from its precursors by complex and non-linear photochemical reaction in presence of sunlight. O_3 concentrations are very difficult to model because of the different interactions between pollutants and meteorological variables (Sousa, 2007).

Concerning the methods of analysis, although several multiple regression models are available to analyze urban air pollution especially surface O_3. It is however difficult to apply these models to deal with the complex cause-effect relationships among meteorological factors, primary pollutants under different wind conditions, and their influences on surface O_3. Therefore, our proposed structural equation model can flexibly represent the aforementioned causal interactions aspects. The development of such models usually involves the choice of appropriate model structures and nonlinear data transformation methods. Then, a spatial and temporal analysis was performed based on our structural equation model with latent variables. A spatial analysis based on spatial pattern is also carried out at two major land use types (i.e., suburban area-SU and central urban area (CA), and a roadside area-RA in central business district in Jakarta City. A temporal analysis was done at roadside station in central Jakarta by considering seasonal and weekly variations.

2. Literature review and methodology

2.1. Relationships between surface ozone and its precursors

In the O_3-NOx system, the dominant chemical reactions in the atmosphere are described below :

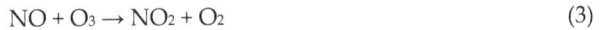

$$NO_2 + hv \rightarrow NO + O \tag{1}$$

$$O + O_2 + M \rightarrow O_3 + M \tag{2}$$

$$NO + O_3 \rightarrow NO_2 + O_2 \tag{3}$$

M represents N_2 or O_2 or another third molecule that absorbs excess energy and consequently stabilizes the O_3 molecule formed (3). The time scale of reaction (2) is very small (~10-6s) relative to the scales of reactions (1) and (3) (~100s and 30s, respectively) (Monoura, 1999). This is the result of O_3 destruction by NO in the nitrogen dioxide photolytic cycle, which is effective at a close distance to NO source due to its short cycle time (about several minutes) (Jenkin, 2000). Since the conversion from NO to NO_2 involving reactive hydrocarbons and the OH radical usually takes several hours, the higher concentration of O_3 is observed in both weekdays and weekend in dry season (Seinfeld & Pandis, 1998).

It is known that O_3 concentration and NO concentration show a logarithmic relationship, and the relationship between O_3 and NO_2 observed at the same time shows a typical linear function. A power function relationship is found between NO and NO_2 observed at the same time (Monoura, 1999). O_3 levels are negatively relevant to nitric oxide and positively to nitrogen dioxide, weakly affected by carbon monoxide (CO) and hardly affected by sulphur dioxide (SO_2) and respirable suspend particles (RSP). A case study in Hong Kong confirms a strong linear relationship between O_3 and NO_2/NO concentration in 1999 and 2000 (Wang, 2003).

High emission of NO from automobile traffic should be the major reason for low O_3 at the curbside (roadside) and lower O_3 at ambient monitoring station. In a city like Bangkok where the emission of NO from traffic is rather uniformly spread over a large area, the processes of O_3 destruction (by NO) and formation should be competing at any locations. Therefore O_3 level is found to be high over the city except for the very heavy traffic center and curbside where the O_3 destruction by NO is significant (Zhang & Kim, 2002).

2.2. Meteorological factors influencing surface ozone

The meteorological conditions of a region (e.g., sunlight, temperature, wind speed, and other factors) also directly affect the formation of O_3. In general, episodes of high O_3 concentration are associated with slow-moving, high barometer pressure weather system. Clear skies, sunshine, and warm conditions usually accompany high-pressure system, accelerating the photochemical formation of O_3 (Rubin, 2001). The relationship between the meteorological variation and daily maximum O_3 concentration can be well represented by a linear function (Gardner & Dorling, 1998).

Solar radiation

O_3 production is dependent on solar radiation, and consequently solar radiation intensity and O_3 concentration usually show positive correlation (Monoura, 1999).

Ambient air temperature

Meteorologically, high temperature is frequently associated with high pressure, stagnant conditions that lead to high O_3 concentration at vertical level (Seinfeld & Pandis, 1998). The rate of photochemical reaction increases as air temperature rises. In many O_3 prediction models, air temperature was found to be the strongest single predictor of O_3 concentration (Boriboonsomsin & Uddin, 2005). In urban and metropolitan areas, paved surface, high-rise building and other constructed surfaces cause air temperature to be higher due to the heat transfer of these surfaces.

Wind speed and direction

Wind speed associated with high-pressure system is typically low. Therefore pollutants stay longer over urban areas and accumulate in the atmosphere (Rubin, 2001). Calm or light winds allow more emissions to accumulate over large area, which result in higher concentration of O_3 precursors. O_3 formation and transport is a complex phenomenon, and O_3 concentration depends on wind speed and direction among others (Hubbard & Cobourn, 1998). The dispersion of air pollutants is roughly inversely related to wind speed (Zhang, 2002). Higher wind speeds promote the dispersion of O_3 concentrations (Sanchez-ccoyllo, 2006). Wind direction is also highly related to O_3 level, for example, downwind locations of precursor emission sources are strongly inclined to high concentration of surface O_3.

Precipitation

Precipitation is one of O_3 destruction mechanisms due to a wet deposition. In this study, precipitation is expressed as relative humidity level. Most tropical rain forest countries such as Indonesia have high relative humidity, especially during night time and wet season.

2.3. Development of surface ozone model in urban areas

2.3.1. Existing model

Various models have been developed to describe the relationship among factors to surface ozone. These models include simple contingency tables, multiple linear and non-linear regression models, time series techniques (Benarie, 1980), artificial neural network approaches and fuzzy logic based methods (Wang, 2003). Linear regression model is a classical and easily applied method. It uses a linear combination of factors to explain the ozone behavior. Artificial neural network approach is capable of modeling complex nonlinear phenomena, but its main drawback is that it results in a 'black box' model which it isn't easy to interpret or justify. Fuzzy logic also allows one to model complex nonlinear phenomena (Peton, 2000). Since fuzzy logic is based on a set of empirical rules, the inherent cause-effect relationships and interactions among factors of the ozone cannot be flexibly incorporated. Time series technique is suitable to capture the temporal change of ozone itself, but they are not capable of incorporating the influential factors into the models.

Multiple regression models have been commonly used for describing the ozone in the last few decades (Boriboonsomsin, 2005). Gardner and Dorling (2000) found that the relationship between meteorological variables analyzed and the daily maximum ozone concentration could be well represented by a linear model. Linear regression gives a first-order approximation of a non-linear function, is easy to calculate and very robust (Geladi, 1999). However, it is quite difficult to apply such linear regression models to properly capture the nonlinear relationships among variables, and to represent the inherent cause-effect relationships and interactions in the model structure. Therefore, it is required to establish an alternative surface ozone model.

2.3.2. Multilevel analysis

Multilevel models are the expansion of classical regression model which data were classified in groups, thus allow coefficients to vary for each group. This has been a popular approach applied in many fields, such as properties and its relation to PM_{10} (Pattenden et al., 2000), pure properties aspect (Gelfanda et al., 2007), and land use fields for crops (Overmars K.P., and Verburg P.H. 2006). The benefits of multilevel models are allows random variations and explanatory variables to be incorporated inside the model at different levels.

Multilevel models are considered as a regression model in which the ultimate power lays on the regression coefficients that are given a probability model (Gelman and Hill, 2007). The second-level has parameters of its own which are estimated from data. Varying coefficients across different levels are a critical difference from classical regression models. Also, those varying coefficients serve as a model as well. Although classical regression models sometimes are also able to accommodate varying coefficients by using explanatory variables, however multilevel models has one ultimate attractive feature that it allows for modeling of the variation between groups, which classical regression is incapable off.

The multilevel model essentially treats multiple hierarchical and cross-classifications unobserved heterogeneities by introducing corresponding variation components. To describe the variations concentration pollutant i, in multilevel analysis, the model buildings strategies can be either top-down and bottom-up (J.J Hox, 2010). In this study, we select bottom-up approach in which analysis starts with a simplest model and proceed by adding parameters. Concretely speaking, first, we start with model without explanatory variables (called *Null* model). This model, the intercept-only model, can be defined as follows:

$$Y_{ij} = \gamma_{00} + \mu_{oj} + \varepsilon_{ij} \tag{4}$$

where γ_{00} is regression intercept and μ_{oj} and ε_{ij} are residuals at group-level and individual-level (Here, "group level" means monitoring sites, and "individual level" means measurements within the same station), following the normal distribution with mean 0 and variances $\sigma_{\mu0}^2$ and σ_e^2, respectively. Using *Null* model, it possible to clarify reason of "why the concentrations are fluctuates?" based on the component of variance. It is also gives estimate of interclass correlation (ρ) among measurements in stations. The interclass correlation (ICC, δ) is estimated as follows:

$$\sigma_{\mu0}^2 / (\sigma_{\mu0}^2 + \sigma_e^2) \tag{5}$$

Second, we analyze a model with all explanatory variables (called as the *Full* model). This model is expressed as follows:

$$Y_{ijk} = \gamma_{00} + \gamma_{10}X_{ijk} + \mu_{oj} + \varepsilon_{ij} \tag{6}$$

Where Y_{ijk} is dependent variable concentration of pollutant i at monitoring station j of measurement k. γ_{00} and $\gamma_{\lambda0}$ are unknown parameters, X_{ijk} indicates explanatory variables including monitoring station` j attributes (e.g., emission intensity which reflected by systematically day-to-day variation, open space area nearby station, etc), atmospheric situations (e.g., presence or concentration of other pollutants), temporal attributes (e.g., annual variation and seasonal variation). Parameters μ_{oj} and ε_{ij} represent random components which indicate inter- monitoring location variation and inter-measurement variation within same location respectively. In this step, we assess the contribution of explanatory variables. The significance of each predictor can be tested and also possible to assess what changes occur in the first-level and second-level variance terms. We use chi-square test based on the deviances of *Null* and *Full* models to test the assumption whether variation across group is significant. Whenever explanatory variables introduced, we expect the variance $\sigma_{\mu0}^2$ and σ_e^2 to go down or in other words the introduced explanatory variables explain part of measurements and part of monitoring station variances.

2.3.3. Structural equation model with latent variables

This paper also proposes to apply a structural equation model with latent variables to capture the complex cause-effect relationships and interactions in photochemical process. Structural equation model (SEM) is a modeling technique that can handle a large number of the observed endogenous and exogenous variables, as well as (unobserved) latent variables specified as linear combinations (weighted averages) of the observed variables (Golob, 2003). The models play many roles, including simultaneous equation systems, linear causal analysis, path analysis, structural equation models, dependence analysis, and cross-legged panel correlation technique (Joreskoq, 1989). It is a confirmatory, rather than explanatory method, because the modeler is required to construct a model in term of a system of unidirectional effects of one variable on another. SEM is used to specify the phenomenon under study in terms of putative cause-effect variables and their indicators. Following the descriptions by Jöreskog and Sörbom (1989), the full model structure can be summarized by the following three equations.

Structural Equation Model:

$$\eta = B\eta + \Gamma\xi + \zeta \tag{7}$$

Measurement Model for y:

$$y = \Lambda_y\eta + \varepsilon \tag{8}$$

Measurement Model for x:

$$x = \Lambda_x\xi + \delta \tag{9}$$

Here, $\eta' = (\eta_1, \eta_2, ..., \eta_m)$ and $\xi' = (\xi_1, \xi_2, ..., \xi_m)$ are latent dependent and independent variables, respectively. Vectors η and ξ are not observed, but instead $y' = (y_1, y_2, ..., y_p)$ and $x' = (x_1, x_2, ..., x_q)$ are observed dependent and independent variables. $\zeta, \varepsilon, \delta$ are the vectors of error terms, and $B, \Gamma, \Lambda_x, \Lambda_y$ are the unknown parameters.

An important feature of SEM is that it can calculate not only direct effects, but also total effect (Golob, 2003). Direct effect is the link between a productive variable and the variable that is the target of the effect, which corresponds to an arrow in a path diagram. These direct effects embody the causal modeling aspect of SEM. Total effects are defined to be the sum of direct effects and indirect effects, where the indirect effects represent the sum of all the effects along paths between two variables that involve intervening variables. Advantages of SEM compared to most other linear-in-parameter statistical methods include the following capabilities: (1) treatment of both endogenous and exogenous variables as random variables with error of measurement, (2) latent variables with multiple indicators, (3) test of a model overall rather than coefficients individually, (4) modeling of mediating variables, (5) modeling of dynamic phenomena such as habit and inertia (Golob, 2003). One can see that SEM has a very flexible model structure to simultaneously represent various interdependent variables. Therefore, in this study, we adopt the SEM to model and analyze surface ozone in Jakarta City.

The model was built using 11 observed variables that consisted of three meteorological factors (SR, T and RH), two wind factors (WS and WD), five primary pollutants (NO, NO_2, CO, SO_2 and PM_{10}) and a surface O_3. The four latent variables $\xi_1, \eta_1, \eta_2, \eta_3$ as shown in Figure 1 represents these four groups of variables respectively. ξ_1 indicates an exogenous latent variable, and η_1, η_2, η_3 are the endogenous latent variables. The latent variable η_3, which is defined by using both O_3 and its precursor NO, describes the photochemical matters in this study.

Since the SEM still possesses a linear model structure, to capture the non-linear relationship between some variables, here several observed variables need to be properly transformed. The empirical observations results of Jakarta air quality data indicates that the relationship between O_3 concentration and NO concentration may be explained by a negative logarithm function and the relationship between NO and NO_2 by a logarithm function. In addition, the existing research (Monoura, 1999) suggests that the relationship between O_3 and NO_2 is best described by a linear function. The non-linear phenomena is represented by a natural logarithm (LN) function, therefore the pollutant NO is transformed into a new variable LN_NO. LN_NO, NO_2, CO, SO_2 and PM_{10} are specified in one-to-one relationships with the latent variables "Primary Pollutants" (η_2). This latent variable η_2 is specified to represent the influence of primary pollutants emitted from both gasoline and diesel vehicles. The latent variable "Photochemical" (η_3) corresponds to several chemical reactions in photochemical process (Seinfeld & Pandis, 1998).

For the structural equation model with multiple endogenous variables, especially with latent variables, model estimation becomes more challenging, and quite a few different methods have been developed (Golob, 2003). The most commonly used estimation methods are maximum likelihood (ML), general least squares (GLS), weighted least squares (WLS), asymptotically distribution free weighted least squares (ADF or ADF-WLS) and elliptical re-weighted least squares (EGLS or ELS). The most often used estimation method is ML, which maximizes joint probabilities that the observed covariance are drawn from a population that has its variance-covariance generated by the process implied by the model, assuming a multivariate normal distribution.

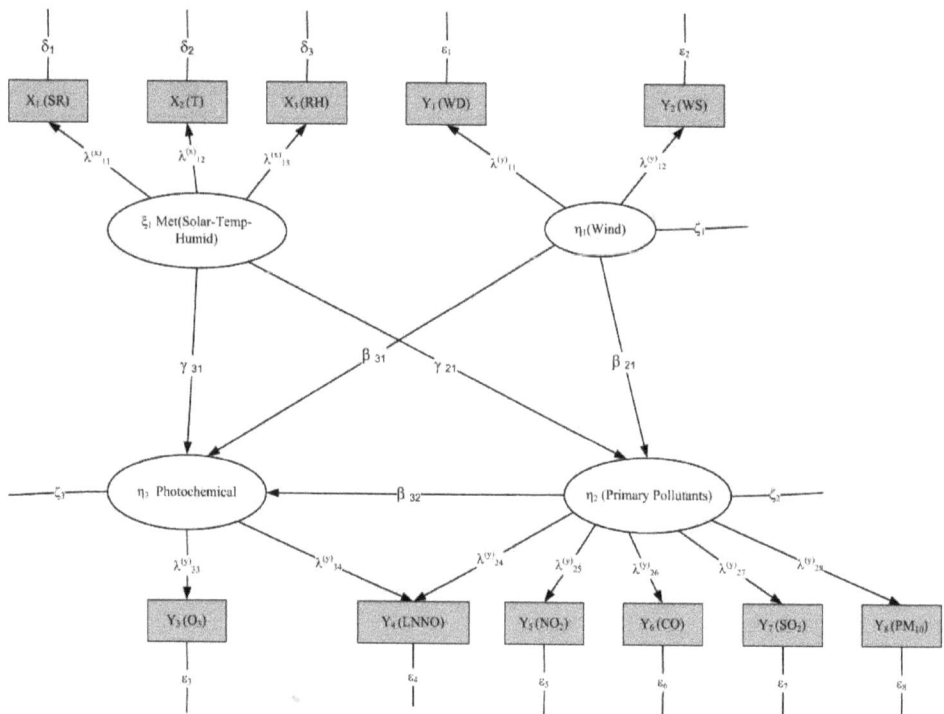

Figure 1. Air Pollutants Interactions Model for Jakarta City

Several criteria have been developed for assessing overall goodness-of-fit of a structural equation model and are used to determine how well one model performs than others. Such model accuracy indices includes: (a) root mean square residual (RMR), (b) standardized RMR (SRMR), (c) the goodness-of-fit index (GFI), (d) adjusted goodness-of-fit index (AGFI) which adjusts GFI for the degree of freedom in the model, and (e) the parsimony-adjusted goodness-of-fit index (PGFI). In this study, the GFI and AGFI are used to assess the models and to compare model results for different areas. Nowadays, there are several software that can estimate the structural equation models. The Analysis of Moment Structure (AMOS) software, which has a very attractive and user-friendly interface is used for this study.

In the work by Boriboonsomsin and Uddin (2005), they incorporated precursor emissions (mobile sources and stationery sources) into the model and found that traffic is highly associated with the change of O_3 concentration. The traffic behaviors are strongly influenced by land use type, which in the behavior of pollutant species are reflected as spatial and temporal variables such as location of stations and systematically day-to-day variation. It assumed that day-to-day variation has linear relationship with traffic data and it is expected lower emission intensity occurs on weekend as result of decreasing vehicle usage on weekend days. Furthermore, we also assumed that variation of emission intensity especially in weekend days will affect simultaneously on concentration of primary pollutants in weekend days. Then, this study examines those impact on secondary pollutants ozone.

3. Study area and data

3.1. Description of study area

Jakarta is comprised of 664 km² land area and stretchs along the coast of the Java Sea. The topography is very flat with a mean elevation of seven meters above sea level. Jakarta is a part of the greater Metropolitan Jabodetabek (Jakarta, Bogor, Depok, Tangerang and Bekasi) area. Jakarta's climate is generally tropical. The 'rainy/wet' season starts from November to March and 'dry' season from May to September. A few weeks in April and October are the transition period between dry and wet seasons, respectively.

The Jakarta Office of Environment (Bapedalda DKI Jakarta and later BPLHD DKI Jakarta) has regularly monitored the air pollution in Jakarta since 1985. At the beginning, twelve manual monitoring stations that are located at housing, industrial, recreation and mixed areas measures sulphur dioxide (SO_2), nitrogen oxides (NOx), and total suspended particulate (TSP) (Haq, 2002). Those stations are operated on a rotational basis, and the parameters are measured for twenty-four hours every eight days at each manual monitoring station (Syahril, 2002). Since 1992, Jakarta has another six continuous monitoring stations which consist of four ambient fix stations and two roadside fix stations. The fix monitoring stations records air quality every 10 minutes. At the end of 2001, another six new monitoring stations were activated which consist of five ambient fix stations and one mobile roadside station. These stations equipped with measurement analyzers to monitor NO, NO_2, NO_x, SO_2, CO, O_3 and PM_{10} every 30 second. The fix stations are centrally connected to data computer at Jakarta Office of Environment and the data are transferred every half an hour.

No	Monitoring Stations	Location	Land-use
A	Ambient Stations (Fixed Station)		
1	Gelora Bung Karno (Senayan)	Central Jakarta	City center-commercial area (CBD)-
2	Kemayoran	North Jakarta	Commercial & Industry-Urban Fringe
3	Kantor Walikota Jakarta Timur	East Jakarta	Residential – Sub urban
4	Pondok Indah	South Jakarta	Residential – Urban fringe
5	Kantor Walikota Jakarta Barat	West Jakarta	Commercial and residential area-Sub Urban
B	Roadside (Mobile) Station		
1	Casablanca	Central Jakarta	Central business district (CBD)

Table 1. Air Quality Monitoring Stations in Jakarta City

Land-use map & Air Quality Monitoring Stations

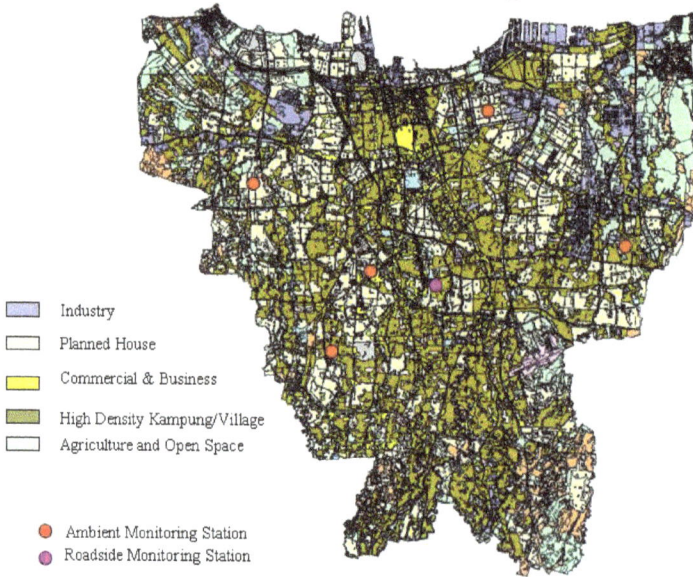

Figure 2. Air Quality Monitoring Stations in Jakarta city

Nowadays, only the latest five fix stations that remains to provide air quality data on daily basis for parameters CO, NO, NO_2, SO_2, PM_{10}, and O_3. The data are used to calculate the Pollutants Standard Index (PSI), which are subsequently published on data displays to the public. In-situ meteorological data i.e. solar radiation (SR), temperature (T), relative humidity (RH), wind speed (WS) and direction (WD) are also recorded using the basic meteorological sensors, which are installed at 10 meter height above the ground. Four data displays are located at Gambir (central Jakarta), Kelapa Gading (east Jakarta), Pondok Indah (south Jakarta) and Grogol (west Jakarta). Figure 2 and Table 1 provides detail information on the stations location.

This study used air quality data for weekday and weekend at wet and dry season in 2001-2003 from five fixed ambient monitoring stations and the roadside street-level ambient monitoring station. The general ambient air quality monitoring stations are located more than 100 meters away from main roads and the roadside street-level ambient air quality monitoring station is located 5-10 meter from the main road. The five of monitoring stations are Senayan (Central Jakarta), Kemayoran (North), Pondok Indah (South Jakarta), Walikota Jakarta Barat and Walikota Jakarta Timur (East station). The West Station (SUW) is located 20 km from city center and represents suburban area at western part of Jakarta. The East Station (SUE) is located 25 km from city center and represents suburban area in eastern part of Jakarta. The Senayan Station (CA) is located at city sport facilities in Jakarta's central business district area. This station is nearby the heaviest traffic roads in Jakarta (Jl Sudirman and Jl Gatot Subroto). The North Station (NUF) and South Station (SUF) are represents urban fringe area non-CBD in north and south Jakarta. Finally, the Roadside Station (RA) is located at the Jakarta Office of Environment on Jl Casablanca, which is also located in central business district area.

These all stations were selected to make a spatial and temporal analysis of the surface O_3 behavior in Jakarta city. Analysis was performed for several set situations as provided in table 2.

No	Type of Analysis	Approach	Data
1	Spatial and temporal variations of daily peak concentration of ozone (analysis of events)	Multilevel Analysis	Events of peak concentration of ozone at five six stations on 2001 to 2003.
2	Spatial and temporal variations of daily average concentration of	Multilevel Analysis	Daily average concentration at five fixed station in 2001-2003. Parameter: PM_{10}, SO_2, CO, O_3, NO_2, and NO
3	Spatial and temporal Analysis of causal interaction among pollutants	Structural Equation Model	Spatial Analysis: Three stations at West Jakarta (SA), Central Jakarta (DA) and mobile station (RA) in Dry season 2003 Temporal Analysis: Seasonal variation and weekly variation at Roadside station (RA) in 2003.

Table 2. Distribution of data in Spatio-Temporal Analysis

3.2. Ambient air quality monitoring data in Jakarta city

Table 3 summarizes the data availability for diurnal analysis from six current monitoring stations in Jakarta. Due to technical failure, the data from North and South Stations were incomplete, therefore only the data from the four remaining stations were used in this diurnal analysis. The weekly variation for dry and wet seasons in year 2003 that start from 00.30 a.m. on Monday and end at 24.00 on Sunday were identified. The data time interval is 30 minutes, therefore 336 average concentration data should be available in a week for each corresponding hour and day in a week. The results of analysis for pollutants O_3 is discussed below.

Locations	Data Avialability			
	Dry Season		Wet Season	
	Weekdays	Weekend	Weekdays	Weekend
East	5520	2208	6240	2496
West	3648	2496	3456	2496
Central	5568	2160	4128	1632
Roadside (Central)	5760	2352	5520	2208
North	NA[1]	NA[1]	NA[1]	NA[1]
South	NA[1]	NA[1]	12 [2]	NA[1]

Note: NA[1]: Not available for NO and NO_2
12[2] : Limited data for NO and NO_2

Table 3. Data availability for diurnal analysis

Figures 3 and 4 show weekly variations of average O_3 concentrations at each station during wet and dry seasons in year 2003, respectively. The concentrations of O_3 increased after the sunrise and reached the highest level at around 10:00-12:00 a.m. in all the locations. We found only a single peak of O_3 occurs in a day. It is obvious that the formation of O_3 was coincided with the abrupt dropped of NO concentrations after sunrise. During the daytime, the O_3 production was faster than the O_3 consumption. During this period, some O_3 might be transported from the upper atmosphere to the ground level accompanied by convection in the mixing layer (Monoura, 1999). The highest average concentration for dry season was identified at the Central Station (CA), but not for wet season. The average concentration of O_3 showed a seasonal variation, which average concentrations for dry season were slightly high. Although the highest daytime O_3 concentration during wet and dry season is measured at the East Station, the lowest concentrations were also measured at the same location.

The findings for O_3 concentration variation seems in agreement with the Hubbard & Cobourn (1998) finding that indicates that unlike primary pollutants, the O_3 concentration does not show obvious weekly cycles. Unlike CO and SO_2 which showed a weekly cycle with lower concentration during the weekend at the Roadside Station (RA), the O_3 concentration remained stable. The findings reveal that the ambient air quality standard for 1-hour O_3 (200 ug/m^3-1hr, Governor Decree of DKI Jakarta no 551/2001) was exceeded several times at all the locations.

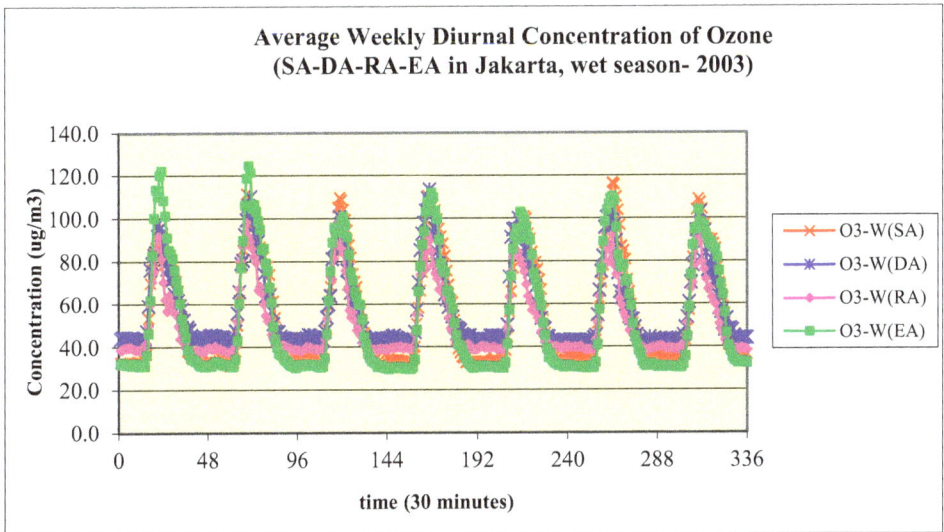

Figure 3. Weekly variations of average O₃ concentrations during wet season in 2003

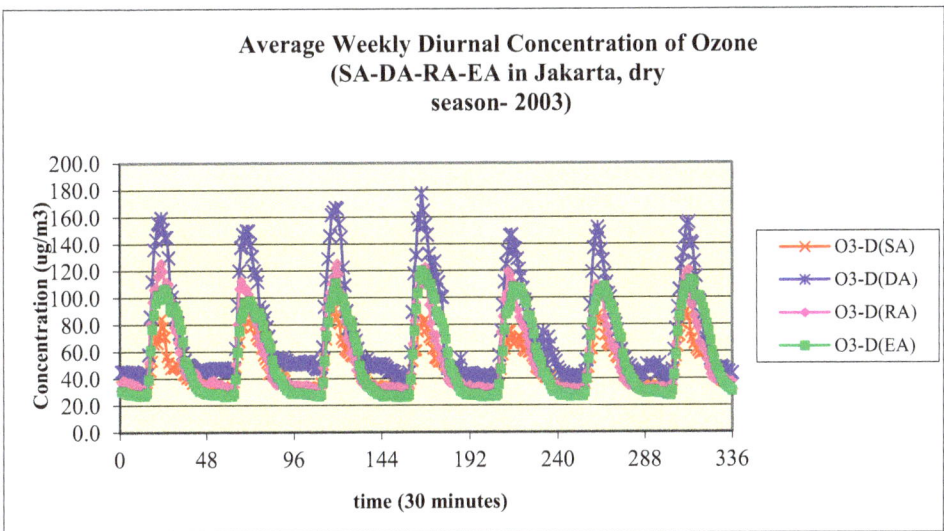

Figure 4. Weekly variations of average O₃ concentrations during dry season in 2003

3.3. Observed causal interaction among pollutants

In order to enhance understanding of the surface O₃ behavior in Jakarta, it is necessary to examine the relationships among O₃ precursors and meteorological factors. Figure 5 shows the relationship between NO and O₃ at the Roadside Station., A logarithmic relationship is observed between O₃ concentration and NO concentration as indicated in solid lines. The highest R^2 0.1319 is obtained for weekday-wet season. O_3 formation is solar radiation (SR) dependent. Figure 6 shows the relationships between O_3 and SR that are linear at three different areas. The highest R^2 value is found for weekday-dry season. Some observed relationships between O_3-NO, NO_2-NO and O_3-SR might be derived from the reactions (1) ~ (3) as mentioned earlier in the paper and follow the basic photochemical cycle of NO, NO_2, CO, O_3 and SR (Seinfeld & Pandis). These observations are helpful to develop and understand the structure of surface O_3 model for urban roadside in Jakarta city.

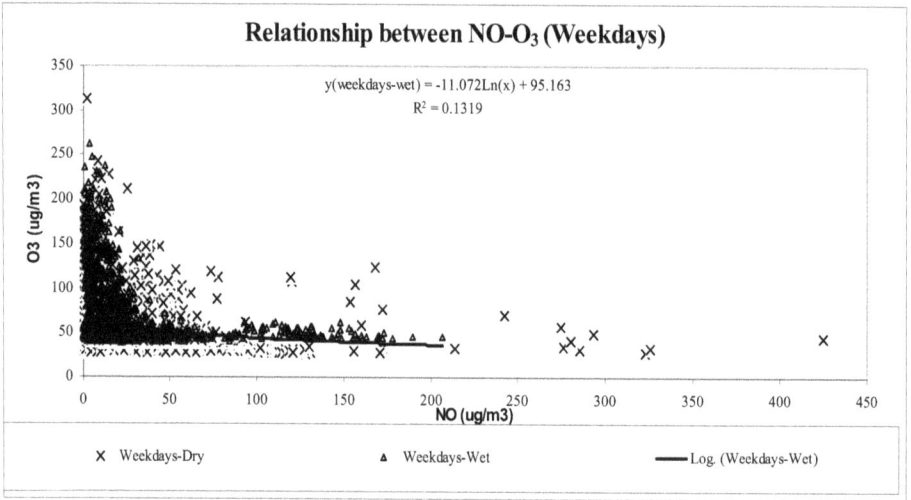

(a) Relationships between O_3 – NO for Weekday Situations

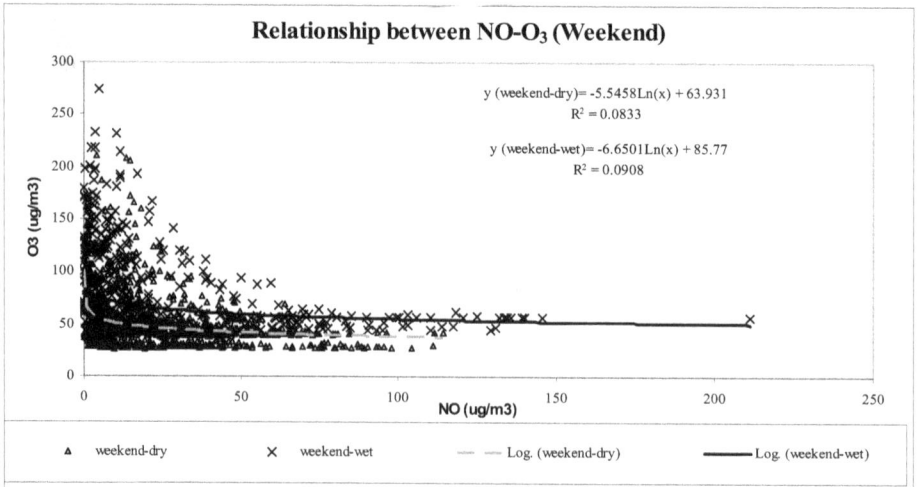

(b) Relationships between O_3 – NO for Weekend Situations

Figure 5. Relationships between O_3 – NO at roadside station in 2003

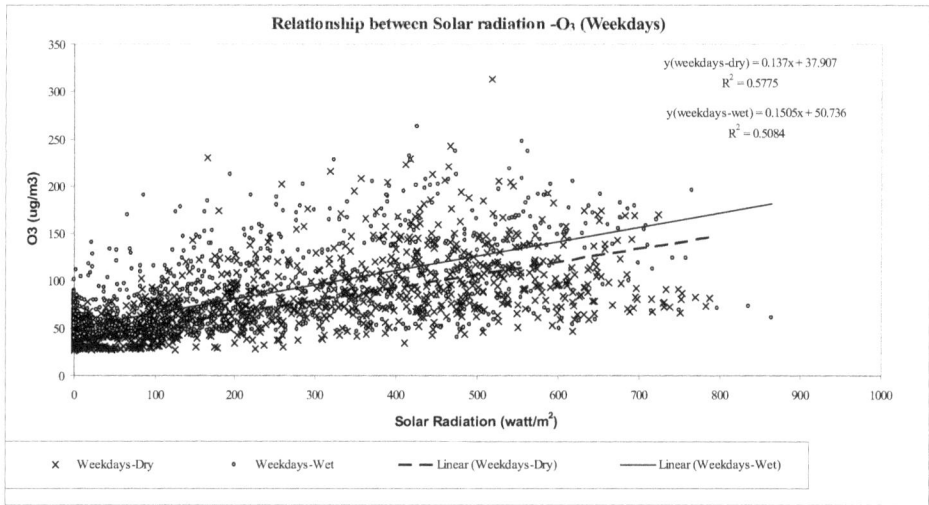

(a) Relationships between O₃ – SR for Weekday Situations

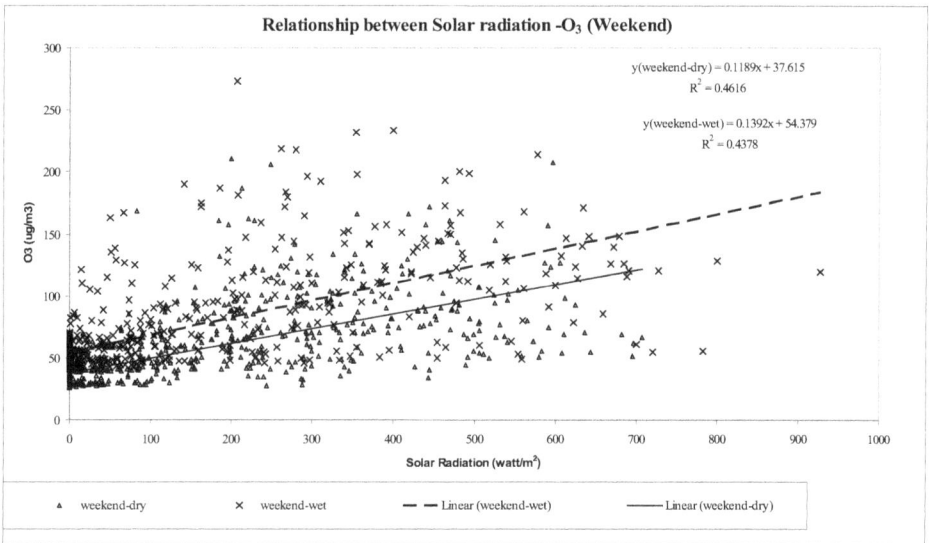

(b) Relationships between O₃ – SR for Weekend Situations

Figure 6. Relationships between O₃ – SR at roadside station in 2003

4. Result and discussion

This section discuss about estimation results for several issues mentioned in above. It is organized as follows. First part discuss about spatial and temporal analysis by multilevel approach and secondly spatial and temporal analysis of causal interaction among factors in urban ambient air pollution. In the first part, there are two main topics to be analyzed which are (a) daily event of peak concentration of ozone, when it happened and (b) analysis of daily average ozone concentration. In the second part, spatial and temporal analysis was done by using the proposed structural equation model.

4.1. Spatial and temporal analysis by multilevel approach

4.1.1. Spatial and temporal variation of Events of Daily Peak Concentration of ozone

The dependent variable, time of daily peak concentration of surface ozone is expressed in minute counted from midnight as zero. First, the *Null* model is estimated for intercept (location) only and the result is presented in Table 4. Estimation result show only small variation (1.7%) of event of daily peak concentration due to different location in Jakarta city. Next step, it is necessary to examine how much of unobserved variance of random component can be explained by observed information. We use half model (spatial and temporal information) and full model (spatial, temporal and systematic day-to-day variation) to examine unobserved variance. Both two models show zero random component of inter-monitoring which means there is no variation among locations. The selected variable of observed information successfully explained all unobserved variance of random component (1.7%) of the *Null* model.

Comparing the *Null*, *Half* and *Full* models as shown in table 4, we could conclude that variation of event when peak concentration of ozone happened mostly caused by locations. The dummy variable of Sub-urban and Urban-fringe show the event of peak concentration ozone in Sub-urban and Urban Fringe usually 38 and 40 minutes later than Central Business District (urban core/central Jakarta) around 688 minutes from midnight or 11:28 am. The temporal variations are insignificant in all temporal variables which are long-term (annual), seasonal and weekly (day-to-day variation). Looking at systematically day-to-day variation, by using event peak on Tuesday, Wednesday and Thursday as the references, we could see there are insignificant different among other days. This result support the findings for O_3 concentration variation seems in agreement with the Hubbard & Cobourn (1998) finding that indicates that unlike primary pollutants, the O_3 concentration does not show obvious weekly cycles.

4.1.2. Spatial and temporal variation of Daily Average Concentration of ozone

The dependent variable, daily average concentration of surface ozone is expressed in ug/m^3 as also measured by automatic ambient air monitoring stations. First, the *Null* model is estimated for intercept (location) only and the result is presented in Table 5. Estimation result shows variation around 22.6% due to different specific characteristic among monitoring station which contribute to the variation of daily average concentration. The rest parts are due to variations inside the boundary nearby stations which influence on ambient air pollution measured at the stations. Next step, it is necessary to examine how much of unobserved variance of random component can be explained by observed information. We use half model (spatial and temporal information) and full model (spatial, temporal and interaction with other pollutants in ambient air) to examine unobserved variance. In the *Half* model we could found there is no significant different among location (spatial impact). The estimation results of dummy variable Sub-urban and Urban fringe are insignificant. As for temporal aspect, long-term aspect (annual impact) shows positive and significant which mean daily average concentration of ozone increase year by year significantly. It shows consistent result (positive and significant) in the *Full* model. In the full model, we also found the positive significant impact of dummy variable wet season on surface ozone concentration.

No	Description	Null Model	Half model: Spatial & Temporal	Full Model: With systematic day-to-day
I	**Fixed Part**			
A	**Intercept (Location)**	706.243(84.95)	688.521(112.17)	687.298(104.22)
B	**Spatial**			
1	Sub-Urban (Dummy)		38.111(6.47)	38.147(6.48)
2	Urban Fringe(Dummy)		40.553(6.34)	40.614(6.35)
C	**Temporal**			
1	Long-term (Year)		-3.013(-1.28)	-3.007(-1.28)
2	Seasonal (Dummy wet season)		-7.862(-1.55)	-7.894(-1.56)
3	Weekly Weekend (Dummy)		0.580(0.100)	
D	**Systematic day-to-day variation**			
1	Monday			-6.660(-0.86)
2	Friday			12.450(1.62)
3	Saturday			-6.644(-0.86)
4	Sunday			10.316(1.33)
II	**Random Part**			
	σ_e^2 (Within monitoring)	21527.55	21518	21489
	$\sigma_{\mu 0}^2$(Inter-monitoring)	374.16	0	0
III	**Model Performance**			
	AIC	43456	16060	14499
	BIC	43474	16104	14570
	-2*Log likelihood	43450	43406	43382
	Degree of freedom	3	8	11
	No of Samples	3390	3390	3390

Note: ( ) → t-statistic

Table 4. Model of Daily Event of Peak Concentration of Ozone (Peak O_3)

Looking at *Full* model, the model performance is increase based on some indicators such as AIC, BIC and log likelihood estimation. The inter-monitoring location` variances also decrease from 22.5 % (*Null* model) to 8.2% in the *Full* model and selected observed variables show meaningful information to explain unobserved variance properties. Instead of spatial and temporal variables, the interaction effect of pollutants on surface ozone is also significant. By using *Full* model, we successfully explore the significant impact of ozone precursors (NO_2 and NO) and PM_{10}. We leave other two parameters (SO_2 and CO) since the estimation results show insignificant effects of these two parameters on daily average concentration of ozone. Daily average concentration of PM_{10} slightly increase ozone concentration while in contrast, NO_2 will decrease ozone concentration. The ratio between NO and NO_2 is crucial factor since it give a negative and significant impact on ozone. This result leads to policy maker to manage the ratio NO and NO_2 to decrease ozone concentration in urban area. Finally, we also found accumulation impact on surface ozone concentration. By using dummy variable of prior day concentration (t-1), this dummy variable significantly shows a positive sign which mean today`s average concentration of ozone is significantly affected by yesterday` concentration, a time series dependent concentration phenomena. We leave systematic day-to-day variation in *Half* and *Full* model since this variables are insignificant. This result also support the findings for O_3 concentration variation seems in agreement with the Hubbard & Cobourn (1998) finding that indicates that unlike primary pollutants, the O_3 concentration does not show obvious weekly cycles. We can preliminary conclude that there is no ozone weekend effect phenomena in Jakarta city.

4.2. Spatial analysis on causal interaction by structural equation model

4.2.1. Spatial analysis

The model for the Sub-urban west (SUW) shows the highest GFI (AGFI) value of 0.787 (0.629), followed by that for the RA with the value of GFI (AGFI) 0.770 (0.600). The model for the CA has the lowest GFI (AGFI) of 0.731 (0.533). Peton (2000) highlights that environmental data usually have some measurement and sampling errors. These errors may due to the disordered operation of measurement equipments, some missing observations, and some very small observed data that fluctuated around the detection limit of monitoring equipments and also sometimes irrelevant measurements. Thus, this kind of measurement issues might influence model performance. Indeed, the calculated GFI and AGFI values for this model imply that the model is statistically acceptable. Among the three models, the sub-urban model performance is the best.

For all of the structural equation models and measurement models, it is found that all the parameters are statistically significant at the 1% or 5% level. This finding indicate the validity of the postulated model structure in this case study. The log-transformed variable LN_NO is also statistically a meaningful parameter. All the signs of the estimated parameters are intuitive and consistent with expectations. It can be imagined that positive parameter indicating the influence of "Primary Pollutants" on "Photochemical" might be also logical, considering that at the SUW, other than the pollutants from mobile sources, stationary sources (e.g., household and industrial emissions) also contribute to the air pollutants. Indeed, this findings need to be further explored when the data is available.

No	Description	Null Model	Half model: Spatial & Temporal	Full Model: With pollutants interactions
I	**Fixed Part**			
A	**Intercept (Location)**	50.125(10.12)	42.627(3.848)	8.722(1.97)
B	**Spatial**			
1	Sub-Urban (Dummy)		-13.255(-0.992)	-4.513(-0.96)
2	Urban Fringe(Dummy)		-18.743(-1.401)	-6.873(-1.45)
C	**Temporal**			
1	Long-term (Year)		9.383(12.333)	3.410(6.27)
2	Seasonal (Dummy wet season)		-1.370(-1.472)	2.186(3.32)
3	Weekly Weekend (Dummy)		1.077(1.054)	0.909(1.370)
D	**Interaction with other pollutants**			
1	PM_{10}			0.129(9.68)
2	NO_2			-0.056(-2.76)
3	NO			0.050(1.50)
E	**Atmospheric Condition**			
	Ratio NO/NO_2			-2.046 (-4.02)
F	**Accumulation Impacts**			
	Prior day concentration			0.669 (40.88)
II	**Random Part**			
	σ_e^2 (Within monitoring)	416.43	384.38	160.559
	$\sigma_{\mu 0}^2$(Inter-monitoring)	121.47	118.34	14.333
III	**Model Performance**			
	AIC	16060	16060	14499
	BIC	16104	16104	14570
	-2*Log likelihood	16213	16044	14473
	Degree of freedom	3	8	13
	No of Samples	1826	1826	1826

Note: ( ) → t-statistic

Table 5. Model of Daily Average Concentration of Ozone (O_3)

The latent variable "Photochemical" consistently receives the largest influence from the latent variable "Meteorology" at all the locations (see Table 6). This is consistent with the scientific evidences about photochemical reactions as described earlier in this chapter. O_3 is the secondary pollutant, which is chemically transformed from the primary pollutants and the dominant driving forces for such chemical transformation are meteorological factors. Among the meteorological factors, humidity has a negative effect on "Photochemical" in contrast to solar radiation and temperature, which have positive effects. It is also found that parameter of wind speed has a negative value and parameter of wind direction (i.e., degree from the north) is positive. Since wind speed is usually slow, and major wind comes from the north direction in Jakarta City, wind speed and direction works in the same way to increase the O_3 production. Primary pollutants, on the one hand, produce the O_3, but on the other, they cause O_3 destruction too. The latent variable "Wind" shows the second largest influence on the "Photochemical", followed by the latent variable "Primary Pollutants". "Primary Pollutants" shows positive influence on the "Photochemical" at the SUW, but negative at CA & RA because major precursors of O_3 are NO, NO_2 and CO, the increase in "Primary Pollutants" usually results in the reduction of O_3 production. Accordingly, negative influence at city center (CA & RA) is intuitive. On the other hand, the higher loading of PM_{10}, then lower loading of major precursors NO, NO_2 and CO at SUW. To verify the influence of PM_{10} on major precursors NO, NO_2 and CO, we also tried to incorporate such influence in the model structure, but we failed to get reasonable estimation results. Then it is difficult to clarify the reason why the influence of "Primary Pollutants" on the "Photochemical" is positive at the SUW. However, because of the negative interaction between PM_{10} and major precursors NO, NO_2 and CO, it seems that the influence of "Primary Pollutants" on the "Photochemical" is also dependent on the relative magnitude of each pollutant. This should be further explored in the future.

Concerning the interactions among the "Meteorology", "Wind" and "Primary Pollutants", it is found that "Meteorology" negatively affects "Primary Pollutants" at all the locations, "Wind" has positive influence on "Primary Pollutants" at the SUW and the RA, but negative at the CA. Looking at the total effects as shown in Table 7, one can see that at the SUW and the RA, influence of "Meteorology" on "Photochemical" is clearly larger than "Wind", however, "Meteorology" and "Wind" have almost equal influence at the CA.

4.2.2. Temporal analysis

Observing the model accuracy indices (i.e., GFI and AGFI), the model for weekdays-wet season shows the highest GFI (AGFI) value 0.845 (0.724), followed by the model for weekend-wet season with the value of GFI (AGFI) 0.822 (0.683) and than followed by the model for weekdays-dry season with the value of GFI (AGFI) 0.783 (0.612). The model for weekend-dry season has the lowest GFI (AGFI) 0.775 (0.599). Despite the possible measurement and sampling errors, the GFI and AGFI values indicate the model is statistically acceptable. Among all models, the model accuracy for the weekday-wet season is the best.

Covariances				Weekdays - Dry Season		
				Sub-Urban (SUW)	CBD (CA)	Roadside (RA)
Primary ($\acute\eta_2$)	<---	Met (ξ_1)	γ_{21}	-0.017	-0.142 ***	-0.080 ***
Primary ($\acute\eta_2$)	<---	Wind ($\acute\eta_1$)	β_{21}	0.547 ***	-0.072 ***	0.180 ***
Photochem ($\acute\eta_3$)	<---	Wind ($\acute\eta_1$)	β_{31}	0.420 ****	0.683 ***	0.156 ***
Photochem ($\acute\eta_3$)	<---	Met (ξ_1)	γ_{31}	0.816 ****	0.759 ***	0.743 ***
Photochem ($\acute\eta_3$)	<---	Primary ($\acute\eta_2$)	β_{32}	0.109 ***	-0.040 **	-0.142 ***
SR (X_1)	<---	Met (ξ_1)	$\lambda^{(x)}_{11}$	0.685 ***	0.796 ***	0.793 ***
T (X_2)	<---	Met (ξ_1)	$\lambda^{(x)}_{12}$	0.972 ***	0.969 ***	0.980 ***
RH (X_3)	<---	Met (ξ_1)	$\lambda^{(x)}_{13}$	-0.967 ***	-0.930 ***	-0.952 ***
WD (Y_1)	<---	Wind ($\acute\eta_1$)	$\lambda^{(y)}_{11}$	0.664 ***	0.494 ***	0.995 ***
WS (Y_2)	<---	Wind ($\acute\eta_1$)	$\lambda^{(y)}_{12}$	-0.977 ***	-0.672 ***	0.617 ***
LN NO (Y_4)	<---	Primary ($\acute\eta_2$)	$\lambda^{(y)}_{24}$	0.548 ***	0.525 ***	0.719 ***
NO$_2$ (Y_5)	<---	Primary ($\acute\eta_2$)	$\lambda^{(y)}_{25}$	0.688 ***	0.659 ***	0.684 ***
CO (Y_6)	<---	Primary ($\acute\eta_2$)	$\lambda^{(y)}_{26}$	0.790 ***	0.831 ***	0.944 ***
SO$_2$ (Y_7)	<---	Primary ($\acute\eta_2$)	$\lambda^{(y)}_{27}$	0.210 ***	0.311 ***	0.368 ***
PM$_{10}$ (Y_8)	<---	Primary ($\acute\eta_2$)	$\lambda^{(y)}_{28}$	0.777 ***	0.469 ***	0.449 ***
O$_3$ (Y_3)	<---	Photochem ($\acute\eta_3$)	$\lambda^{(y)}_{33}$	0.795 ***	0.879 ***	0.979 ***
LN NO (Y_4)	<---	Photochem ($\acute\eta_3$)	$\lambda^{(y)}_{34}$	-0.660 ***	-0.642 ***	-0.231 ***
GFI				0.787	0.731	0.770
AGFI				0.629	0.533	0.600
df				37	37	37
Sample Size				1916	3179	2145
Notes : *** Significant at 1 %; ** Significant at 5%						

Table 6. Estimation Results of Spatial Analysis (comparison among locations)

Components	Dry Season											
	Sub-Urban (West Jakarta-SUW)				CBD (Central-CA)				Roadside (JAM/Mobile-RA)			
	Met (ξ_1)	Wind ($\acute\eta_1$)	Primary ($\acute\eta_2$)	Photochem ($\acute\eta_3$)	Met (ξ_1)	Wind ($\acute\eta_1$)	Primary ($\acute\eta_2$)	Photochem ($\acute\eta_3$)	Met (ξ_1)	Wind ($\acute\eta_1$)	Primary ($\acute\eta_2$)	Photochem ($\acute\eta_3$)
Primary ($\acute\eta_2$)	-0.017	0.547	0.000	0.000	-0.142	-0.072	0.000	0.000	-0.080	0.180	0.000	0.000
Photochem ($\acute\eta_3$)	0.814	0.480	0.109	0.000	0.765	0.686	-0.040	0.000	0.754	0.131	-0.142	0.000
O$_3$ (Y_3)	0.647	0.382	0.086	0.795	0.673	0.603	-0.035	0.879	0.738	0.128	-0.139	0.979
PM$_{10}$ (Y_8)	-0.013	0.425	0.777	0.000	-0.066	-0.034	0.469	0.000	-0.036	0.081	0.449	0.000
SO$_2$ (Y_7)	-0.003	0.115	0.210	0.000	-0.044	-0.022	0.311	0.000	-0.029	0.066	0.368	0.000
LN NO (Y_4)	-0.547	-0.018	0.476	-0.660	-0.566	-0.478	0.550	-0.642	-0.232	0.099	0.752	-0.231
NO$_2$ (Y_5)	-0.011	0.376	0.688	0.000	-0.093	-0.047	0.659	0.000	-0.055	0.123	0.684	0.000
CO (Y_6)	-0.013	0.432	0.790	0.000	-0.118	-0.060	0.831	0.000	-0.075	0.170	0.944	0.000
WS (Y_2)	0.000	-0.977	0.000	0.000	0.000	-0.672	0.000	0.000	0.000	0.617	0.000	0.000
WD (Y_1)	0.000	0.664	0.000	0.000	0.000	0.494	0.000	0.000	0.000	0.995	0.000	0.000
RH (X_3)	0.685	0.000	0.000	0.000	0.796	0.000	0.000	0.000	0.793	0.000	0.000	0.000
T (X_2)	0.972	0.000	0.000	0.000	0.969	0.000	0.000	0.000	0.980	0.000	0.000	0.000
SR (X_1)	-0.967	0.000	0.000	0.000	-0.930	0.000	0.000	0.000	-0.952	0.000	0.000	0.000

Table 7. Estimated Standardized Total Effects of spatial analysis

For all of the structural equation models and measurement models, it is found that all the parameters are statistically significant at the 1% or 5% level. This findings indicate that the the postulated model structure in this case study is valid. In addition, the log-transformed variable NO (LN_NO) is also statistically a meaningful parameter. All the signs of the estimated parameters are intuitive and consistent with expectations. It can be imagined that positive parameter indicating the influence of "Primary Pollutants" on "Photochemical" might be also logical, considering weather/meteorological situations, also contribute to the reaction of air pollutants in roadside. Needless to say, this findings need to be further explored when the data is available.

The latent variable "Photochemical consistently receives the largest effect from the latent variable "Meteorological" at all the situations (see Table 8). This is consistent with the scientific evidences about photochemical reactions as described earlier in this chapter. O_3 is the secondary pollutant which is chemically transformed from the primary pollutants and the dominant driving forces for such chemical transformation are meteorological factors. Among the meteorological factors, humidity has negative effect on "Photochemical", in contrast to solar radiation and temperature that have a positive effect. The signs of these parameters seem in agreement with the photochemical process described earlier in this chapter. It is also found that latent variable "Wind" has a negative value during wet season, in contrast to a positive value during dry season, since the wind direction are on the opposite direction seasonally. The wind comes from South East (57 %) and North West (47.4%) during dry season and wet season, respectively.

The Roadside Station is located in the south part of the nearest pollutants source (Casablanca Road) , we preliminary identify that during wet season the wind direction from North West carry the "Primary Pollutants" more intensive than during in dry season. On the one hand, primary pollutants produce the O_3, but on the other hand also cause O_3 destruction. The latent variable "Wind" shows the second largest influence on the "Photochemical", followed by the latent variable "Primary Pollutants" during wet season. On the contrary, "Primary Pollutants" shows the second largest influence on the "Photochemical", followed by the latent variable "Wind" during dry season period. The "Primary Pollutants" shows negative influence on the "Photochemical" for weekday-dry, weekday-wet and weekend-dry season, because major precursors of O_3 are NO, NO_2 and CO. The increase in "Primary Pollutants" usually reduces O_3 production. Accordingly, negative influences for weekday-wet, weekdays-dry and weekend-dry season are intuitive. The "Primary Pollutants" shows positive influence on the "Photochemical" for weekend-wet season, but not significant for all confidence level (see Table 8). Therefore, the data for weekend-wet season in particular should be further explored to explain the positive value. The load of CO is the highest among other pollutants SO_2, NO, NO_2 and CO for all situations. The influence of CO has been incorporated into the model structure to verify its effect to the model especially for weekend-wet season, but all the estimation results are below the reasonable confidence level, despite the fact that .the emission source (road) is relatively close to the monitoring station. The influence of meteorological factors seems more dominant than primary pollutants. Indeed, this should be further explored in the future.

Concerning the interactions among the "Meteorological", "Wind" and "Primary Pollutants", it is found that "Meteorological" and "Wind" positively affects "Primary Pollutants" for all data sets. The influence of "Meteorological" on "Photochemical" is obviously larger than the "Wind" and "Primary Pollutants" for all situations as depicted in Table 9 and 10.

Estimated Free Structural Parameter			Weekdays		Weekend	
			Dry season	Wet Season	Dry Season	Wet season
Wind ($\acute{η}_1$)	<---	Met ($ξ_1$) $γ_{11}$	-0.156 ***	0.679 ***	-0.237 ***	-0.129
Primary ($\acute{η}_2$)	<---	Met ($ξ_1$) $γ_{21}$	0.02	0.027	0.117	0.005
Primary ($\acute{η}_2$)	<---	Wind ($\acute{η}_1$) $β_{21}$	0.363 ***	0.479	0.305 ***	0.538 ***
Photochem ($\acute{η}_3$)	<---	Wind ($\acute{η}_1$) $β_{31}$	0.118 ****	-0.315 ***	0.075 *	-0.054
Photochem ($\acute{η}_3$)	<---	Met ($ξ_1$) $γ_{31}$	0.769 ****	0.971 ***	0.777 ***	0.761 ***
Photochem ($\acute{η}_3$)	<---	Primary ($\acute{η}_2$) $β_{32}$	-0.17 ****	-0.142 ***	-0.163 ***	0.022
SR (X_1)	<---	Met ($ξ_1$) $λ^{(x)}_{11}$	0.795 ***	0.724 ***	0.775 ***	0.796 ***
T (X_2)	<---	Met ($ξ_1$) $λ^{(x)}_{12}$	0.975 ***	1 ***	0.978 ***	0.989 ***
RH (X_3)	<---	Met ($ξ_1$) $λ^{(x)}_{13}$	-0.949 ***	-0.95 ***	-0.963 ***	-0.958 ***
WD (Y_1)	<---	Wind ($\acute{η}_1$) $λ^{(y)}_{11}$	0.979 ***	0.441 ***	0.724 ***	0.453 ***
WS (Y_2)	<---	Wind ($\acute{η}_1$) $λ^{(y)}_{12}$	0.473 ***	0.855 ***	0.525 ***	0.383 ***
LN NO (Y_4)	<---	Primary ($\acute{η}_2$) $λ^{(y)}_{24}$	0.742 ***	0.551 ***	0.629 ***	0.525 ***
NO₂ (Y_5)	<---	Primary ($\acute{η}_2$) $λ^{(y)}_{25}$	0.737 ***	0.786 ***	0.711 ***	0.94 ***
CO (Y_6)	<---	Primary ($\acute{η}_2$) $λ^{(y)}_{26}$	0.91 ***	0.936 ***	0.991 ***	0.962 ***
SO₂ (Y_7)	<---	Primary ($\acute{η}_2$) $λ^{(y)}_{27}$	0.206 ***	0.673 ***	0.239 ***	0.329 ***
PM₁₀ (Y_8)	<---	Primary ($\acute{η}_2$) $λ^{(y)}_{28}$	0.411 ***	0.512 ***	0.4 ***	0.563 ***
O₃ (Y_3)	<---	Photochem ($\acute{η}_3$) $λ^{(y)}_{33}$	0.962 ***	0.946 ***	0.967 ***	0.93 ***
LN NO (Y_4)	<---	Photochem ($\acute{η}_3$) $λ^{(y)}_{34}$	-0.254 ***	-0.408 ***	-0.243 ***	-0.317 ***
Goodness-of-fit index (GFI)			**0.783**	**0.845**	**0.775**	**0.822**
Adjusted Goodness-of-fit Index (AGFI)			**0.612**	**0.724**	**0.599**	**0.683**
df			37	37	37	37
Estimation Method : Maximum Likelihood						
Notes : * Significant at 1 % ; * significant at 10%**						

Table 8. Estimation Results of Temporal Variations at Roadside of Jakarta City

Variables	Weekdays							
	Dry Season				Wet Season			
	Met ($ξ_1$)	Wind ($\acute{η}_1$)	Primary ($\acute{η}_2$)	Photochem ($\acute{η}_3$)	Met ($ξ_1$)	Wind ($\acute{η}_1$)	Primary ($\acute{η}_2$)	Photochem ($\acute{η}_3$)
Wind ($\acute{η}_1$)	-0.156	0	0	0	0.679	0	0	0
Primary ($\acute{η}_2$)	-0.037	0.363	0	0	0.352	0.479	0	0
Photochem ($\acute{η}_3$)	0.757	0.056	-0.17	0	0.708	-0.383	-0.142	0
O₃ (Y_3)	0.728	0.054	-0.164	0.962	0.669	-0.362	-0.135	0.946
PM₁₀ (Y_8)	-0.015	0.149	0.411	0	0.18	0.245	0.512	0
SO₂ (Y_7)	-0.008	0.075	0.206	0	0.237	0.322	0.673	0
LN NO (Y_4)	-0.219	0.255	0.786	-0.254	-0.095	0.42	0.609	-0.408
NO₂ (Y_5)	-0.027	0.268	0.737	0	0.277	0.376	0.786	0
CO (Y_6)	-0.033	0.33	0.91	0	0.33	0.448	0.936	0
WS (Y_2)	-0.074	0.473	0	0	0.58	0.855	0	0
WD (Y_1)	-0.152	0.979	0	0	0.3	0.441	0	0
RH (X_3)	-0.949	0	0	0	-0.95	0	0	0
T (X_2)	0.975	0	0	0	1	0	0	0
SR (X_1)	0.795	0	0	0	0.724	0	0	0

Table 9. Estimated standardized total effects of surface O₃ model for Jakarta City (weekday)

Variables	Weekend							
	Dry Season				Wet Season			
	Met (ξ_1)	Wind ($\acute{\eta}_1$)	Primary ($\acute{\eta}_2$)	Photochem ($\acute{\eta}_3$)	Met (ξ_1)	Wind ($\acute{\eta}_1$)	Primary ($\acute{\eta}_2$)	Photochem ($\acute{\eta}_3$)
Wind ($\acute{\eta}_1$)	-0.237	0	0	0	-0.129	0	0	0
Primary ($\acute{\eta}_2$)	0.045	0.305	0	0	-0.065	0.538	0	0
Photochem ($\acute{\eta}_3$)	0.752	0.026	-0.163	0	0.766	-0.042	0.022	0
O_3 (Y_3)	0.727	0.025	-0.158	0.967	0.713	-0.039	0.021	0.93
PM_{10} (Y_8)	0.018	0.122	0.4	0	-0.036	0.303	0.563	0
SO_2 (Y_7)	0.011	0.073	0.239	0	-0.021	0.177	0.329	0
LN NO (Y_4)	-0.154	0.186	0.668	-0.243	-0.277	0.296	0.518	-0.317
NO_2 (Y_5)	0.032	0.217	0.711	0	-0.061	0.506	0.94	0
CO (Y_6)	0.044	0.303	0.991	0	-0.062	0.518	0.962	0
WS (Y_2)	-0.124	0.525	0	0	-0.05	0.383	0	0
WD (Y_1)	-0.171	0.724	0	0	-0.059	0.453	0	0
RH (X_3)	-0.963	0	0	0	-0.958	0	0	0
T (X_2)	0.978	0	0	0	0.989	0	0	0
SR (X_1)	0.775	0	0	0	0.796	0	0	0

Table 10. Estimated standardized total effects of surface O_3 model for Jakarta City (weekend)

5. Conclusion

Surface ozone is potentially high in Jakarta, serious problem and getting worse every year. In this paper, a spatial and temporal analysis of surface ozone related issues were done by two major approach multilevel analysis and structural equation model. A spatial and temporal analysis was conducted by using time series data, which were collected at the existing air quality monitoring stations in Jakarta city from 2001 to 2003.

This paper first applied a multilevel analysis to examine the variation properties affect on event of daily peak ozone concentration. Secondly, we analyze variations properties on daily average surface ozone concentration by introducing observed information related to spatial aspect and temporal aspect. The year of measurement, seasonal and weekly variables were selected to represent long-term, medium/seasonal-term and day-to-day (short term) variation of daily average ozone concentration. Finally, we established a structural equation model, which can endogenously incorporate various cause-effect relationships and interactions among meteorological factors, wind, and primary pollutants, which affect on a half-hour concentration of surface ozone. The established model also incorporated non-linear relationships existing in the observed variables. Using the data collected from the above-mentioned fixed monitoring stations in Jakarta City, the effectiveness of the established model is empirically confirmed. The best model for spatial analysis, that it has the highest goodness-of-fit index, is the one for the suburban area. As for temporal analysis, the model effectiveness was empirically tested using the air quality data from Roadside Station in Central Jakarta. The best model indicated with the highest goodness-of-fit index, was the one for the weekdays during wet season.

The event of daily peak ozone concentration is singular and usually occurred at 11.28 am in central business district of Jakarta city. These events will be slightly late at sub-urban monitoring stations and urban fringe around 38 to 40 minutes later than central Jakarta. The events of daily peak concentration of ozone are almost stable in all measurement period. We couldn't found variations among year of measurement, among dry and wet seasonal variations and also among days in a week. In contrast, by using daily average concentration we couldn't find significant impact of location which mean location properties are minor factor on daily average concentration of surface ozone occurs in Jakarta city. The main factors affects on daily average concentration are temporal aspects and the presence of other pollutants. The medium and long-term variations are significantly increase ozone concentration. In contrast, short-term (day-to-day) variation is insignificant. This analysis shows the tendency of daily average surface ozone concentration in Jakarta city are increase year by year and getting worse. The expected washing phenomena caused by rain are smaller than the emission increase due to traffic jam or chaotic traffic situation on the rainy situation in Jakarta city. As results, daily average concentration of surface ozone concentration measured at wet season is slightly high than dry season. The influence of precursor pollutants on surface ozone concentration shows the logical reason and accumulation process of daily average surface ozone concentration was exist in the urban ozone atmospheric conditions.

The establishment of causal interaction in urban ozone atmospheric condition was successfully captured by proposed structural equations model. The proposed structural equation model also examine by empirical data for very short term concentration of ozone in Jakarta city. The structural equation model incorporates various cause-effect relationships and interactions among meteorological variables, wind, and primary pollutants, which affect the surface O_3. The model also incorporated the existing non-linear relationships in the observed variables. The model effectiveness was empirically tested and the best model was defined for the one that has the highest goodness-of-fit index, which was the one for the suburban area and weekdays-wet season` model. In micro urban environment studies, all models used in this study showed that meteorological variables consistently had the largest influence on photochemical, followed by the wind conditions and lastly the primary pollutants. Among the meteorological variables, relative humidity had a negative influence while solar radiation and temperature had positive influences. The model estimations demonstrated that the influence of meteorological factors on photochemical was definitely larger than the wind conditions at all situations.

Primary pollutants had a negative influence for all temporal situations in roadside area except for the weekend during wet season. It seems that PM_{10} behaved quite differently compared to the other primary pollutants at the suburban area and city center, i.e. the higher the PM_{10} load, the lower the major precursors NO, NO_2 and CO loads. On the roadside area in the city center, It is found that CO concentration was the highest among the other primary pollutants for all situations. In addition, the higher the CO load, the lower the other major precursors (NO and NO_2) loads.

Further study should be carried out to combine both spatial and temporal issues and causal interaction among factors on surface ozone concentration at urban areas. A study based on multilevel structural equation model should be conducted to solve these issues.

This understanding can assist the policy maker in the developing O_3 pollution control strategies.

Author details

S. B. Nugroho, A. Fujiwara and J. Zhang

Transportation Engineering Laboratory, Graduate School for International Development and Cooperation, Hiroshima University, Japan,
Kagamiyama, Higashi Hiroshima, Japan

Acknowledgement

This research is partially supported by Global Environmental Leadership Program at Graduate School of International Development and Cooperation, Hiroshima University, Japan.

6. References

[1] Benarie, M.M. (1980) *Urban Air Pollution Modeling*. The Macmillan Press Ltd, London.

[2] Boriboonsomsin, K. and Uddin, W. (2005) Tropospheric ozone modeling considering vehicle emission and point & aviation source, validation, and implementation in rural areas. The 84th Annual Meeting of Transportation Research Board, Washington D.C. (CD-ROM)

[3] Gao, H.O and Debbie A Niemeier. (2007) The impact of rush hour traffic and mix on ozone weekend effect in southern California, Transportation research part D, 12, pp83-89.

[4] Gardner, M.W. and Dorling, S.R. (1998) Artificial Neural networks (the multilayer perceptron)-a review of applications in the atmospheric sciences. Atmospheric Environment, Vol. 32, 2627-2636.

[5] Geladi, P., Hadjiiski, L. and Hopke, P. (1999) Multiple regression for environmental data: Nonlinearitas and prediction bias. *Chemometrics and Intelligent Laboratory Systems*, 47, 165-173.

[6] Gelfanda A.E., Banerjee S., Sirmans C.F., Tu Y., Ong S.E. 2007. Multilevel modeling using spatial processes: Application to the Singapore housing market. Computational Statistics & Data Analysis 51(7), Pages 3567–3579

[7] Gelman A., and Hill J. 2007. Data Analysis Using Regression and Multilevel/Hierarchical Models. Cambridge University Press, New York, America

[8] Golob, T. F. (2003) Structural equation modeling for travel behavior research, Transportation Research Part B, Vol. 37, 1-25

[9] Haq G, W.J Han, C. Kim, H Vallack (2002) Benchmarking Urban Air Quality Management and Practice in Major and Megacities of Asia-Stage I. APMA project, Korea Environment Institute.

[10] Hubbard, M.C. and Cobourn, W.G. (1998) Development of regression model to forecast ground-level ozone concentration in Louisville, KY. Atmospheric Environment, Vol. 32, 2637-2647

[11] Jenkin, M.E. and Chemitshaw, K.C. (2000) Ozone and other secondary photochemical pollutants: chemical processes governing their formation in the planetary boundary layer. Atmospheric Environment, Vol. 34, 2499-2527

[12] Jöreskog, K.G. and Sörbom, D. (1989) LISREL7, A Guide to the Program and Application, 2nd Edition. SPSS Inc., Chicago

[13] Joop J Hox (2010) Multilevel Analysis, Techniques and Applications. Routledge, New York (2010) pp 155.

[14] Monoura, H. (1999) Some characteristics of surface ozone concentration observed in an urban atmosphere. Atmospheric Research, Vol. 51, 153-169.

[15] Overmars K.P., and Verburg P.H. 2006. Multilevel modelling of land use from field to village level in the Philippines. Agricultural Systems 89(2–3), Pages 435–456

[16] Pattenden S., Armstrong B.G., Houthuijs D., Leonardi G.S., Dusseldorp A., Boeva B., Hruba F., Brunekreef B., and Fletcher T. 2000. Methodological approaches to the analysis of hierarchical studies of air pollution and respiratory health - examples from the CESAR study. Journal of Exposure Analysis and Environmental Epidemiology 10, 420-426

[17] Peton, N., Dray, D., Pearson, D., Mesbah, M. and Vuillot, B. (2000) Modeling and analysis of ozone episodes. Environmental Modelling and Software, Vol. 15, 647-652.

[18] Rubin, E.S. (2001) Introduction to Engineering and the Environment. McGraw-Hill International Edition, Singapore.

[19] Sanchez-ccoyllo, O.R., R.Y. Ynoue, L.D. Martins, Maria de Fatima Andrade (2006) Impacts of O_3 precursor limitation and meteorological variables on O_3 concentration in Sao Paulo, Brazil. Atmospheric Environment 40, pp 552-562.

[20] Seinfeld, J.H. and Pandis, S.N. (1998) Atmospheric Chemistry and Physics. John wiley and sons, Inc., New York.

[21] Sousa, S.I.V., F.G. Martins, M.C. Pereira, M.C.M. Alvim-ferraz, (2007) Multiple linear regression and artificial neural network based on principal components to predict O_3 concentrations. Environmental Modeling and Software, 22, pp 97-103.

[22] Stern, A.C., Wohlers, H.C., Boubel, R.W. and Lowry, W.P. (1973) Fundamentals of Air Pollution. Academic Press, Inc., New York.

[23] Syahril, S., Budi, P.R., Haryo, S.T. (2002) 'Study on Air Quality in Jakarta, Indonesia', ADB Report, Manila 2002.

[24] Wang, W., Lu, W., Wang, X. and Leung, A.Y.T. (2003) Prediction of maximum daily ozone level using combined neural network and statistical characteristics. Environmental International, Vol. 29, 555-562.

[25] Zhang, B.N. and Oanh, N.T.K. (2002) Photochemical smog pollution in the Bangkok Metropolitan Region of Thailand in relation to O_3 precursor concentrations and meteorological conditions. Atmospheric Environment, Vol. 36, 4211-4222.

Chapter 6

Time Series Analysis of Surface Ozone Monitoring Records in Kemaman, Malaysia

Marzuki Ismail, Azrin Suroto and Nurul Ain Ismail

Additional information is available at the end of the chapter

1. Introduction

Tropospheric ozone is known as environmental air pollutants that arise from photochemical reaction among various natural and anthropogenic precursors that are volatile organic compounds (VOCs) and organic nitrogen (NO_x). Accumulation of the ozone may highly happen under favorable meteorological conditions and will have an adverse effect on human health and ecosystem [1]. Chan & Chan, 2001 concluded that people in Asia also cannot escape from the adversely impact ozone pollution as there were elevated ozone level being detected. Nevertheless, the long-term ozone trend has been less researched, especially in Malaysia.

The time series analysis is one of the best tool in understanding cause and effect relationship of environmental pollution [3, 4,5]. Its applications in many studies were done to describe the past movement of particular variable with respect to time. However, there were several different techniques applied by researcher so that the change of air pollution behavior through time period can be determined [6, 7]. A study by Kuang-Jung Hsu,2003 was done by using autoregression variation (VAR) in order to establish interdependence between primary and secondary air pollutants in area of Taipei. Besides, Omidravi et al., 2008 had applied the time series analysis in their investigation in order to find the answer that relate to extreme high ozone concentrations for each season in Ishafan by using Fast Fourier Transform. Therefore, this study aims to determine qualitative and quantitative aspect of the tropospheric ozone concentrations so that prediction on future concentration of the anthropogenic air pollutant can be achieved in the study area, i.e. Kemaman, Malaysia.

2. Material and method

This study was conducted in Kemaman (04°12'N, 103°18'E), a developing Malaysian town located in between the industrializing of Kertih Petrochemical Industrial Area in the north and industrializing and urbanizing of Gabeng Industrial Area in the South (Figure 1). In this area, there are dominant sources of ozone precursors related to industrial activities and road traffic.

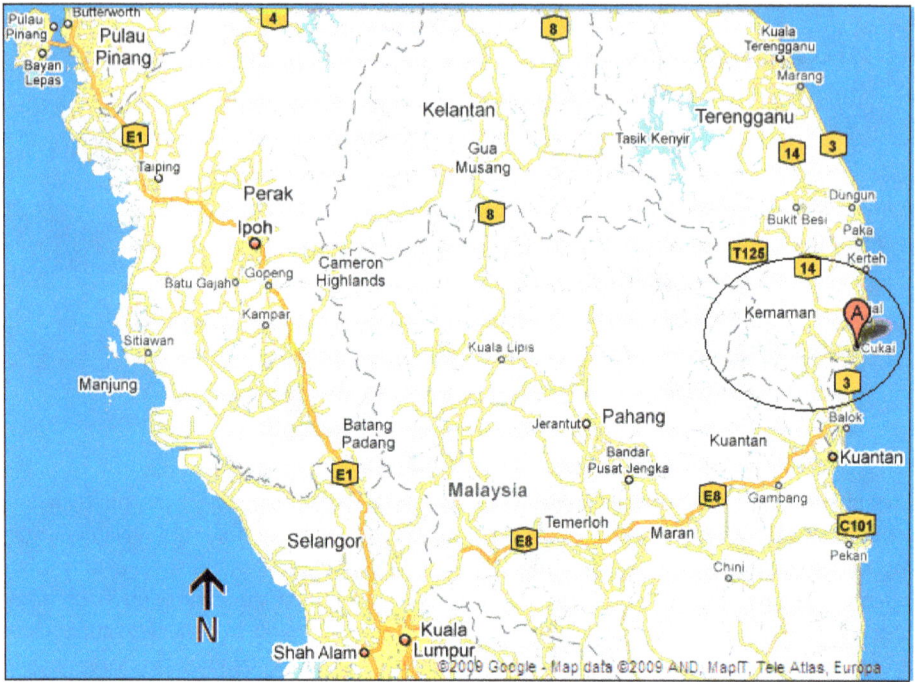

Figure 1. Locations of air monitoring station in Kemaman

In this study, ozone trend was examined using ozone data consisting of 144 monthly observations from January 1996 to December 2007 acquired from the Air Quality Division of ASMA for Sekolah Rendah Bukit Kuang station located in Kemaman district; one of the earliest operational stations in Malaysia. The monitoring network was installed, operated and maintained by Alam Sekitar Malaysia Sdn. Bhd. (ASMA) under concession by the Department of Environment Malaysia [10]. Tropospheric ozone concentrations data was recorded using a system based on the Beer-Lambert law for measuring low ranges of ozone in ambient air manufactured by Teledyne Technologies Incorporated (Model 400E). A 254 nm UV light signal is passed through the sample cell where it is absorbed in proportion to the amount of ozone present. Every three seconds, a switching valve alternates measurement between the sample stream and a sample that has been scrubbed of ozone. The result is a true, stable ozone measurement [11].

Time series analysis was implemented using STATGRAPHICS® statistical software package. A time series consists of a set of sequential numeric data taken at equally spaced intervals, usually over a period of time or space. This study provides statistical models for two time series methods: trend analysis and seasonal component which are both in time scale.

The seasonal decomposition was used to decompose the seasonal series into a seasonal component, a combined trend and cycle component, and a short-term variation component, i.e,

$$O_t = T_t \ x \ S_t \ x \ I_t \tag{1}$$

where O_t is the original ozone time series, T_t is the long term trend component, S_t is the seasonal variation, and I_t is the short-term variation component or called the error component. As the seasonality increase with the level of the series, a multiplicative model was used to estimate the seasonal index. Under this model, the trend has the same units as the original series, but the seasonal and irregular components are unitless factors, distributed around 1. As the underlying level of the series changes, the magnitude of the seasonal fluctuations varies as well. The seasonal index was the average deviation of each month's ozone value from the ozone level that was due to the other components in that month.

In trend analysis, Box-Jenkins Autoregressive Integrated Moving Average (ARIMA) model was applied to model the time series behavior in generating the forecasting trend. The methodology consisting of a four-step iterative procedure was used in this study. The first step is model identification, where the historical data are used to tentatively identify an appropriate Box-Jenkins model followed by estimation of the parameters of the tentatively identified model. Subsequently, the diagnostic checking step must be executed to check the adequacy of the identified model in order to choose the best model. A better model ought to be identified if the model is inadequate. Finally, the best model is used to establish the time series forecasting value.

In model identification (step 1), the data was examined to check for the most appropriate class of ARIMA processes through selecting the order of the consecutive and seasonal differencing required to make series stationary, as well as specifying the order of the regular and seasonal auto regressive and moving average polynomials necessary to adequately represent the time series model. The Autocorrelation Function (ACF) and the Partial Autocorrelation Function (PACF) are the most important elements of time series analysis and forecasting. The ACF measures the amount of linear dependence between observations in a time series that are separated by a lag k. The PACF plot helps to determine how many auto regressive terms are necessary to reveal one or more of the following characteristics: time lags where high correlations appear, seasonality of the series, trend either in the mean level or in the variance of the series. The general model introduced by Box and Jenkins includes autoregressive and moving average parameters as well as differencing in the formulation of the model.

The three types of parameters in the model are: the autoregressive parameters (p), the number of differencing passes (d) and moving average parameters (q). Box-Jenkins model are summarized as ARIMA (p, d, q). For example, a model described as ARIMA (1,1,1) means that this contains 1 autoregressive (p) parameter and 1 moving average (q) parameter for the time series data after it was differenced once to attain stationary. In addition to the non-seasonal ARIMA (p, d, q) model, introduced above, we could identify seasonal ARIMA (P, D, Q) parameters for our data. These parameters are: seasonal autoregressive (P), seasonal differencing (D) and seasonal moving average (Q). Seasonality is defined as a pattern that repeats itself over fixed interval of time. In general, seasonality can be found by identifying a large autocorrelation coefficient or large partial autocorrelation coefficient at a seasonal lag. For example, ARIMA $(1,1,1)(1,1,1)^{12}$ describes a model that includes 1 autoregressive parameter, 1 moving average parameter, 1 seasonal autoregressive parameter and 1 seasonal moving average parameter. These parameters were computed after the series was differenced once at lag 1 and differenced once at lag 12.

The general form of the above model describing the current value Z_t of a time series by its own past is:

$$(1 - \phi_1 B)(1 - \alpha_1 B^{12})(1 - B)(1 - B^{12}) Z_t = (1 - \theta_1 B)(1 - \gamma_1 B^{12}) \varepsilon_t \qquad (2)$$

Where:

$1 - \phi_1 B$	= non seasonal autoregressive of order 1
$1 - \alpha_1 B^{12}$	= seasonal autoregressive of order 1
Z_t	= the current value of the time series examined
B	= the backward shift operator $BZ_t = Z_{t-1}$ and $B^{12}Z_t = Z_{t-12}$
1-B	= 1st order non-seasonal difference
$1-B^{12}$	= seasonal difference of order 1
$1 - \theta_1 B$	= non seasonal moving average of order 1
$1 - \gamma_1 B^{12}$	= seasonal moving average of order 1

For the seasonal model, we used the Akaike Information Criterion (AIC) for model selection. The AIC is a combination of two conflicting factors: the mean square error and the number of estimated parameters of a model. Generally, the model with smallest value of AIC is chosen as the best model [12].

After choosing the most appropriate model, the model parameters are estimated (step 2) - the plot of the ACF and PACF of the stationary data was examined to identify what autoregressive or moving average terms are suggested. Here, values of the parameters are chosen using the least square method to make the Sum of the Squared Residuals (SSR) between the real data and the estimated values as small as possible. In most cases, nonlinear estimation method is used to estimate the above identified parameters to maximize the likelihood (probability) of the observed series given the parameter values [13].

In diagnose checking step (step 3), the residuals from the fitted model is examined against adequacy. This is usually done by correlation analysis through the residual ACF plots and the goodness-of-fit test by means of Chi-square statistics χ^2. If the residuals are correlated, then the model should be refined as in step one above. Otherwise, the autocorrelations are white noise and the model is adequate to represent our time series.

The final stage for the modeling process (step 4) is forecasting, which gives results as three different options: - forecasted values, upper, and lower limits that provide a confidence interval of 95%. Any forecasted values within the confidence limit are satisfactory. Finally, the accuracy of the model is checked with the Mean-Square error (MS) to compare fits of different ARIMA models. A lower MS value corresponds to a better fitting model.

3. Results and discussion

The first step in time series analysis is to draw time series plot which provide a preliminary understanding of time behavior of the series as shown in Figure 2. Trend of the original series appear to be slightly increasing. Nonetheless, this needs to be tested and conformed through descriptive analysis and trend modeling.

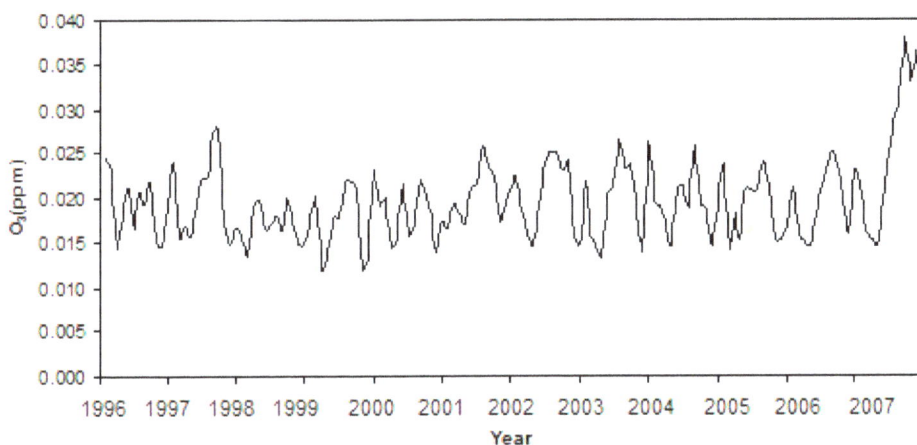

Figure 2. Original monthly ozone concentration for Kemaman

In seasonality of ozone, a well-defined annual cycle was consistent with the highest ozone means occurring in August, and the lowest ozone means in November (Figure 3). Table 1 show the seasonal indices range from a low of 80.047 in November to a high of 122.058 in August. This indicates that there is a seasonal swing from 80.047% of average to 122.058% of average throughout the course of one complete cycle i.e. one year. The seasonal variation pattern in Kemaman differed from other countries, such as United States, United Kingdom, Italy, Canada, and Japan, in that the peak ozone concentration did not correspond to maximum photochemical activity in summer [14,15,16].

For the purpose of forecasting the trend in this study, the first 132 observations (January 1996 to December 2006) were used to fit the ARIMA models while the subsequent 12 observations (from January 2007 to December 2007) were kept for the post sample forecast accuracy check. Ozone concentrations data has been adjusted in the following way before the model was fit: - simple differences of order 1 and seasonal differences of order 1 were taken. The model with the lowest value (-11.8601) of the Akaike Information Criterion (AIC)

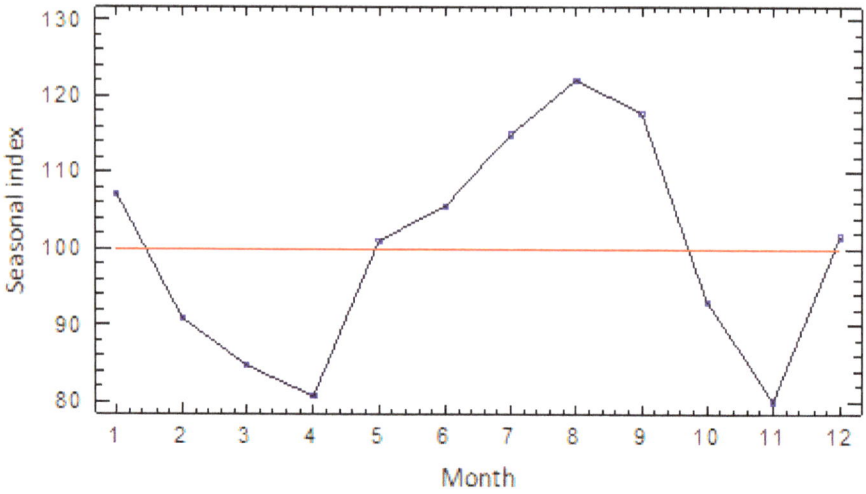

Figure 3. Annual variation of monthly ozone means

Month	Seasonal Index
January	107.199
February	90.8259
March	84.7179
April	80.7204
May	101.135
June	105.618
July	115.073
August	122.058
September	117.771
October	93.0941
November	80.0473
December	101.741

Table 1. Seasonal Index of Ozone

is (ARIMA) (0, 1, 1) x (1, 1, 2)^{12} was selected and has been used to generate the forecasts (Figure 4). This model assumes that the best forecast for future data is given by a parametric model relating the most recent data value to previous data values and previous noise. As shown in Table 2, The P-value for the MA (1) term, SAR (1) term, SMA (1) term and SMA (2) term, respectively are less than 0.05, so they are significantly different from 0. Meanwhile, the estimated standard deviation of the input white noise equals 0.00277984. Since no tests are statistically significant at the 95% or higher confidence level, the current model is adequate to represent the data and could be used to forecast the upcoming ozone concentration. Therefore, we can assume that the best model for ground level ozone in Kemaman is the mathematical expression:

$$Z(t) = a(t) + 0.53a(t-12) - 0.82(t-1) - 1.67a(t-12) + 0.73a(t-24)$$
$$+0.82(1.67)a(t-13) - 0.82(0.73)a(t-25)$$

(3)

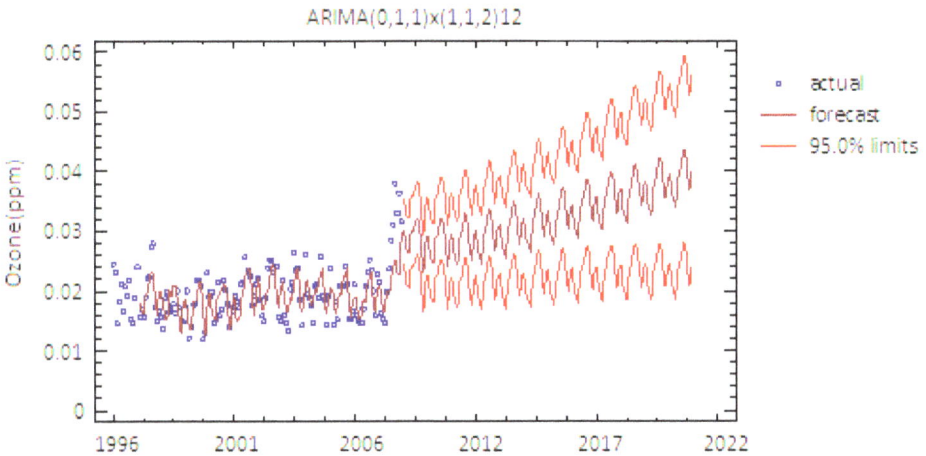

Figure 4. Model predicted plot of ozone concentration with actual and 95% confidence band

Parameter	Estimate	Stnd. Error	T	P-value
MA(1)	0.818786	0.0478133	17.1246	0.000000
SAR(1)	0.531745	0.146213	3.63678	0.000400
SMA(1)	1.67374	0.092474	18.0996	0.000000
SMA(2)	-0.728689	0.081741	-8.91461	0.000000

Table 2. ARIMA (0, 1, 1) x (1, 1, 2)^{12} model parameter characteristics

Model*	RMSE	MAE	MAPE	ME	MPE	AIC
(A)	0.00337	0.00249	13.526	0.000004	-1.5017	-11.2253
(B)	0.00271	0.00206	11.086	0.000002	-1.8498	-11.6431
(C)	0.00269	0.00201	10.786	0.000002	-1.8157	-11.6409
(H)	0.00267	0.00198	10.707	0.000003	-1.7185	-11.6712
(I)	0.00271	0.00201	10.870	-0.000050	-1.9817	-11.6423
(J)	0.00270	0.00199	10.671	0.000206	-0.5469	-11.6286
(M)	0.00258	0.00206	11.250	0.000031	-1.5638	-11.8601
(N)	0.00257	0.00204	11.192	-0.00009	-2.0636	-11.8478
(O)	0.00258	0.00206	11.298	-0.000053	-1.9673	-11.8392
(P)	0.00259	0.00207	11.260		-1.7473	-11.8382
(Q)	0.00259	0.00207	11.267	0.000030	-1.5789	-11.8335

*Models
(A) Random walk; (B) Constant mean = 0.0190056; (C) Linear trend = 0.0184806 + 0.00000789502 t
(H) Simple exponential smoothing with α = 0.109; (I) Brown's linear exp. smoothing with α = 0.0572
(J) Holt's linear exp. smoothing with α = 0.1291 and β = 0.0301; (M) ARIMA(0,1,1)x(1,1,2)12
(N) ARIMA(1,0,1)x(1,1,2)12; (O) ARIMA(0,1,1)x(1,1,2)12 with constant
(P) ARIMA(0,1,1)x(2,1,2)12; (Q) ARIMA(0,1,2)x(1,1,2)12

Table 3. Model Comparison

According to plots of residual ACF (Figure 5) and PACF (Figure 6), residuals are white noise and not-auto correlated. Furthermore, as shown in Figure 7 of normal probability plot, residuals of the model are normal.

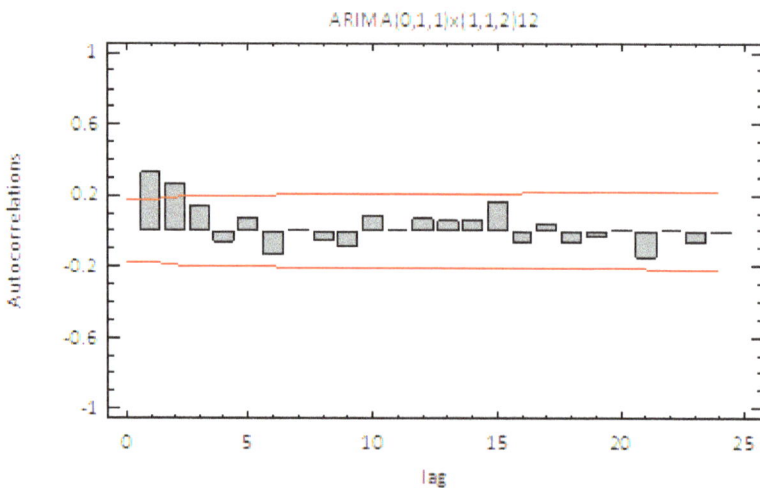

Figure 5. Residual autocorrelation functions (ACF) plot

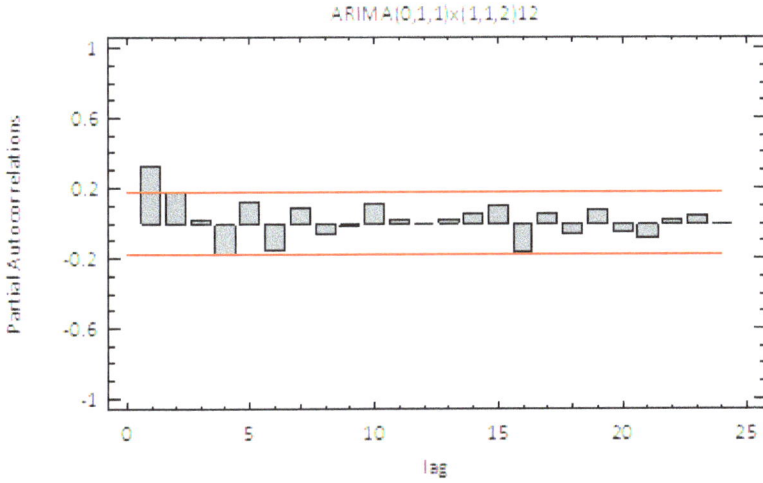

Figure 6. Residual partial autocorrelation (PACF) functions plot

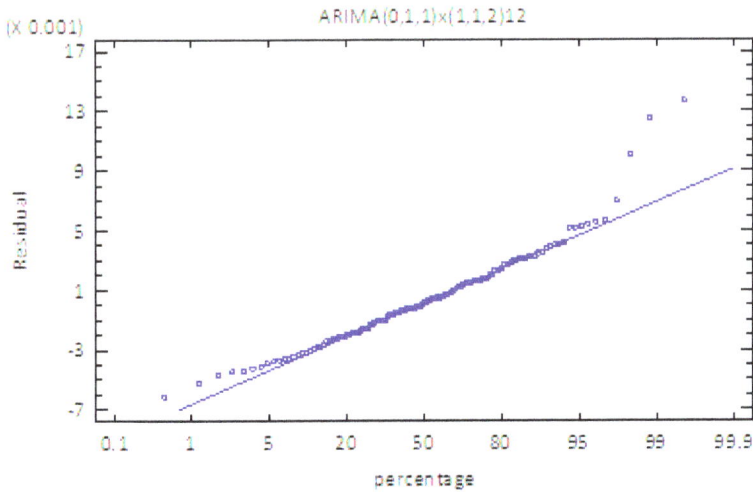

Figure 7. Residual normal probability plot

Based on the prediction for ozone concentration (Figure 4), there is a statistical significant upward trend at Kemaman station. The detection of a steady statistical significant upward trend for ozone concentration in Kemaman is quite alarming. This is likely due to sources of ozone precursors related to industrial activities from nearby areas and the increase in road traffic volume.

4. Conclusion

Time series analysis is an important tool in modeling and forecasting air pollutants. Although, this piece of information was not appropriate to predict the exact monthly ozone concentration, ARIMA $(0, 1, 1) \times (1, 1, 2)^{12}$ model give us information that can help the decision makers establish strategies, priorities and proper use of fossil fuel resources in Kemaman. This is very important because ground level ozone (O_3) is formed from NO_x and VOCs brought about by human activities (largely the combustion of fossil fuel). In summary, the ozone level increased steadily in Kemaman area and is predicted to exceed 40 ppb by 2019 if no effective countermeasures are introduced.

Author details

Marzuki Ismail*, Azrin Suroto and Nurul Ain Ismail
Department of Engineering Science, Faculty of Science and Technology, Universiti Malaysia Terengganu, Kuala Terengganu, Malaysia

Acknowledgement

The researchers would like to thank DOE Malaysia for providing pollutants data from 1996-2007 and the Ministry of Higher Education (MOHE) for allocating a research grant to accomplish this study.

5. References

[1] Wu, H.W.Y. & Chan, L.Y. 2001. Surface Ozone Trends in Hong Kong in 1985-1995. *Journal Of Environmental International* 26: 213-222
[2] Chan, C.Y. & Chan, L.Y. 2000. The effect of Meteorology and Air Pollutant Transport on Ozone Episodes at a Subtropical Coastal Asian City, Hong Kong. *Journal of Geophysical Research* 105: 20707- 20719.
[3] Kyriakidis, P.C. & Journal, A. G. 2001. Stochastic Modelling of Atmospheric Pollution: A Special Time Series Framework, Part II: Application to Monitoring Monthly Sulfate Deposition Over Europe. *Journal of Atmospheric Environment* 35: 2339-2348
[4] Salcedo, R.L.R., Alvim, F.M., Alves, C. & Martins, F. 1999. Time Series Analysis of Air Pollution data. *Journal of Atmospheric Environment* 33 : 2361-2372.

---

* Corresponding Author

[5] Schwartz, J. & Marcus, A. 1990. Mortality and Air Pollution in London: A Time Series Analysis. *American Journal of Epidemiology* 131: 85-194

[6] Hies, T., Treffeisen, R., Sebald, L. & Reimer, E. 2003. Spectral Analysis of Air Pollutants. Part I: Elemental Carbon Time Series. *Journal of Atmospheric Environment* 34: 3495-3502

[7] Kocak, K., Saylan, L. & Sen, O. 2000. Nonlinear Time Series Prediction of O_3 Concentration in Istanbul. *Journal of Atmospheric Environment* 34: 1267-1271

[8] Kuang-Jung Hsu. 2003. Time Series Analysis of the Independence Among Air Pollutants. Atmospheric Environment. Vol 26B, No.4,pp, 491-503,1992.

[9] M. Omidravi, S. Hassanzadah, F. Hossicinibalam. 2008. Time Series Analysis of Ozone Data in Isfahan. Physica A. Statistical Mechanics and Its Applications, 387 (16-17),pp 4393-4403.

[10] Afroz, R., Hassan, M.N., Ibrahim, N.A. 2003. Review of Air Pollution and Health Impacts in Malaysia. *Journal of Environmental Research* 92 (2): 71-77

[11] ASMA., 2008. Alam Sekitar Malaysia Sdn Bhd. http://www.enviromalaysia.com.my 23/12/09.

[12] Hong, W. 1997. A Time Series Analysis of United States Carrots Exports to Canada. Msc. Thesis, North Dakota State University.

[13] Naill P.E. & Momani M., 2009. Time Series Analysis Model for Rainfall Data in Jordan: Case Study for Using Time Series Analysis. *American Journal of Environmental Sciences* 5 (5): 599-604

[14] Angle, R.P. & Sandhu, H.S. 1989. Urban and Rural Ozone Concentrations in Alberta, Canada. *Journal of Atmospheric Environment* 23: 215- 221

[15] Colbeck, I., MacKenzie, A.R. Air Quality Monographs, Air Pollution by Photochemical Oxidants, Vol.1. Amsterdam: Elsevier,1994 pp107-71, 232-326.

[16] Lorenzini, G., Nali, C. & Panicucci, A. 1994. Surface Ozone in Pisa (Italy): A Six Year Study. *Journal of Atmospheric Environment* 28:3155-3164.

[17] Bencala, K.E. & Seinfield, J.H. 1979. On Frequency distribution of Air Pollutant Concentrations. *Journal of Atmospheric Environment* 10: 941-950

[18] Gouveia, N. & Fletcher, T. 2000. Time Series Analysis of Air Pollution and Mortality : Effects by Cause, Age and Socioeconomic Status. *Journal of Epidemiology and Community Health* 54: 750-755

[19] Hertzberg, A.M. & Frew, L. 2003. Can Public Policy be Influenced? *Environmetrics* 14 (1): 1- 10

[20] Lee, C.K. 2002. Multiracial Characteristics in Air Pollutant Concentration Time Series. *Journal of Water Air Soil Pollution* 135: 389-409

[21] Liu, C.M., Hung, C.Y., Shieh, S.L., & Wu, C.C. 1994. Important Meteorological Parameters For Ozone Episodes Experienced In The Taipei Basin. *Journal of Atmospheric Environment* 28: 159-173

[22] Roberts, S. 2003. Combining data From Multiple Monitors in Air Pollution Mortality Time Series Studies. *Journal of Atmospheric Environment* 35: 2339-2348.

[23] Touloumi, G., Atkinson, R. & Terte, A.L. 2004. Analysis of Health Outcome Time Series Data in Epidemiological Studies. *Journal of Environmetrics* 15: 101-117

[24] Yee, E. & Chen, R. 1997. A Simple Model for Probability Density Functions of Concentration Fluctuations in Atmospheric Plumes. *Journal of Atmospheric Environment* 31: 991-1002

Air Pollution and Death Due to Cardiovascular Diseases (Case Study: Isfahan Province of Iran)

Masoumeh Rashidi, Mohammad Hossein Rameshat and Hadi Gharib

Additional information is available at the end of the chapter

1. Introduction

Complex interaction between anthropogenic activities, air quality and human health in urban areas, sustains the need for the development of an interdisciplinary and integrated risk-assessment methodology. Such study would help for the establishment of a sustainable development in urban areas that can maintain the integrity of air quality and preserve human health. For the last century, the worldwide development of anthropogenic activities as well as modifications of spatial management and occupational uses in urban areas have lead to considerable degradation of air quality through the production of a large number of pollutants[1]. Sustainable development introduced during the 1980 represents a sure mean to withstand deleterious effects of pollutants observed in most large cities. However, making effective such concept in urban areas requires the validation of a risk-assessment methodology that can integrate and connect anthropogenic uses of urban areas, air pollution and the occurrence of some pathologies. One actual and major challenge is how to apprehend complexity of systems due to the interaction of multiple parameters at each level of organization (anthropogenic or biological, individual or population) and scale (regional or local). Such challenge can be facilitated by the development of a multidisciplinary and integrative approach using tools from biology and geography that can allow the analysis of complex systems For example, the introduction of biomarkers at the cellular and molecular levels in the detection of early biological events induced by pollutants constitutes promising tools in estimating exposure of human population. Cardiovascular disease is one of the most prevalent diseases in the world and it is expected to be the main cause of death by 2020[2]. Nowadays, the cardiovascular diseases are one of the important issues in health care. The prevalence of this disease in more countries has been rising as the third leading cause of death and the first group of chronic diseases and concerns the health and treatment in the Iran. New eating habits, increased smoking, increasing air population, and the older demographic composition are the predisposing factors in increasing the cases of cardiovascular diseases. It is estimated that one third of cases of cardiovascular disease is preventable and a third contingent on early diagnosis, are potentially treatable. Scientific advances and progress in many cases of cardiovascular diseases have been caused to disease containment and control of its causes and have provided the increase long-term survival for patients with a wide range of types of invasive diseases(3). Despite the lack of attention to air pollution, one of the main reasons is the occurrence of cardiovascular diseases (4). For example, nearly one million ton of Plumb is added to the globe soils annually in which large quantities of atmospheric dust, scattering ash, and chemical fertilizers used in agriculture, industry and urban wastes are included. In many cases, air pollution factors affecting disease is less under consideration (5). Isfahan province, with an area of about 107,045 square kilometers, equivalent to 6.3% of the total area of Iran is located between 30 degrees 43 minutes and 34 degrees 27 minutes north latitude and 49 degrees 38 minutes and 55 degrees 32 minutes east of the Greenwich meridian(6,7). The province is 1550 meters above the sea level altitude (Figure1). This study aimed at mapping the distribution of death due to Cardiovascular Diseases and its relationship with Air Pollution in this province.

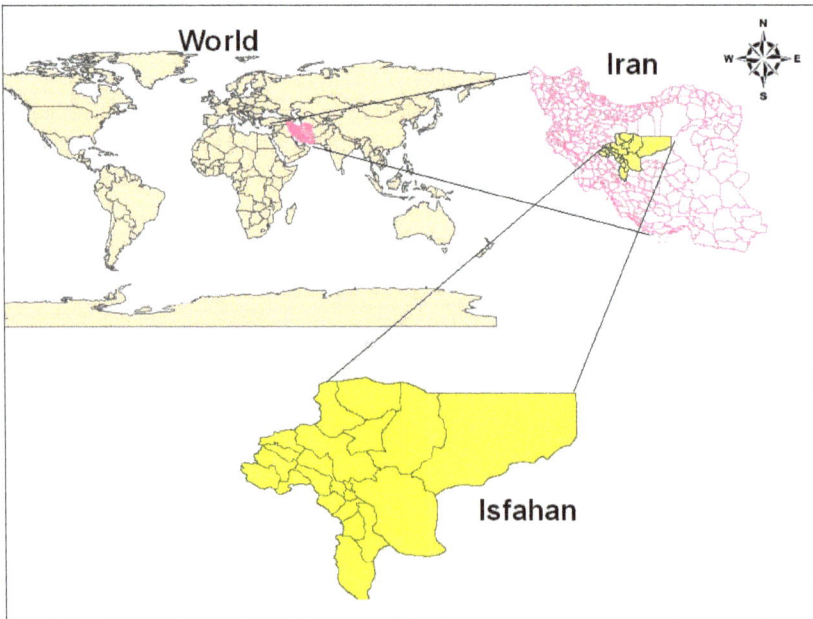

Figure 1. Geographical situation of Isfahan province

2. Method used

The software of geographic information system (GIS) was used after entering data in the mapping information table; spatial distribution was mapped and distribution of Geographical Epidemiology of Death Due to Cardiovascular Diseases in the was determined this case study, the rate of all the deaths in Isfahan province (Iran) within 2005 to 2009 was provided. The collected data was used to find out the rate of deaths due to cardiovascular diseases and preparing geographical distribution maps. Then, by putting down the death rates for different sexes (men and women), the geographical distribution map for deaths with regards to cardiovascular diseases was drawn.

3. Air pollution

Any visible or invisible particle or gas found in the air that is not part of the original, normal composition. Generally any substance that people introduce into the atmosphere that has damaging effects on living things and the environment is considered air pollution. Carbon dioxide, a greenhouse gas, is the main pollutant that is warming Earth. Though living things emit carbon dioxide when they breathe, carbon dioxide is widely considered to be a pollutant when associated with cars, planes, power plants, and other human activities that involve the burning of fossil fuels such as gasoline and natural gas. In the past 150 years, such activities have pumped enough carbon dioxide into the atmosphere to raise its levels higher than they have been for hundreds of thousands of years. Other greenhouse gases include methane—which comes from such sources as swamps and gas emitted by livestock—and chlorofluorocarbons (CFCs), which were used in refrigerants and aerosol propellants until they were banned because of their deteriorating effect on Earth's ozone layer. Another pollutant associated with climate change is sulfur dioxide, a component of smog (9). Sulfur dioxide and closely related chemicals are known primarily as a cause of acid rain. But they also reflect light when released in the atmosphere, which keeps sunlight out and causes Earth to cool. Volcanic eruptions can spew massive amounts of sulfur dioxide into the atmosphere, sometimes causing cooling that lasts for years. In fact, volcanoes used to be the main source of atmospheric sulfur dioxide; today people are(10). Industrialized countries have worked to reduce levels of sulfur dioxide, smog, and smoke in order to improve people's health. But a result, not predicted until recently, is that the lower sulfur dioxide levels may actually make global warming worse. Just as sulfur dioxide from volcanoes can cool the planet by blocking sunlight, cutting the amount of the compound in the atmosphere lets more sunlight through, warming the Earth. This effect is exaggerated when elevated levels of other greenhouse gases in the atmosphere trap the additional heat.

3.1. Major classes of air pollution

- Carbon Oxides (CO and CO2)
- Sulfur Oxides (SO2)
- Nitrogen Oxides (NO and NO2)
- Volatile Organic Compounds (VOCs – CFCs)
- Suspended Particulate Matter (soot, dust, asbestos, lead etc).
- Photochemical Oxidants (ozone O3)
- Radioactive Substances (Radon)
- Hazardous Air Pollutants (carcinogens, etc) (11).

3.2. Where do these pollutants come from?

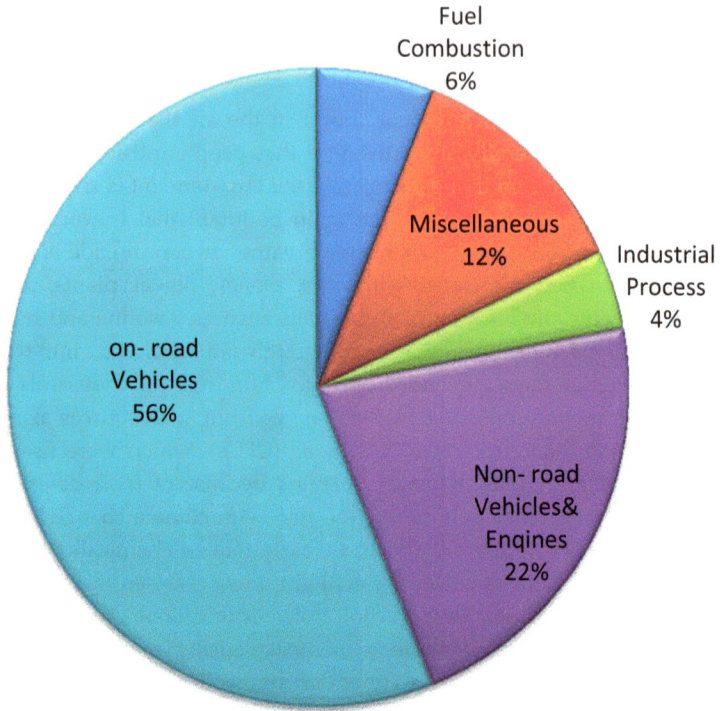

Figure 2. Source of Air Pollution (12)

Health and Effects of Air Pollution	
Pollutant	**Health Effects**
Ozone (O3)	Decreases lung function and causes respiratory symptoms, such as coughing and shortness of breath; aggravates asthma and other lung diseases leading to increased medication use, hospital admissions, emergency department (ED) visits, and premature mortality.
Particulate Matter (PM)	Short-term exposures can aggravate heart or lung diseases leading to symptoms, increased medication use, hospital admissions, ED visits, and premature mortality; long-term exposures can lead to the development of heart or lung disease and premature mortality.
Lead (Pb)	Damages the developing nervous system, resulting in IQ loss and impacts on learning, memory, and behavior in children. Cardiovascular and renal effects in adults and early effects related to anemia.

Oxides of Sulfur (SOx)	Aggravate asthma, leading to wheezing, chest tightness and shortness of breath, increased medication use, hospital admissions, and ED visits; very high levels can cause respiratory symptoms in people without lung disease.
Oxides of Nitrogen (NOx)	Aggravate lung diseases leading to respiratory symptoms, hospital admissions, and ED visits; increase susceptibility to respiratory infection.
Carbon Monoxide (CO)	Reduces the amount of oxygen reaching the body's organs and tissues; aggravates heart disease, resulting in chest pain and other symptoms leading to hospital admissions and ED visits.
Ammonia (NH3)	Contributes to particle formation with associated health effects.
Volatile Organic Compounds (VOCs)	Some are toxic air pollutants that cause cancer and other serious health problems. Contribute to ozone formation with associated health effects.
Mercury (Hg)	Causes liver, kidney, and brain damage and neurological and developmental damage.
Other Toxic Air Pollutants	Cause cancer; immune system damage; and neurological, reproductive, developmental, respiratory, and other health problems. Some toxic air pollutants contribute to ozone and particle pollution with associated health effects.

Table 1. Effects of air pollution on human health (13)

4. Cardiovascular disease and air pollution

Diseases of the heart or blood vessels, or cardiovascular disease, and in particular coronary heart disease (harm to the heart resulting from an insufficient supply of oxygenated blood) are leading causes of death in the Iran (14). Prevention of these killers has traditionally focused on controlling hypertension, cholesterol levels, and smoking and making healthy choices in regard to diet, exercise, and avoiding second-hand smoke. However, accumulating evidence indicates that air pollutants contribute to serious, even fatal damage to the cardiovascular system – and air pollution is a factor that you can't control just through healthy lifestyle. Harmful air pollutants lead to cardiovascular diseases such as artery blockages leading to heart attacks (arterial occlusion) and death of heart tissue due to oxygen deprivation, leading to permanent heart damage (infarct formation). The mechanisms by which air pollution causes cardiovascular disease are thought to be the same as those responsible for respiratory disease: pulmonary inflammation and oxidative stress.

5. Finding in case study

The population includes 35273 records from death due to Cardiovascular Diseases in the province. The period studied, according to the number of samples is sufficiently reliable and, over 5 years (from 2005 until early 2009) was considered .Impaired synthesis of hemoglobin and anemia, Cardiovascular Diseases, Respiratory Diseases, malignant disease, hypertension, kidney damage, miscarriages and premature infants, nervous system disorders, brain damage, male infertility, loss of learning and behavioral disorders in children are from the negative effects of high concentrations of the Air Pollution in the body. Air Pollution exists naturally in the environment, but in most cases the increase in quantity, is the result of human activities. There were 19614 men (i.e. 56%) and 15659 women (44%) regarding the mentioned mortality rate, showing more men than women. Analysis of the mortality conforming rate of cardiovascular diseases in men shower to be highest in Isfahan, Najafabad, Borkhar&Meimeh, Fereidan, Natanz, Ardestan, Mobarakeh, Lenjan&Naein, respectively and the lowest rates were in Golpayegan, Tiran &Karvand, Falavarjan&Chadegan, that means that mortality rates were higher in central counties of the province.This is observed for the total population, and men and women separately. It was significant in most of the central counties of the province.

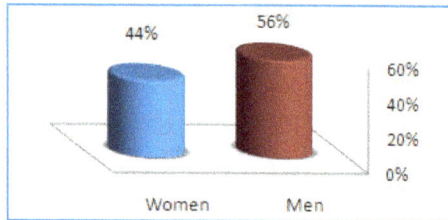

Figure 3. Graph of statistical comparison death due to cardiovascular diseases in Men& Women

After drafting the diagram for distribution of death due to cardiovascular diseases in Men& Women (Figure 3). Death rate was higher in men than women.

6. Cities with higher air pollution

Cities such as Isfahan, Najafabadf, Borkhar&Meimeh, Ardestan and Natanz,… are Population centers and air pollution in these cities, according to survey is more than the other cities because human activity is higher in these cities, The following map shows the cities with high pollution in Isfahan province(Figure4).

Air pollution levels in the study province are increased and with increased air pollution also death due to cardiovascular disease has gone up (Figure5, 6).

Most of the mortality rates with regards to cardiovascular diseases in women was in Isfahan, Najafabadf, Borkhar&Meimeh, Ardestan and Natanz that is much less in comparison to men (Figure7, 8). This means that men have gone ahead of women in this respect. The role of weather pollution in emergence of cardiovascular diseases in urban communities is considered as an effective factor that could not be modified, such that comforting environmental pollution has been considered relative to different cardiovascular effects including angina, heart stroke and hearty failures. Heart disease is the bitter achievement of advance technology. The useful role of technology somewhat allows people to have longer life and the harmful role of technology provides the change in life style and immobility.

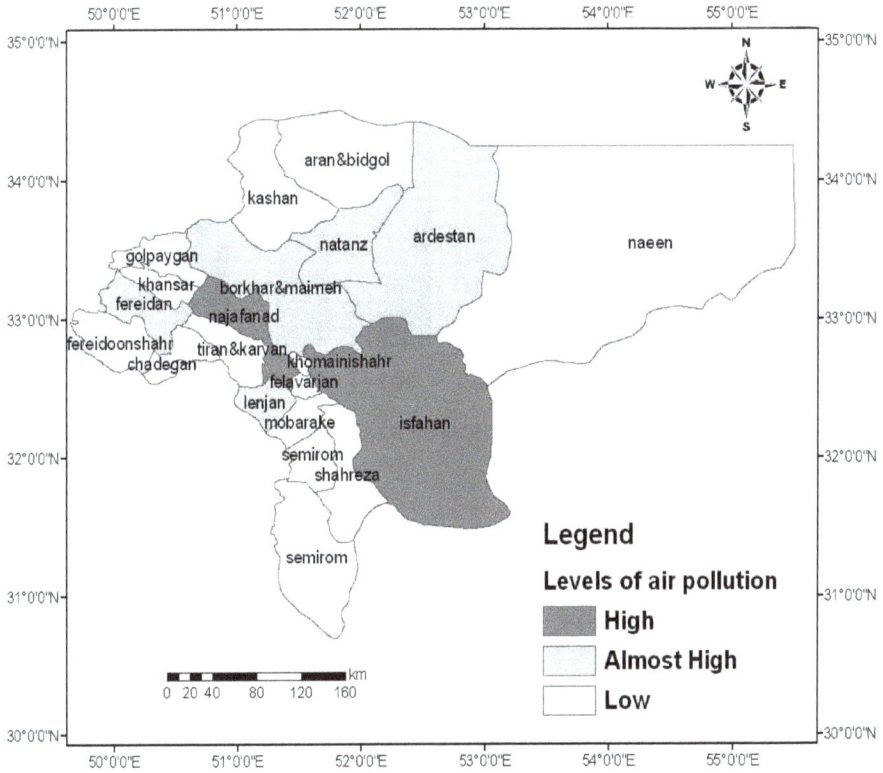

Figure 4. Levels of Air Pollution in Isfahan province

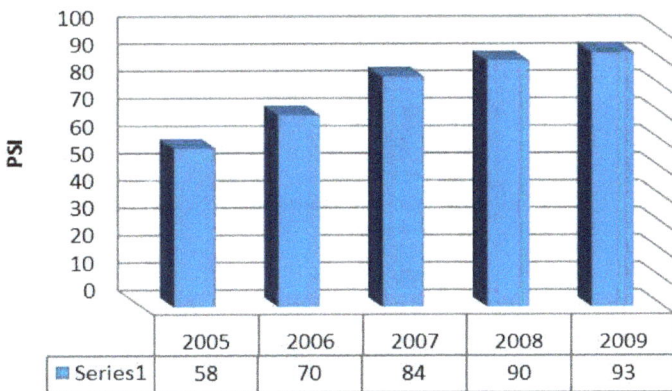

Figure 5. A comparison chart increase air pollution in the years 2005-2009

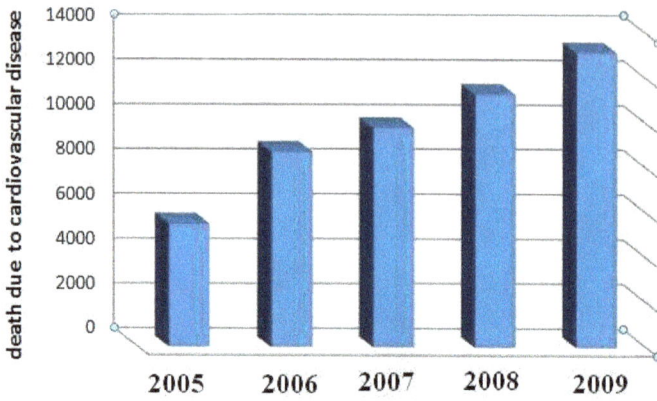

Figure 6. A comparison chart increase death due to cardiovascular disease in the years 2005-2009

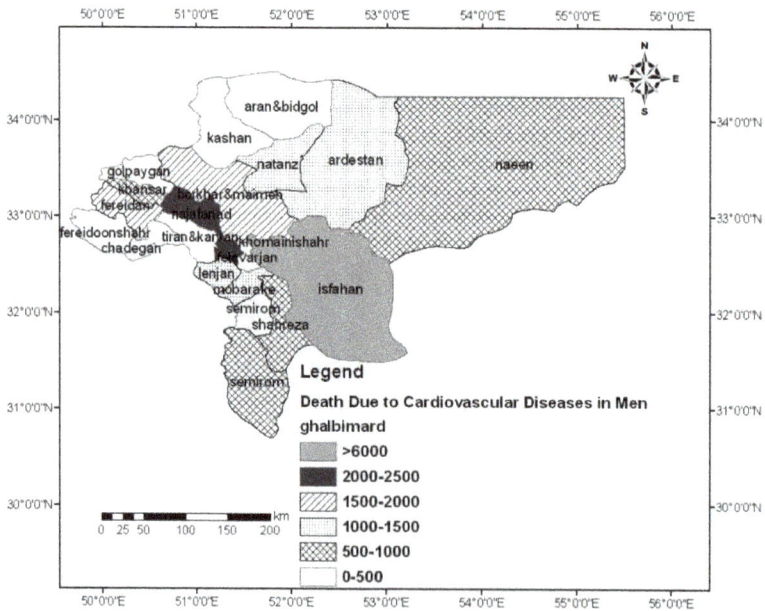

Figure 7. Spatial distribution of death due to cardiovascular diseases in Men

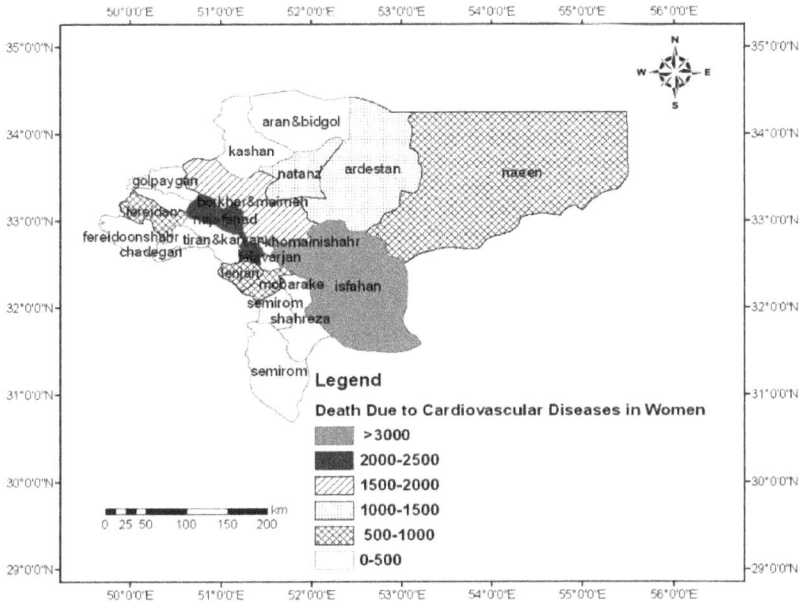

Figure 8. Spatial distribution of death due to cardiovascular diseases in Women

7. Conclusion

By drawing the geographical distribution of mortality due to cardiovascular reasons (by the use of GIS software), it was observed that the rate of mortality I higher in control and main counties of the province, which could be due to two reasons: 1) Due to existence Air pollution in the main cities of the province including Isfahan, Najafabad, Borkhar&Meimeh, More vehicles are movement, in these cities than the other places in the province. 2) Improper diet including saturated fat due to mechanized life and better welfare in these regions and immobility, use of new technologies, and environmental pollutions including the existence of some specific elements & hard urban life all express the verification of the hypothesis. Also, it was observed that mortality in men is higher in the province than women and there could be different reasons for that, which may include Most men work outside the home in , male hormones, some social factors, increasing fat around stomach in men, stimulating behaviors and sometimes offensive behavior, not observing the weight, stress in work places and smoking. Since the basis for campaign against non-epidemic diseases, including cardiovascular disease is changing the people's life style, it seems that it can be achieved by instructions and training people, making required polices and enacting laws and necessary regulations to provide on environment that is suitable for promoting healthy behaviors in life. By proper intervention in the society the effects of risk factors could be totally eliminated or reduced. Even partial changes could be very useful. Prevention is possible by intervening the risk factors in cardiovascular diseases such as identification of some elements in the environment and finding the place of their distribution, avoiding the use of air pollutants or using them as little as possible, proper use of technology, changing diets, behaviors, physical habits, reducing anxieties and mental stresses and other environmental diseases.

Author details

Masoumeh Rashidi
Department of Geography, and Medicine Geography Researchers, the University of Isfahan, Iran

Mohammad Hossein Rameshat
Department of Geography, Isfahan University, Iran

Hadi Gharib
Iran Space Agency, Iran

8. References

Azizi, F, (2001), *"Epidemiology and control of common diseases in Iran"*, Volume II, Tehran University Press, 2001.

Braunwald E. (2005), *Approach to the Patient with Cardiovascular Disease*. In: Kasper DL, Branwald E, Favci AS, Havser SL, Longo DL, Jameson JL. Harrison's Principles of Internal Medicine. McGraw-Hill. New York, 1301-4.

Reddy KS, (2004), *Cardiovascular Disease in Non-Western Countries* Engl J Med. 350: 2438-40.

Nagavi, M. (2005), the pattern of mortality within 23 provinces of Iran in 2003, Health Deputy, Iranian Ministry of Health. Tehran.

Rezaeian M, Dunn G. St. Leger, S. Appleby L. (2007), *Geographical epidemiology, spatial analysis and geographical information systems: a multidisciplinary glossary*. J Epidemiol Community Health; 61: 98-102.

Rezaeian, M. Dunn, G. St. Leger, S. Appleby L. (2004), *the production and interpretation of disease maps: A methodological case study. Soc Psychiatry* Psychiatr Epidemiol.; 39: 947-54.

Rezaeian, M. (2004), *an introduction to the practical methods for mapping the geographical morbidity and mortality rates*.Tollo-e-behdasht. 2: 41-51.

Isfahan Health Center, *Center for death Statistics, 2009.*

Samet, J. M., Zeger, S. L., Kelsall, J., Xu, J., and Kalkstein, L. (1997), *"Air Pollution, Weathera nd Mortalityi n Philadelphia,"in ParticulateA ir Pollution and Daily Mortality: Analyses of the Effects of Weather and Multiple* Air Pollutants, The Phase IB report of the Particle Epidemiology Evaluation Project, Cambridge, MA: Health Effects Institute.

Schwartz, J. (1994), *"Air Pollution and Daily Mortality: A Review and Meta Analysis,"* EnvironmentaRl esearch, 64, 36-52.

Zeger, S. L., Dominici, F., and Samet, J. M. (1999), *"Harvesting-Resistant Estimates of Pollution Effects on Mortality,"*E pidemiology, 8 9, 171-175.

Research Centre for Atmospheric Chemistry, Ozone *and Air Pollution in Isfahan province.*

WHO (2007). *Health risks of heavy metals from long-range transboundary air pollution.* Copenhagen, World Health Organization Regional Office for Europe.

Rashidi M, Ramehsat M.H, Ghias, M(2011), *Geographical Epidemiology of Death Due to Cardiovascular Diseases in Isfahan Povince, Iran*; Journal of Isfahan Medical School, Vol 29, No 125, 1st week, April.

Chapter 8

Particulate Air Pollutants and Respiratory Diseases

An-Soo Jang

Additional information is available at the end of the chapter

1. Introduction

Air pollution is composed of a mixture of toxins, consisting of particles and gases emitted in large quantities from many different combustion sources, including cars and industries. A variety of anthropogenic and natural particle sources are present in ambient air. Throughout the past decade, the composition of air pollution has changed in developed countries from classical type 1 pollution, consisting of sulfur dioxide and large dust particles, to modern type II pollution, characterized by nitrogen oxides, organic compounds, ozone, and ultra-fine particles (Schäfer & Ring, 1997).

Particulate matter (PM) is the principal component of indoor and outdoor air pollution. PM is a complex, multi-pollutant mixture of solid and liquid particles suspended in gas (Ristovski et al., 2011). PM originates from a variety of manmade and natural sources. Natural sources include pollen, spores, bacteria, plant and animal debris, and suspended materials. Human-made sources include industrial emissions and combustion byproducts from incinerators, motor vehicles, and power plants. Indoor sources include cigarette smoking, cooking, wood and other materials burned in stoves and fireplaces, cleaning activities that resuspend dust particles, and the infiltration of outdoor particles into the indoor environment (2003, McCormack et al., 2008).

Vehicle emissions are the predominant source of fine PM (2.5 μ PM with an aerodynamic diameter <2.5 μm) in urban areas, where most people live globally (Ristovski et al., 2011). Airborne PM less than 10 μm in aerodynamic diameter (PM 10) is a complex mixture of materials with a carbonaceous core and associated materials such as organic compounds, acids, and fine metal particles (Pagan et al., 2003).

The physical properties of PM including the mass, surface area, and number/size/ distribution of particles, as well as their physical state, influence respiratory health in different ways (Ristovski et al., 2011). The primary exposure mechanism to PM and other particle sources is by inhalation (Ristovski et al., 2011).

Growing epidemiologic evidence indicates that inhalation of airborne PM increases respiratory and cardiac mortality and morbidity, and produces a range of adverse respiratory health outcomes such as asthma, lung function decline, lung cancer, and chronic obstructive pulmonary disease (COPD) (Ayres et al., 2008, Ristovski et al., 2011). Epidemiologic data indicate that air pollution also aggravates asthma, with the exacerbation correlating with levels of environmental particles (Schwartz et al. 1993). Likewise, the rate of decline seen in COPD patients correlates with the level of air pollution where the patients live (Pope & Kanner, 1993).

PM induces inflammation, innate and acquired immunity, and oxidative stress. It also increases innate and adaptive immune responses in both animals and humans. That derived from traffic and various industries is associated with allergic airway disorders, including asthma. Understanding the mechanisms of lung injury from PM will enhance efforts to protect at-risk individuals from the harmful respiratory effects of air pollutants. PM functions as an adjuvant inducing lung inflammation to allergens or respiratory viruses. Inhalation of PM aggravates respiratory symptoms in patients with chronic airway diseases, but the mechanisms underlying this response remain poorly understood. This review focuses on the adverse effects of exposure to ambient PM air pollution on the exacerbation, progression, and development of asthma, COPD, and respiratory diseases. It also attempts to offer insights into the mechanisms by which particles may influence airway inflammation, and several mechanisms that may explain the relationship between particulate air pollutants and respiratory diseases are discussed.

2. Adverse effects of PM on respiratory diseases identified in epidemiologic studies (Figure 1)

PM is a mixture of organic and inorganic solid and liquid particles of different origins, size, and composition. It is a major component of urban air pollution and greatly effects health. Penetration of the tracheobronchial tract is related to particle size and the efficiency of airway defense mechanisms (D'Amato et al., 2010). Particles smaller than 10 μm can get into the large upper branches, just below the throat, where they are caught and removed (by coughing and spitting or swallowing). Particles smaller than 5 μm can get into the bronchial tubes at the top of the lungs, while particles smaller than 2.5 μm in diameter can penetrate the deepest (alveolar) portions of the lung. If these particles are soluble in water, they pass directly into the blood in the alveolar capillaries. If they are insoluble in water, they are retained deep within the lungs for extensive periods of time. About 60% of PM10 particles (by weight) have a diameter of 2.5 μm or less.

According to the World Health Organization, 24% of the global disease burden and 23% of all deaths are attributable to environmental factors (Pruss-Ustun & Corvalan, 2006). The cause of, and route of exposure that lead to, disease and death is often complex and poorly

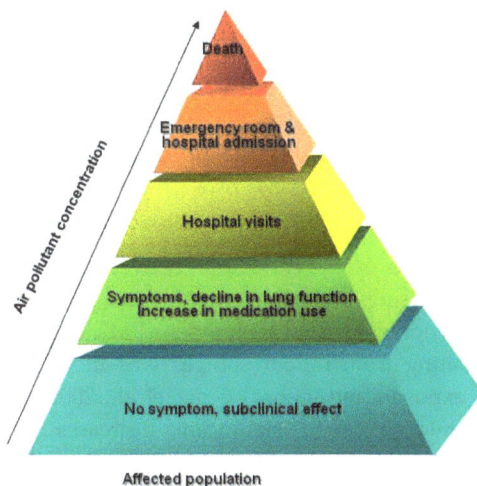

Figure 1. Particulate matter effect on respiratory diseases.

understood. The increased air pollution emanating from traffic and various industries has caused an increase in the incidence of allergic diseases. In children, acute exposure to air pollution is associated with increased respiratory symptoms and decreased lung function. Chronic exposure to increased levels of inhalable particles is associated with up to a threefold increase in non-specific respiratory symptoms, such as chronic cough, asthma, and chronic airway diseases (Nicolai, 1989). Exposure to heavy traffic leads to significant increases in respiratory symptoms, while a direct effect of traffic on asthma initiation has been documented (Nicolai, 1989). Indeed, outdoor air pollution levels have been associated with adverse health in asthma subjects (Nicolai, 1989). Exposure to traffic-related air pollution, in particular diesel exhaust particles (DEP), may lead to reduced lung function in children living near major motorways (Brunekreef et al., 1997). The prevalence of airway hyper-responsiveness (AHR) has increased over the last few decades, potentially because of environmental factors. Air pollution is convincingly associated with many signs of asthma aggravation, including pulmonary function decrease, increased AHR, additional visits to emergency departments, increased hospital admissions, increased medication use, and more reported symptoms. It is also associated with inflammatory changes, interactions between air pollution and allergen challenges, and changes in immune response (Koenig, 1999). There is a significant association between traffic-related air pollution and wheezing in children (Hisch et al., 1999), and exposure to DEPs may reduce lung function in children living near motorways. DEPs account for most airborne PM in the world's largest cities (Rield & Diaz-Sanchez, 2005), and are composed of fine (2.5–0.1µm) and ultra-fine (<0.1µm) particles, although primary DEPs can coalesce to form aggregates of varying sizes. Acute exposure to DEPs causes irritation of the nose and eyes, headache, lung function abnormalities, fatigue, and nausea, while chronic exposure is associated with cough, sputum production, and diminished lung function (McCreanor et al., 2007).

There is strong evidence that episodes of air pollution aggravate respiratory disease, especially asthma. A study of the relationship between fine PM and emergency room visits for asthma in the metropolitan Seattle area was designed to substantiate that air pollution was a risk factor for asthma (Mortimer et al., 2002). Using Poisson regression analyses that controlled for weather, season, time trends, age, hospital, and day of the week, a significant association was found between fine particles measured at the monitoring station and visits to emergency departments in eight nearby participating hospitals (Mortimer et al., 2002).

There are relatively few studies on the correlation between indoor PM and asthma. A sub-group of 10 children not using inhaled corticosteroids in Seattle were found to have decreased pulmonary function associated with indoor PM2.5 exposure (Koenig et al., 2005). Moreover, PM2.5 originating from indoor sources was more potent in decreasing lung function than was PM exposure outdoors (Koenig et al., 2005). A California study of 19 predominantly white children found significant decreases in lung function (FEV1) associated with indoor PM. While this study found associations between ambient PM and lung function, a stronger association was found with indoor central site PM concentrations than outdoor PM (McConnellet et al., 2003). Significant determinants of indoor PM concentrations include smoking, sweeping, and stove use (McCormack et al., 2008), activities that are modifiable and provide opportunities for exposure reduction. Smoking has been a major source of indoor particulates over the last several decades, with more than 30% of all U.S. children exposed to secondhand smoke (Winickoff et al., 2005).

Asthma symptoms are associated with indoor coarse PM. For example, in a previous study, every 10 mg/m^3 increase in indoor PM 2.5–10 concentration led to a 6% increase in the number of days of coughing, wheezing, or chest tightness, after adjusting for age, race, sex, socioeconomic status, season, indoor fine PM, and ambient fine and coarse PM concentrations (Breysse et al., 2010). This study also found that higher indoor coarse PM concentrations were also associated with increased incidences of symptoms severe enough to slow a child's activity, cause wheezing that limited speaking ability, nocturnal symptoms, and rescue medication use; and although outdoor coarse PM was not associated with increased asthma symptoms or rescue medication use, fine PM was positively associated with respiratory symptoms and rescue medication use (Breysse et al., 2010). These findings demonstrate that both indoor coarse and fine PM distinctly affect respiratory health in children with asthma.

Although fine PM may be capable of reaching the alveoli, the regions responsible for gas exchange, the deposition of coarse PM in upper airways and subsequent bronchial hyper-reactivity may be responsible for the symptomatic response measured in preschool children (Breysse et al., 2010).

In asthmatic children attending school in urban Amsterdam, black smoke was the most important air pollution indicator associated with acute changes in lung function, respiratory symptoms, and medication use (Gielen et al., 1997). In one polluted area (Jang et al., 2003), 670 schoolchildren (100%) had normal pulmonary function, while 257 (38.3%) had AHR. A significantly greater proportion of children had AHR in the polluted area (45.0% [138/306], 6.50±0.48) than in rural (31.9% [52/163], 9.84±0.83) or coastal (33.3% [67/201], 7.17±0.68) areas.

Schoolchildren with atopy had lower PC20 levels than those without (5.98± 0.60 vs. 8.15±0.45, p < 0.001). In a multiple logistic regression model, a positive allergy skin test and living in the polluted area near a chemical factory were independently associated with AHR (odds ratio for location=2.4875, CI 1.6542-3.7406, P < 0.01; odds ratio for allergy skin test=1.5782, CI 1.1130 - 2.2379, p < 0.05), when adjusted for sex, parents' smoking habits, age, body mass index, nose symptoms, and lung symptoms. This suggests that air quality near the polluted area contributes to the development of AHR, and that controlling air pollution is important for preventing the development of asthma. Asthma, a complex disease influenced by both environmental and genetic factors, is common and the prevalence is increasing worldwide (Holgate, 1999). Indoor environmental factors thought to modify asthma severity include pollutants such as PM, nitrogen oxides, secondhand smoke, and allergens from pests, pets, and molds (Diette, 2008). In contrast to the outdoors, individuals have a greater ability to modify indoor environmental exposure risks, making indoor air pollution an attractive target for disease prevention (Breysse et al., 2010).

DEP plays a role in increasing asthma prevalence, although a causal relationship has yet to be established. In a modification of the classical ovalbumin sensitization and challenge model, mice were exposed to intranasal DEP and challenged with aerosolized DEP on days 6–8 (Song et al., 2008). Delivery of aerosolized DEP, following exposure with intranasal DEP, induced a significant increase in methacholine-induced airway hyper-responsiveness. Pope and Dockery (Pope & Dockery, 2006) suggested that there is a 0.6–2.2% increase in respiratory mortality risk for a 10 µg/m³ increase in ambient PM. Indeed, many cohort studies have demonstrated that airborne PM, of which PM is a major contributor (Robinson et al., 2010), causes respiratory mortality and morbidity (Pope & Dockery, 2006).

A cross-sectional study of 20,000 children between 6 and 12 years old found a weak association between decreased pediatric lung function and secondhand smoking (Moshammer et al., 2006). Also, children living in homes that use organic fuels for cooking, heating, and lighting are exposed to much higher levels of PM than children living in homes where parents smoke and use clean fuels (e.g., a mean indoor level of 200 mg/m³ PM per 24 h; Jiang & Bell, 2008).

There are many sources of air pollution in the home environment. Air pollution inside homes consists of a complex mixture of agents penetrating from ambient outdoor air, and agents generated by indoor sources. Indoor pollutants can vary in their potential health hazard and intensity, as well as in their distribution across geographic areas, cultural backgrounds, and socioeconomic status (Breysse et al., 2010).

In a British cohort of 4,400 preschool children, a significant association was found between exposure to primary PM10 at the home address and prevalence of coughing without a cold (Pierse et al, 2006). Data from the Third U.S. National Health and Nutrition Examination Survey (1988–1994) found that exposure to environmental tobacco smoke is associated with increased prevalence of pediatric asthma, wheezing, and chronic bronchitis (Gergen et al., 1998).

Entering adulthood with impaired lung function is a non-specific risk factor for respiratory disease in adulthood. Lower lung function predisposes children to further structural damage to the developing lung (Grigg, 2009). COPD is the non-specific terminology commonly used to describe the spectrum of diseases limiting respiratory airflow, e.g., asthma, chronic bronchitis, and emphysema (Matthay, 1992). There are several reasons why environmental exposures in childhood are relevant to the pathogenesis of COPD (Grigg, 2009). First, attenuation of lung growth due to air pollution in childhood is a risk factor for adult-onset respiratory disease. Second, there may be common cellular and molecular mechanisms underlying impaired pulmonary innate host defenses in children exposed to air pollution, and susceptibility to infection in COPD. Third, lung damage initiated in childhood may contribute to an emerging global health issue, namely, COPD due to smoke exposure.

Studies showing an association between lifelong organic smoke and the development of COPD in nonsmoking women provides a direct link between exposure of children to PM and increased vulnerability to respiratory disease in adulthood (Grigg, 2009). Chronic exposure to PM (Grigg, 2009) likely interferes with maximal lung function attainment in childhood, accelerates lung function decline in adulthood, stimulates airway mucus production, and impairs pulmonary innate immunity. Similar associations between air particulate pollution (PM10 or PM2.5) and hospital admissions for COPD have been reported for a variety of urban areas (Yang et al., 2005). The strong association between respiratory hospital admissions and PM10 pollution (Pope, 1991) supports the role of PM10 in the incidence and severity of respiratory disease.

Long-term studies usually use a cohort design when comparing mortality across populations, and vary in their long-term exposure to air pollution. An overall reduction in PM 2.5 levels over time results in reduced long-term risk of death due to cardiovascular and/or respiratory disease (Laden et al., 2006). A large European cohort study of mortality and air pollution showed smaller effective estimates, which were significant only for all-cause and respiratory mortality (Beelen et al., 2008). Epidemiological studies from controlled human exposure to toxins have identified characteristics of populations that may be more susceptible to PM-related health issues (Sacks et al., 2011): children and older adults with preexisting cardiovascular and respiratory diseases, populations with lower income and less education, and the presence of genetic polymorphisms. In addition, PM-related health effects are sometimes observed in individuals with diabetes, COPD, and increased body mass index. A cohort study of Swiss adults demonstrated that a decrease in ambient PM10 was associated with reduced respiratory symptoms (Schindler et al., 2009).

Given the increasing evidence that air pollution has both short- and long-term effects on health, the public health impact of reducing pollutant levels has gained attention. A large study across 211 U.S. counties demonstrated significant improvements in life expectancy related to reductions in PM2.5 concentrations (Pope et al., 2009).

3. Molecular mechanisms in *in vitro* and *in vivo* studies (Figure 2)

Because the lung interfaces with the external environment and is frequently exposed to air pollutants, such as PM, it is prone to oxidant-mediated cellular damage (Nel et al., 2006).

The adverse health effects of particulate pollutants may be explained by several mechanisms, including innate immunity, adaptive immunity, and the production of reactive oxygen species (Nel et al., 2006).

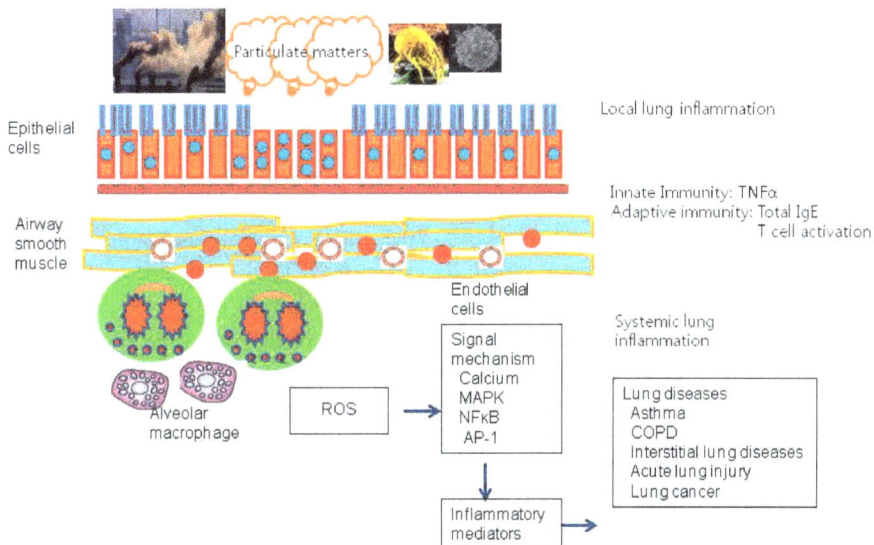

Figure 2. Proposed mechanism of lung diseases by PM.

Innate immunity

The pathways associated with acute inflammation in response to particle exposure involve an orchestrated sequence of events, mediated in part by chemokines and cytokines (Seagrave, 2008). Particles larger than 10 μm generally get caught in the nose and throat, and never enter the lungs (Yang & Omaye, 2009). After inhalation of PM, phagocytic cells including neutrophils and macrophages are recruited to the foreign particle by cytokines and chemokines, and transported by the mucociliary escalator for removal (Donaldson and Tran, 2002). PM induces the release of inflammatory cytokines, such as IL-6, IL-8, GM-CSF, and TNF-α (Stone et al., 2007) from immune cells (e.g., macrophages) as well as structural airway cells (Totlandsdal et al., 2010).

DEPs exert their effect through agents such as polyaromatic hydrocarbons (PAHs). The particles are deposited on the airway mucosa; their hydrophobic nature allows them to diffuse easily through cell membranes and to bind to cytosolic receptor complexes. Through subsequent nuclear activity, PAHs can modify both cell growth and cell differentiation programs.

Experimental studies have shown that DEP-PAHs can modify the immune response in animals and humans and modulate airway inflammatory processes. In other words, DEPs exert an adjuvant immunological effect on IgE synthesis in atopic subjects, thereby causing sensitization to airborne allergens (Diaz Sanchez et al., 1997). They also cause respiratory symptoms and modify the immune response in atopic subjects (Rield & Diaz-Sanchez, 2005, Diaz Sanchez et al., 1997), and can interact with aeroallergens to enhance antigen-induced responses, with the result that allergen-specific IgE levels are up to 50-fold greater in allergic patients stimulated with DEPs and allergens than in patients treated with allergen alone (Diaz Sanchez et al., 1997). A combined challenge of DEPs and ragweed allergen markedly increases the expression of human nasal ragweed-specific IgE *in vivo* and skews cytokine production to a type 2 helper T-cell pattern (Diaz Sanchez et al., 1997).

Chitin is commonly found in organisms including parasites, fungi, and bacteria, but does not occur in mammalian tissues (Guo et al., 2000), allowing for selective antimicrobial activity of chitinase. Macrophage-synthesized Ym1 and Ym2 are homologous to chitinase, and have chitinase activity (Sun et al., 2001, Jin et al., 1998). Through the IL-4/STAT 6 signal transduction pathway, Ym1 was implicated in allergic peritonitis (Welch et al., 2002). Acid mammalian chitinase may also be an important mediator of IL13-induced responses in Th2 disorders, such as asthma (Zhu et al., 2004). Indeed, polymorphisms in acid mammalian chitinase are associated with asthma, further supporting the involvement of acid mammalian chitinase in asthma development (Bierbaum et al., 2005). DEP induces airway hyper-responsiveness as well as Ym mRNA expression, a Th2 cell-biased response by activated macrophages. The chitinase Ym1 is expressed in the spleen and lungs, with lower expression in the thymus, intestine, and kidney, whereas Ym2 is expressed at high levels in the stomach, with lower levels in the thymus and kidney (Ward et al., 2001). Conserved STAT6 sites probably account for the similar, striking induction of Ym1 and Ym2 expression in Th2-type environments. In a murine model of DEP exposure, with BALB/c mice exposed intranasally to DEP followed by a DEP challenge, upregulation of lung-specific expression of Ym1 and Ym2 transcripts was seen relative to mice that were not exposed nor similarly challenged (Song et al., 2008). Alveolar macrophages play an important role in particle-induced airway and lung inflammation via direct production of IL-13. Treatment of epithelial cells with bovine serum albumin-coated titanium dioxide particles led to 20 altered proteins on two-dimensional gels, which were further analyzed by nano-LC-MS/MS. These proteins included defense-related, cell-activating, and cytoskeletal proteins implicated in responses to oxidative stress (Kang et al., 2005). Titanum dioxide (TiO2) treatment increased macrophage migration-inhibitory factor (MIF) mRNA levels. MIF was expressed primarily in the epithelium and was elevated in lung tissues and bronchoalveolar lavage (BAL) fluids of TiO2-treated rats, compared to sham-treated rats. Carbon and DEPs also induce the expression of MIF protein in epithelial cells. The regulation and function of chitinase has not been well explored in air pollution asthma models. However, in one study, Ym1 was one of the most highly induced IL-4 target genes, exhibiting at least a 70-fold increase in macrophage populations (Kang et al., 2005). Nitric oxide (NO) was shown to be a short-lived molecule that causes vasodilation and bronchodilation (Moncada et al., 1991). In that study, the nitrite concentration in BAL fluids, indicative of the *in vivo* generation of NO in the airways, was significantly greater in DEP-exposed animals than in the control group. In another study, alveolar macrophages produced nitrite during *in vitro* exposure to DEP particles (50 µg/ml), with maximal induction 4 h after exposure (Song et al., 2008). The inflammatory effects of PM 10 were demonstrated in experimental animal studies following direct instillation into the lung, prior to human studies that showed the pulmonary effects after experimental exposure to PM (Ghio & Devlin, 2001). Clinically, PM 10 particles likely provoke airway inflammation via the release of mediators that exacerbate lung disease in susceptible individuals (Seaton et al., 1995); even a single exposure compromises a host's ability to handle ongoing pulmonary infections (Zelikoff et al., 2003). Fine and ultra-fine particles directly stimulate macrophages and epithelial cells to produce inflammatory cytokines such as TNF-α, TGF-β1, GM-CSF, PDGF, IL-6, and IL-8 (Fugii et al., 2001), and reactive oxygen species are responsible for acute and chronic lung inflammation (Li et al., 2003).

Adaptive immunity

PM induces a Th2-like environment, with the overproduction of IL-4 and IL-13 (Kang et al., 2005). We found that IL-13 mRNA levels in lung tissue extracts were significantly increased 24 h after treatment with TiO2 particles, compared to sham-treated rats (Kang et al., 2005). IL-13 levels were also significantly increased in the BAL fluids of TiO2-treated rats 72 h after treatment (n=8), relative to sham-treated rats (n=8). To investigate the time- and dose-dependency of macrophage IL-13 production, purified alveolar macrophages were stimulated with 1, 10, and 40 µg/ml TiO2 for 24, 48, and 72 h (n=6 in each experiment). The control group (n=6) consisted of untreated alveolar macrophages. IL-13 levels in the supernatants of the macrophage cultures were measured by ELISA. Macrophages that were cultured for 48 h with TiO2 produced IL-13 in a dose-dependent manner. In addition, 10 µg/ml TiO2 significantly enhanced IL-13 production relative to controls. IL-13 protein production increased in a time-dependent manner, and peaked 48 h after TiO2 exposure. Using immunohistochemical staining, we also found that TiO2-engulfing macrophages were the main source of IL-13 in TiO2-particle-induced lung inflammation. Taken together, our results suggest that alveolar macrophages may be major effectors of innate immunity by modulating inflammatory responses towards a Th2-phenotype by producing IL-13, as seen in the adaptive immune response (Figure 3). Proteomics offers a unique means of analyzing expressed proteins, and was successfully used to examine the effects of oxidative stress at the cellular level. In addition to revealing protein modifications, this approach can also be used to look at changes in protein expression levels (Blackford et al., 1997). In a previous study, 20 proteins were identified (Table 1) whose expression levels in the human bronchial epithelial cell line BEAS-2B changed in response to TiO2 particle exposure (Cha et al., 2007). These proteins included defense-related, cell-activating, and cytoskeletal proteins implicated in the response to oxidative stress, and can be classified into four groups according to the pattern of their TiO2-induced change in expression over time (Figure 4). One protein, MIF, was induced at the transcriptional level by stimulation of cells with any of three different particulate molecules; expression of MIF increased in lungs of TiO2-instilled rats. These results indicate that some of these proteins may serve as mediators of, or markers for, airway disease caused by exposure to PM.

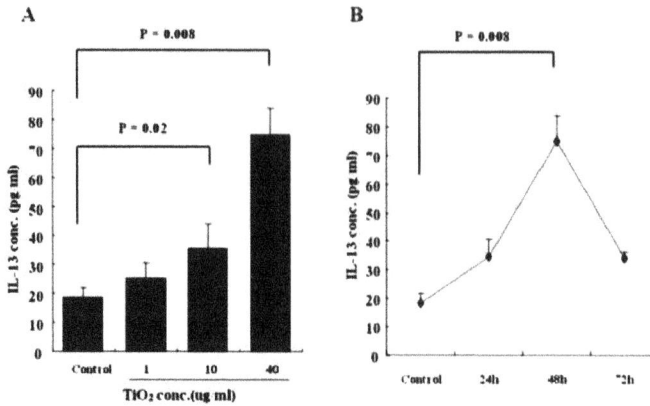

Figure 3. Time and dose responses of IL-13 production by macrophages exposed to TiO2 particles. Purified alveolar macrophages stimulated with 1, 10, and 40 g/ml TiO2 for 24, 48, and 72 h (n=6 in each experiment). The control group (n=6) consisted of unstimulated alveolar macrophages. The IL-13 in the 48-h culture supernatants is produced in a dosedependent manner after TiO2 treatment (A). TiO2 concentrations 10 g/ml significantly enhance IL-13 production when compared with the control group. The production of IL-13 protein is increased in a time-dependent manner and peaks 48 h after TiO2 stimulation (B). The results are expressed as means ± SEM. * Significant difference (P< 0.05) when compared with the control group.

No.	Protein name	Abbreviation	Accession no.	Amino acid sequence	pI/molecular mass (Da)
Group 1					
1	ATPase, H$^+$-transporting	ATP6V1B2	19913428	(K)AVVQVFEGTSGIDAK(K)	5.4/55,401.6
2	Keratin 6A	K6A	15559584	(K)ADTLTDEINFLR(A)	8.09/60,018.2
3	Macrophage migration-inhibitory factor	MIF	30583135	(K)LLCGLLAER(L)	7.73/12,476.4
Group 2					
4	Heat-shock 60-kDa protein 1	HSPD1	31542947	(K)VGEVIVTKDDAmLLK(G)	5.7/61,055.0
5	RuvB-like 2	RUVBL2	5730023	(R)ALESDmAPVLIMATNR(G)	5.49/51,156.8
6	Proliferating cell nuclear antigen	PCNA	33239451	(R)DLSHIGDAVVISCAK(D)	4.57/28,769.0
7	Transaldolase 1	TALDO1	16307182	(K)ALAGCDFLTISPK(L)	9.07/35,329.0
8	Chloride intracellular channel 1	CLIC1	14251209	(K)LAALNPESNTAGLDIFAK(F)	5.09/259,229
Group 3					
9	Replication licensing factor MCM7	MCM7	20981696	(R)TQRPADVIFATVR(E)	6.08/81,281.4
10	Calpain 1	CAPN1	12408656	(R)DMETIGFAVYEVPPELVGQPAVHLKR(D)	5.49/81,890.5
11	Ribonuclease/angiogenin inhibitor	RNH1	15029922	(K)ELSLAGNELGDEGAR(L)	4.83/48,368.0
12	Vimentin	VIM	62414289	(K)FADLSEAANR(N)	5.00/52,438.6
13	26 S proteasome subunit 9	PSMD9	2150046	(R)DIQENDEEAVQVK(E)	6.08/47,448.0
14	Actin-related protein 2	ACTR2	15778930	(K)HIVLSGGSTMYPGLPSR(L)	6.29/44,761.0
15	26 S proteasome-associated pad1 homologue	PSMD14	5031981	(R)AVAVVVDPIQSVK(G)	6.06/34,577.3
Group 4					
16	PRP19/PSO4 pre-mRNA processing factor 19 homologue	PRP19	7657381	(R)QELSHALYQHDAACR(V)	6.14/55,181.1
17	Ribosomal protein P0	RPLP0	12654583	(K)TSFFQALGITTK(I)	5.42/34,274
18	Heat-shock 27-kDa protein 1	HSPB1	4504517	(R)LFDQAFGLPR(L)	5.98/22,782.6
19	Phosphoglycarate mutase 1	PGAM1	38566176	(R)VLIAAHGNSLR(G)	6.68/28,520.1
20	Platelet-activating factor acetylhydrolase	PAFAH1B3	4505587	(R)VVVLGLLPR(G)	6.33/25,734.4

Table 1. List of proteins identified by LC-MS/MS analysis

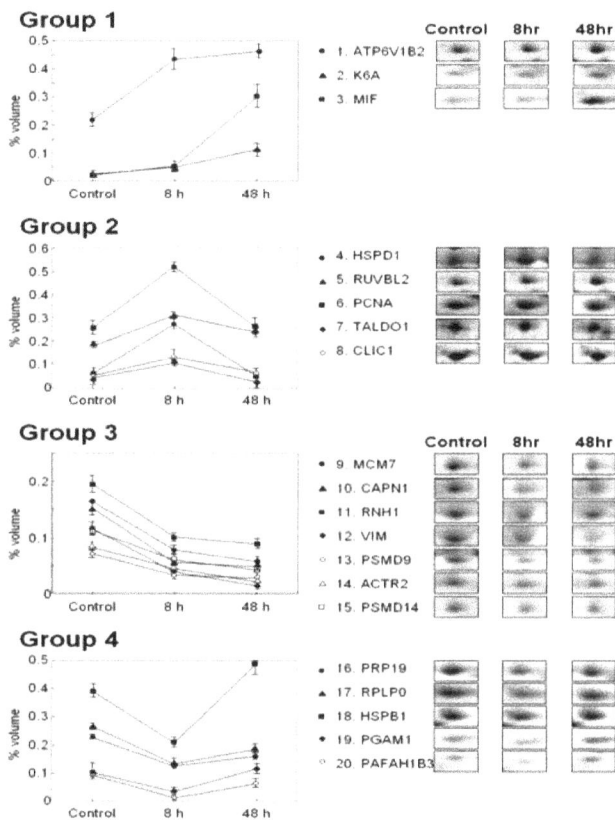

Figure 4. Cluster analysis of 20 proteins with significant differential expression (>2-fold change) at 8 or 48 h caused by TiO2 treatment of BEAS-2B epithelial cells. The expression profiles of the individual proteins were classified by cluster analysis. Protein names (National Center for Biotechnology Information (NCBI)) are displayed for each cluster.

However, there is a lack of evidence showing a direct relationship between particulates and the induction of Th2-like cytokines, including IL-4 and IL-13. TiO2 particles are a component of PM 10 found in dusty workplaces in industries involved in the crushing and grinding of the mineral ore rutile (Templeton, 1994). Garabrant et al. (1987) reported that 50% of TiO2-exposed workers have respiratory symptoms accompanied by reduced pulmonary function. Because acute and chronic exposures to TiO2 particles also induce inflammatory responses in the airways and alveolar spaces of rats (Ahn et al., 2005, Kang et al., 2005, Schapira et al., 1995, Waheit et al., 1997), TiO2-treated rats are good models for studying epithelial responses to PM10 particles. Proteomics has been successfully used to examine oxidative stress at the cellular level (Xiao et al., 2003). PM10 or DEP increase lung inflammation by inhalant allergens or respiratory viral infection by acting as adjuvants. The response may enhance already existing allergies or IgE responses to neo-allergens and susceptibility to respiratory infection. This adjuvant effect is exerted by the enhanced production of inflammatory Th2 and/or Th1 cytokines (Diaz-Sanchez et al., 1997). In animal experiments and human studies, several cytokines and CC chemokines including IL-4, IL-5, IL-13, GM-CSF, RANTES, MCP-3, MIP-1 were increased when lymphocytes and macrophages/monocytes were co-stimulated with particulates in the presence of specific allergens (Hamilton et al., 2004). The immune system responds in different ways depending on the type of particulate. DEP favors a Th2 response, while asbestos fiber and carbon particles upregulate both Th1 and Th2 cytokines produced by autologous lymphocyte stimulated by antigen (Hamilton et al., 2004). In addition to the adjuvant effects, inhaled inert particles cause a spectrum of pulmonary responses, ranging from minimal changes to marked acute and chronic inflammation.

Oxidative stress

ROS production and the generation of oxidative stress are relevant to lung diseases. Oxygen is readily reduced with an electron to form oxygen free radicals, such as superoxide (Bast, et al., 2010, Finkel, 2011, Comhair & Erzyrum, 2010). Superoxide takes up a second electron, leading to hydrogen peroxide, which will generate the extremely reactive hydroxyl radical in the presence of iron ions. Hydroxyl radicals react very quickly with biomolecules, such as proteins, fatty acids, and DNA (Bast, et al., 2010, Finkel, 2011, Comhair & Erzyrum, 2010). All molecules in the direct vicinity of the hydroxyl radical will react with this reactive form of oxygen (Bast, et al., 2010, Finkel, 2011, Comhair & Erzyrum, 2010). The various forms of oxygen are called ROS (Bast, et al., 2010). Formation of ROS takes place constantly in every cell during normal metabolic processes (Bast, et al., 2010, Finkel, 2011, Ballaban, et al., 2005, Comhair & Erzyrum, 2010). Cellular sites for production of ROS include mitochondria,

microsomes, and enzymes (e.g., xanthine oxidase, P450 monooxygenase, cyclooxygenase, lipoxygenase, indole amine dioxygenase, monoamine oxidase) (Nadeem, et al., 2008). One of the most dangerous forms of PM pollution is diesel exhaust particles. Diesel exhaust particles consist of polyaromatic hydrocarbons, hydrophobic molecules that can diffuse easily through cell membranes. As free radicals cause oxidative damage to biological macromolecules, such as DNA, lipids, and protein, they are believed to be involved in the pathogenesis of many diseases (Tredaniel, et al., 1994). The particles are able to induce the generation of free radicals, which may lead to an increase in oxidative stress, exacerbating some respiratory symptoms. Metals present on the particle surface, including Fe, Co, Cr, and V, undergo redox cycling, while Cd, Hg, and Ni, as well as Pb, deplete glutathione and protein-bound sulfhydryl groups resulting in ROS production (Stohs, et al., 2001, Valko, et al., 2005). Metal-induced oxidative stress has been shown to subsequently affect the immune system, by causing neutrophilic lung injury and release of inflammatory mediators by several lung cell types (Ghio & Delvin, 2001), and to act as the cornerstone for subsequent particle-induced inflammation (Dye et al., 1999). Another mechanism involves phagocytosis, characterized by the removal of microorganisms and pollutant particles (Forman & Toress, 2001), and an essential element of the immune defense system, which may mediate alveolar macrophage binding of certain inert and environmental particulate matter, such as Fe_2O_3, silicates, TiO_2, quartz, and iron oxide (Cha et al., 2007). Redox reactions regulate signal transduction as important chemical processes. The response of a cell to a reactive oxygen-rich environment often involves the activation of numerous intracellular signaling pathways, which can cause transcriptional changes and allow the cells to respond appropriately to the perceived oxidative stress (Finkel, 2011, Comhair & Erzyrum, 2010). Nuclear factor-κB (NF-κB) and activation protein-1 (AP-1) are regulated and influenced by the redox status and have been implicated in the transcriptional regulation of a wide range of genes involved in oxidant stress and cellular response mechanisms (Beamer & Holian, 2005). In the nucleus, redox affects histone acetylation and deacetylation status, which at least partly regulates inflammatory gene expression by activation of redox-sensitive transcription factors (Liu, et al., 2005). NF-κB is activated in epithelial cells and inflammatory cells during oxidative stress, leading to the upregulation of a number of proinflammatory genes (Beamer & Holian, 2005). NF-κB is a protein heterodimer made up of p65 and p50 subunits. There is evidence of activation of NF-κB in bronchial mucosa and sputum inflammatory cells in asthmatic patients (Rhaman, et al., 1996). Many of the inflammatory genes responsible for the pathogenesis of asthma are regulated by NF-kB. AP-1 is a protein dimer composed of a heterodimer of Fos and Jun proteins. AP-1 regulates many of the inflammatory and immune genes in oxidant-mediated diseases. Gene expression of gamma-glutamylcysteine synthetase, the rate-limiting enzyme for GSH synthesis, is induced by the activation of AP-1. In addition, the family of mitogen-activated protein kinases (MAPKs) is directly or indirectly altered by redox changes (Ciencewicki, et al., 2008). Oxidative stress and other stimuli, such as cytokines, activate various signal transduction pathways leading to activation of transcription factors, such as NF-kB and AP-1 (Rahman & Adcock, 2006). Binding of transcription factors to DNA elements leads to recruitment of CREB-binding protein (CBP) and/or other co-activators to the transcriptional initiation complex on the promoter regions of various genes (Rahman & Adcock, 2006). Activation of CBP leads to acetylation of specific core histone lysine residues by an intrinsic histone acetyltransferase activity (Rahman & Adcock, 2006). Redox changes also can activate members of the MAPK, such as extracellular signal-regulated kinase, c-jun N-terminal kinase, p38 kinase, and phosphoinositol-3 kinase, all of which may ultimately promote inflammation (Carvalho, et al., 2004). Both STAT1 and STAT3 activation are regulated by redox (Carvalho, et al., 2004). NF-E2-related factor 2, a basic leucine zipper transcription factor, involved in induction of the antioxidant element (ARE)-mediated transcriptional response is known to play an important role and binds to the ARE and upregulates the expression of several antioxidant genes in response to a variety of stimuli (Nguyen, et al., 2003). ROS (Nadeem, et al., 2008) can influence airway cells and reproduce many of the pathophysiological features associated with asthma by initiating lipid peroxidation, altering protein structure, enhancing release of arachidonic acid from cell membranes, contracting airway smooth muscle, increasing airway reactivity and airway secretions, increasing vascular permeability, increasing the synthesis and release of chemoattractants, inducing the release of tachykinins and neurokinins, decreasing cholinesterase and neutral endopeptidase activities, and impairing the responsiveness of ß-adrenergic receptors (Barnes, et al., 1998). Asthma attacks and experimental allergen challenge are associated with immediate formation of superoxide that persists throughout the late asthmatic response (Calhoun, et al., 1992). Allergen challenge in the airways of atopic individuals caused a twofold increase in superoxide generation (Calhoun, et al., 1992). Spontaneous and experimental allergen-induced asthma attacks lead to eosinophil and neutrophil activation, during which NADPH oxidase is activated and ROS such as superoxide and its dismutation product H_2O_2 are rapidly formed (Klebanoff, 1980). ROS production by asthmatics correlates with the severity of airway reactivity (Calhoun, et al., 1992). Asthma is characterized by oxidative modifications (Sansers, et al., 1995). Increased levels of EPO and MPO parallel the numbers of eosinophils and neutrophils, respectively, and are found at higher than normal levels in asthmatic peripheral blood, induced sputum, and BAL fluid (Sansers, et al., 1995). Malondialdehyde and thiobarbituric acid-reactive substances have also been detected in urine, plasma, sputum, and BAL fluid in relation to the severity of asthma (Mondino, et al., 2004, Wood, et al., 2005) In addition, 8-isoprostane, a biomarker of lipid peroxidation, is also elevated in exhaled breath condensate from adults and children with asthma (Mondino, et al., 2004, Wood, et al, 2005). Generation of reactive oxygen and nitrogen species is markedly increased during acute asthma attacks (MacPherson, et al., 2001, Wu, et al., 2000). The loss of SOD contributes to oxidative stress during acute episodes of asthma exacerbation (MacPherson et al, 2001, Wu et al, 2000). Oxidative modification of MnSOD is present in asthmatic airway epithelial cells (Malik & Storey, 2011).The loss of SOD activity reflects the increased oxidative and nitrative stress in asthmatic patients, suggesting that SOD may serve as a surrogate marker of oxidant stress and asthma severity (Takaku, et al., 2011). ROS, such as superoxide and hydrogen peroxide, enhance vascular endothelial growth factor (VEGF) expression (Kuroki, et al., 1996), while exogenous SOD prevents VEGF expression (Kuroki, et al., 1996), suggesting that the increased vascularization found in asthma may be due to the involvement of oxidative stress via effects on hypoxia-inducible factors (Ghosh, et al., 2003). The catalase activity was found to be 50% lower in BAL fluid of asthmatic lungs than that in healthy controls (Ghosh, et al., 2003). Tyrosine oxidant modifications of catalase occur in asthma, such as chlorination of tyrosine by peroxidase-catalyzed halogenation, and oxidative cross-linking of tyrosine as monitored by dityrosine, a product of tyrosyl radicals (Ghosh, et al., 2003). The most extensive modification found in asthmatic lungs is tyrosine chlorination, which is 20-fold more extensive than tyrosine nitration (Ghosh, et al., 2003). In contrast to SODs and catalase, extracellular GPx is present at higher than normal levels in the lungs of individuals with asthma (Comhair, et al., 2002). This increase is due to induction of GPx mRNA and protein expression by bronchial epithelial cells in response to increased intracellular or extracellular ROS [94]. During asthma exacerbation in humans, the levels of serum TRX1 increase and are inversely correlated with airflow (Yamada, et al., 2003). Cigarette smoke can induce increased oxidant burden and cause irreversible changes to the antioxidant protective effects in the airways (van Der Troorn, et al., 2007). The smoke-derived oxidants damage airway epithelial cells inducing direct injury to membrane lipids, proteins, carbohydrates, and DNA, leading to chronic inflammation (Foronjy, et al., 2008). Cigarette smoking delivers and generates oxidative stress within the lungs (Lin & Thoma, 2010) These imbalances of oxidant burden and antioxidant capacity have been implicated as important contributing factors in the pathogenesis of COPD (Lin & Thoma, 2010) However, smoking also causes the depletion of antioxidants, which further contributes to oxidative tissue damage (Lin & Thoma, 2010) The downregulation of antioxidant pathways has also been associated with acute exacerbations of COPD (Lin & Thoma, 2010). Disruption of the oxidant/antioxidant balance is important in the pathogenesis of acute lung injury and acute respiratory distress syndrome. Different cytokines and growth factors play a role in the pathogenesis of lung fibrosis (Hecker, et al., 2009). ROS mediate the formation of TGF-β in lung epithelial cells (Hecker, et al., 2009). Fibroblasts of patients with idiopathic pulmonary fibrosis produce H_2O_2 upon stimulation with TGF-β. This interplay between H_2O_2 and TGF-β leads to deterioration of re-epithelialization and fibrosis (Hecker, et al., 2009).

4. Conclusions

Epidemiological surveys and animal studies together suggest that air pollutants are involved in the pathogenesis of airway inflammation and aggravate respiratory symptoms. Avoidance of harmful exposures is a key component of national and international guideline recommendations for the management of respiratory diseases. Controlling air pollution is important for the prevention of airway diseases. Finally, *in vitro* and *in vivo* studies are needed to further delineate the role of particulate air pollutants in airway diseases and the molecular mechanisms involved.

Author details

An-Soo Jang
Division of Allergy and Respiratory Medicine, Department of Internal Medicine, Soonchunhayg University Hospital, Bucheon, Korea

Acknowledgement

This subject is supported by Korea Ministry of Environment (2012001360001) as "The Environmental Health Action Program".

5. References

Ahn M.H., Kang C.M., Park C.S., Park S.J., Rhim T., Yoon P.O., Chang H.S., Kim S.H., Kyono H., Kim K.C. *(2005)*. Titanium dioxide particle-induced goblet cell hyperplasia: association with mast cells and IL-13. *Respir Res*, 6, 34.

Ayres, J.G., Borm, P., Cassee, F.R., Castranova, V., Donaldson, K., Ghio, A., Harrison, R.M., Hider, R., Kelly, F., Kooter, I.M., Marano, F., Maynard, R.L., Mudway, I., Nel, A., Sioutas, C., Smith, S., Baeza-Squiban, A., Cho, A., Duggan, S., Froines, J. (2008). Evaluating the toxicity of airborne particulate matter and nanoparticles by measuring oxidative stress potential - a workshop report and consensus statement. *Inhal. Toxicol* ,20, 75-99.

Baldi, I., Tessier, J.F., Kauffmann, F., Jacqmin-Gadda, H., Nejjari, C., Salamon, R. (1999). Prevalence of asthma and mean levels of air pollution: results from the French PAARC survey. Pollution Atomosphérique et Affections Respiratoires Chroniques. *Eur Respir J,* 14,132-138.

Barnes, P.J., Chung, K.F., Page, C.P. (1998). nflammatory mediators of asthma: an update. *Pharmacol Rev,* 50, 515–596.

Bast, A., Weseler, A. R., Haenen, G. R., den Hartog, G.J. (2010) Oxidative stress and antioxidants in interstitial lung disease. *Curr Opin Pulm Med,* 16, 516-520.

Beamer, C.A., Holian, A. (2005). Scavenger receptor class A type I/II (CD204) null mice fail to develop fibrosis following silica exposure. *Am J Physiol Lung Cell Mol Physiol,* 289, 186–195.

Beelen, R., Hoek, G., van den Brandt, P.A., Goldbohm, R.A., Fischer, P., Schouten, L.J., Jerrett, M., Hughes, E., Armstrong, B., Brunekreef, B. (2008). Long-term effects of traffic-related air pollution on mortality in a Dutch cohort (NLCS-AIR study). *Environ Health Perspect,* 116, 196–202.

Bierbaum S., Nickel R., Koch A., Lau S., Deichmann K.A., Wahn U., et al.(2005). Polymorphisms and haplotypes of acid mammalian chitinase are associated with bronchial asthma. *Am J Respir Crit Care Med,* 172, 1505-1509.

Blackford, J.A., Jr, Jones, W., Dey, R.D., Castranova, V. (1997). Comparison of inducible nitric oxide synthase gene expression and lung inflammation following intratracheal instillation of silica, coal, carbonyl iron, or titanium dioxide in rats. *J Toxicol Environ Health,* 51, 203-218.

Breysse, P.N., Diette, G.B., Matsui, E.C., Butz, A.M., Hansel, N.N., McCormack, M.C. (2010). Indoor air pollution and asthma in children. *Proc Am Thorac Soc,* 7, 102-106.

Brunekreef, B., Janssen, A.H., de Hartog, J., Harssema, H., Knape, M., van Vliet, P. (1997). Air pollution from truck traffic and lung function in children living near motoways. *Epidemiology ,* 8, 298-303.

Carvalho, H., Evelson, P., Sigaud, S., González-Flecha, B. (2004). Mitogen-activated protein kinases modulate H(2)O(2)-induced apoptosis in primary rat alveolar epithelial cells. *J Cell Biochem,* 92, 502–513.

Cha, M.H., Rhim, T., Kim, K.H., Jang, A.S., Paik, Y.K., Park, C.S. (2007). Proteomic identification of macrophage migration-inhibitory factor upon exposure to TiO2 particles. *Mol Cell Proteomics,* 6, 56-63.

Ciencewicki, J., Trivedi, S., Kleeberger, S.R. (2008). Oxidants and the pathogenesis of lung diseases. *J Allergy Clin Immunol,* 122, 456–468.

Comhair, S.A., Erzurum, S.C. (2010). Redox control of asthma: molecular mechanisms and therapeutic opportunities. *Antioxid Redox Signal,* 12, 93-124.

Comhair, S.A., Erzurum, S.C. (2002). Antioxidant responses to oxidant-mediated lung diseases. *Am J Physiol Lung Cell Mol Physiol,* 283, 246–255.

D'Amato, G., Cecchi, L., D'Amato, M., Liccardi, G. (2010). Urban air pollution and climate change as environmental risk factors of respiratory allergy: an update. *J Investig Allergol Clin Immunol,* 20, 95-102.

Diaz-Sanchez, D., Tsien, A., Fleming, J., Saxon, A. (1997). Combined diesel exhaust particulate and ragweed allergen challenge markedly enhances human in vivo nasal ragweed-specific IgE and skews cytokine production to a T helper cell 2-type pattern. *J Immunol,* 158, 2406-2413.

Donaldson, K., Tran, C.L. (2002). Inflammation caused by particles and fibers. *Inhal. Toxicol,* 14, 5-27.

Dye, J.A., Adler, K.B., Richards, J.H., Dreher, K.L. (1999). Role of soluble metals in oil fly ash-induced airway epithelial injury and cytokine gene expression. *Am J Physiol,* 277, L498–510.

Finkel, T. (2011). Signal transduction by reactive oxygen species. *J Cell Biol,* 194, 7-15.

Garabrant, D.H., Fine, L.J., Oliver, C., Bernstein, L., Peters, J.M. (1987). Abnormalities of pulmonary function and pleural disease among titanium metal production workers. *Scand J Work Environ Health,* 13, 47-51.

Folinsbee, L.J. (1992). Does nitrogen dioxide exposure increase airways responsiveness? *Toxicol Indust Health,* 8, 273-283.

Forman, H.J., Torres, M. (2001). Redox signaling in macrophages. *Mol Aspects Med,* 22, 189–216.

Foronjy, R., Alison, W., D'Aarmiento, J. (2008). The pharmokinetic limitations of antioxidant treatment for COPD. *Pulm Pharmacol Ther,* 21, 370–379.

Fujii, T., Hayashi, S., Hogg, J.C., Vincent, R., Van Eeden, S.F. (2001). Particulate matter induces cytokine expression in human bronchial epithelial cells. *Am J Respir Cell Mol Biol,* 25, 265-271.

Gergen, P.J., Fowler, J.A., Maurer, K.R., Davis, W.W., Overpeck, M.D. (1998). The burden of environmental tobacco smoke exposure on the respiratory health of children 2 months through 5 years of age in the United States: Third National Health and Nutrition Examination Survey, 1988 to 1994. *Pediatrics,* 101, E8.

Ghio, A.J., Devlin, R.B. (2001). Inflammatory lung injury after bronchial instillation of air pollution particles. *Am J Respir Crit Care Med,* 164, 704-708.

Ghosh, S., Masri, F., Comhair, S. et al.(2003). Nitration of proteins in murine model of asthma. Am J Respir Crit Care Med, 167, 889.

Gielen, M.H., van der Zee, S.C., van Wijnen, J.H., van Steen, C.J., Brunekreef, B. (1997). Acute effects of summer air pollution on respiratory health of asthmatic children. *Am J Respir Crit Care Med,* 155, 2105-2108.

Grigg, J. (2009). Particulate matter exposure in children: relevance to chronic obstructive pulmonary disease. *Proc Am Thorac Soc,* 6, 564-569.

Guo, L., Johnson, R.S., Schuh, J.C. (2000). Biochemical characterization of endogenously formed eosinophilic crystals in the lungs of mice. *J Biol Chem,* 275, 8032-8037.

Hamelmann, E., Schwarze, J., Takeda, K., Oshiba, A., Larsen, G.L., Irvin, C.G., Gelfand, E.W. *(1997).* Noninvasive measurement of airway responsiveness in allergic mice using barometric plethysmography. *Am J Respir Crit Care Med,* 156, 766-775.

Hamilton, R.F. Jr, Holian, A., Morandi, M.T. (2004). A comparison of asbestos and urban particulate matter in the in vitro modification of human alveolar macrophage antigen-presenting cell function. *Exp Lung Res,* 30, 147-162.

Hecker, L., Vittal, R., Jones, T., Jagirdar, R., Luckhardt, T.R., Horowitz, J.C., Pennathur, S., Martinez, F.J., Thannickal, V.J. (2009). NADPH oxidase-4 mediates myofibroblasts activation and fibrogenic responses to lung injury. *Nat Med*, 15, 1077–1081.

Hirsch, T., Weiland, S.K., von Mutius, E., Safeca, A.F., Gräfe, H., Csaplovics, E., Duhme, H., Keil, U., Leupold, W. (1999). Inner city air pollution and respiratory health and atopy in children. *Eur Respir J*, 14, 669-677.

Holgate, S.T. (1999). The epidemic of allergy and asthma. *Nature,* 402, B2–B4.

Diette, G.B., McCormack, M.C., Hansel, N.N., Breysse, P.N., Matsui, E.C.(2008). Environmental issues in managing asthma. *Respir Care*, 53,602–615, discussion 616–617.

Jang, A.S., Choi, I.S., Lee, J.U., Park, S.W., Lee, J.H., Park, C.S. (2004). Changes in the expression of NO synthase isoforms after ozone: the effects of allergen exposure. *Respir Res*, 5, 5.

Jang, A.S., Yeum, C.H., Son, M.H. (2003). Epidemiologic evidence of a relationship between airway hyperresponsiveness and exposure to polluted air. *Allergy*, 58, 585-588.

Jiang, R., Bell, M.L. (2008). A comparison of particulate matter from biomassburning rural and non-biomass-burning urban households in northeastern China. *Environ Health Perspect*, 116, 907–914.

Jin, H.M., Copeland, N.G., Gilbert, D.J., Jenkins, N.A., Kirkpatrick, R.B., Rosenberg, M. (1998). Genetic characterization of the murine Ym1 gene and identification of a cluster of highly homologous genes. *Genomics*, 54, 316-322.

Kang, C.M., Jang, A.S., Ahn, M.H., Shin, J.A., Kim, J.H., Choi, Y.S., Rhim, T.Y., Park, C.S. (*2005*). Interleukin-25 and interleukin-13 production by alveolar macrophages in response to particles. *Am J Respir Cell Mol Biol*, 33, 290-296.

Koenig, J.Q., Mar, T.F., Allen, R.W., Jansen, K., Lumley, T., Sullivan, J.H., Trenga, C.A., Larson, T., Liu, L.J. (2005). Pulmonary effects of indoor- and outdoor-generated particles in children with asthma. *Environ Health Perspect*, 113, 499–503.

Koenig, J.Q. (1999). Air pollution and asthma. *J Allergy Clin Immunol*, 104, 717-722.

Klebanoff, S.J. (1980). Oxygen metabolism and the toxic properties of phagocytes. *Ann Intern Med*, 93, 480–489.

Kreit, J.W., Gross, K.B., Moore, T.B., Lorenzen, T.J., D'Arcy, J., Eschenbacher, W.L. (1989). Ozone-induced changes in pulmonary function and bronchial responsiveness in asthmatics. *J Appl Physiol*, 66, 217-222.

Kuipers, I., Guala, A.S., Aesif, S.W., Konings, G., Bouwman, F.G., Mariman, E.C., Wouters, E.F., Janssen-Heininger, Y.M., Reynaert, N.L. (2011). Cigarette smoke targets glutaredoxin 1, increasing s-glutathionylation and epithelial cell death. *Am J Respir Cell Mol Biol*, 45, 931-937.

Kuroki, M., Voest, E.E., Amano, S., Beerepoot, L.V., Takashima, S., Tolentino, M., Kim, R.Y., Rohan, R.M., Colby, K.A., Yeo, K.T., Adamis, A.P. (1996). Reactive oxygen intermediates increase vascular endothelial growth factor expression in vitro and in vivo. *J Clin Invest*, 98, 1667–1675.

Laden, F., Schwartz, J., Speizer, F.E., Dockery, D.W. (2006). Reduction in fine particulate air pollution and mortality: Extended follow-up of the Harvard Six Cities study. *Am J Respir Crit Care Med*, 173, 667–672.

Laumbach, R.J. (2010). Outdoor air pollutants and patient health. *Am Fam Physician*, 81, 175-180.

Li, J.J., Muralikrishnan, S., Ng, C.T., *Yung, L.Y., Bay, B.H. (2010).* Nanoparticle-induced pulmonary toxicity. *Exp Biol Med*, 235, 1025-1033

Li, N., Sioutas, C., Cho, A., Schmitz, D., Misra, C., Sempf, J., Wang, M., Oberley, T., Froines, J., Nel, A. *(2003).* Ultrafine particulate pollutants induce oxidative stress and mitochondrial damage. *Environ Health Perspect*, 111, 455-460.

Liu, H., Colavitti, R., Rovira, I.I., Finkel, T. (2005). Redox-dependent transcriptional regulation. *Circ Res*, 97, 967–974.

MacPherson, J.C., Comhair, S.A., Erzurum, S.C., Klein, D.F., Lipscomb, M.F., Kavuru, M.S., Samoszuk, M.K., Hazen, S.L. (2001). Eosinophils are a major source of nitric oxide-derived oxidants in severe asthma: Characterization of pathways available to eosinophils for generating reactive nitrogen species. *J Immunol*, 166, 5763–5772.

Malik, A. I., Storey, K. B. (2011). Transcriptional regulation of antioxidant enzymes by FoxO1 under dehydration stress, *Gene*, 485, 114-119.

Matthay, R.A. (1992). Chronic airway diseases, in: J.B. Wyngaarden (Ed.), Cecil Textbook of Medicine, 19th ed.,W.B. Saunders, *Philadelphia*, 386–394.

McConnell, R., Berhane, K., Gilliland, F., Molitor, J., Thomas, D., Lurmann, F., Avol, E., Gauderman, W.J., Peters, J.M. (2003). Prospective study of air pollution and bronchitic symptoms in children with asthma. *Am J Respir Crit Care Med*, 168, 790–797.

McCormack, M.C., Breysse, P.N., Hansel, N.N., Matsui, E.C., Tonorezos, E.S., Curtin-Brosnan, J., Williams, D.L., Buckley, T.J., Eggleston, P.A., Diette, G.B. (2008). Common household activities are associated with elevated particulate matter concentrations in bedrooms of inner-city Baltimore pre-school children. *Environ Res*, 106, 148–155.

McCreanor, J., Cullinan, P., Nieuwenhuijsen, M.J., Stewart-Evans, J., Malliarou, E., Jarup L., Harrington, R., Svartengren, M., Han, I. K., Ohman-Strickland, P., Chung, K.F., Zhang, J. (2007), Respiratory effects of exposure to diesel traffi c in persons with asthma. *N Engl J Med*, 357, 2348-2358.

Mortimer, K.M., Neas, L.M., Dockery, D.W., Redline, S., Tager, I.B. (2002). The effect of air pollution on inner-city children with asthma. *Eur Respir J*, 19, 699–705.

McDonnell, W.F., Abbey, D.E., Nishino, N., Lebowitz, M.D. (1999). Long-term ambient ozone concentration and the incidence of asthma in nonsmoking adults: the AHSMOG study. *Environ Res*, 80, 110-121.

Moncada, S., Palmer, R.M., Higgs, E.A. (1991). Nitric oxide: physiology, pathophysiology and pharmacology. *Pharmacol Rev*, 143, 109-142.

Moshammer, H., Hoek, G., Luttmann-Gibson, H., Neuberger, M.A., Antova, T., Gehring, U., Hruba, F., Pattenden, S., Rudnai, P., Slachtova, H., Zlotkowska, R., Fletcher, T. (2006). Parental smoking and lung function in children: an international study. *Am J Respir Crit Care Med*, 173, 1255–1263.

Mondino, C., Ciabattoni, G., Koch, P., Pistelli, R., Trové, A., Barnes, P.J., Montuschi, P. (2004). Effects of inhaled corticosteroids on exhaled leukotrienes and prostanoids in asthmatic children. *J Allergy Clin Immunol*, 114, 761–767.

Nadeem, A., Masood, A., Siddiqui, N. (2008). Oxidant--antioxidant imbalance in asthma: scientific evidence, epidemiological data and possible therapeutic options. *Ther Adv Respir Dis*, 2, 215-235.

Nel, A., Xia, T., Madler, L., Li, N. (2006). Toxic potential of materials at the nanolevel. *Science*, 311, 622–627.

Nicolai, T. (1999). Air pollution and respiratory disease in children: what is the clinically relevant impact? *Pediatr Pulmonol Suppl*, 18, 9-13.

Nguyen, T., Sherratt, P.J., Pickett, C.B. (2003). Regulatory mechanisms controlling gene expression mediated by the antioxidant response element. *Annu Rev Pharmacol Toxicol*, 43, 233–260.

Pagan, I., Costa, D.L., McGee, J.K., Richards, J.H., Dye, J.A. (2003). Metal mimic airway epithelial injury induced by in vitro exposure to Utah Valley ambient particulate matter extracts. *J Toxicol Environ Health A*, 66, 1087-1112.

Peters, J.M., Avol, E., Gauderman, W.J., Linn, W.S., Navidi, W., London, S.J., Margolis, H., Rappaport, E., Vora, H., Gong, H. Jr, Thomas, D.C. *(1999)*. A study of twelve Southern California communities with differing levels and types of air pollution. II. Effects on pulmonary function. *Am J Respir Crit Care Med*, 159, 768-775.

Pierse, N., Rushton, L., Harris, R.S., Kuehni, C.E., Silverman, M., Grigg, J. (2006). Locally generated particulate pollution and respiratory symptoms in young children. *Thorax*, 61, 216–220.

Pope, C.A. 3rd, Kanner, R.E. (1993). Acute effects of PM10 pollution on pulmonary function of smokers with mild to moderate chronic obstructive pulmonary disease. *Am Rev Respir Dis*, 147, 1336-1340.

Pope, C.A. III, Ezzati, M., Dockery, D.W. (2009). Fine-particulate air pollution and life expectancy in the United States. *N Engl J Med*, 360, 376–386.

Pope III, C.A. (1991). Respiratory hospital admissions associated with PM10 pollution in Utah, Salt lake, and Cahe Valleys. *Arch Environ Health*, 46, 90–97

Pope, C.A., Dockery, D.W. (2006). Health effects of fine particulate air pollution: lines that connect. *J. Air Waste Manage Assoc*, 56, 709-742.

Pruss-Ustun, A., Corvalan, C. (2006). Preventing disease through healthy environments: towards an estimate of the environmental burden of disease. Geneva, Switzerland: World Health Organization.

Rahman, I., Adcock, I.M. (2006). Oxidative stress and redox regulation of lung inflammation in COPD. *Eur Respir J*, 28, 219–242.

Rahman, I., Smith, C.A., Lawson, M.F., Harrison, D.J., MacNee, W. (1996). Induction of gamma-glutamylcysteine synthetase by cigarette smoke is associated with AP-1 in human alveolar epithelial cells. *FEBS Lett*, 396, 21–25.

Ristovski, Z.D., Miljevic, B., Surawski, N.C., Morawska, L., Fong, K.M., Goh, F., Yang, I.A. (2011). Respiratory health effects of diesel particulate matter. *Respirology*, 29. doi: 10.1111/j.1440-1843.2011.02109.x. [Epub ahead of print]

Sacks, J.D., Stanek, L.W., Luben, T.J., Johns, D.O., Buckley, B.J., Brown, J.S., Ross, M. (2011). Particulate matter-induced health effects: who is susceptible? *Environ Health Perspect*, 119, 446-454.

Sanders, S.P. Zweier, J.L. Harrison, S.J., Trush, M.A., Rembish, S.J., Liu, M.C. (1995). Spontaneous oxygen radical production at sites of antigen challenge in allergic subjects. *Am J Respir Crit Care Med*, 151, 1725–1733.

Schäfer, T., Ring, J. (1997). Epidemiology of allergic diseases. *Allergy*, 52(38 Suppl), S14-22.

Schapira, R.M., Ghio, A.J., Effros, R.M., Morrisey, J., Almagro, U.A., Dawson, C.A., Hacker, A.D. *(1995)*. Hydroxyl radical production and lung injury in the rat following silica or titanium dioxide instillation in vivo. *Am J Respir Cell Mol Biol*, 12, 220-226.

Schwartz, J., Slater, D., Larson, T.V., Pierson, W.E., Koenig, J.Q. (1993). Particulate air pollution and hospital emergency room visits for asthma in Seattle. *Am Rev Respir Dis*, 147, 826-831.

Schindler, C., Keidel, D., Gerbase, M.W., Zemp, E., Bettschart, R., Brändli, O., Brutsche, M.H., Burdet, L., Karrer, W., Knöpfli, B., Pons, M., Rapp, R., Bayer-Oglesby, L., Künzli, N., Schwartz, J., Liu, L.J., Ackermann-Liebrich, U., Rochat, T.; SAPALDIA Team. (2009). Improvements in PM10 exposure and reduced rates of respiratory symptoms in a cohort of Swiss adults (SAPALDIA). *Am J Respir Crit Care Med*, 179, 579–587.

Seaton, A., MacNee, W., Donaldson, K., Godden, D. (1995). Particulate airpollution and acute health effects. *Lancet*, 345, 176-178.

Seagrave, J. (2008). Mechanisms and implications of air pollution particle associations with chemokines. *Toxicol App. Pharmacol*, 232, 469-477.

Song, H.M., Jang, A.S., Ahn, M.H., Takizawa, H., Lee, S.H., Kwon, J.H., Lee, Y.M., Rhim, T.Y., Park, C.S. *(2008)*. Ym1 and Ym2 expression in a mouse model exposed to diesel exhaust particles. *Environ Toxicol*, 23, 110-116.

Sun, Y.J., Chang, N.C., Hung, S.I., Chang, A.C., Chou, C.C., Hsiao, C.D. (2001). The crystal structure of a novel mammalian lectin, Ym1, suggests a saccharide binding site. *J Biol Chem*, 276, 17507-17514.

Stohs, S.J., Bagci, D., Hassoun, E., Bagchi, M. (2001). Oxidative mechanisms in the toxicity of chromium and cadmium ions. *J Environ Pathol Toxicol Oncol*, 20, 77–88.

Stone, V., Johnston, H., Clift, M.J.D. (2007). Air pollution, ultrafine and nanoparticle toxicology: Cellular and molecular interactions. *IEEE T Nanobiosci*, 6, 331-340.

Takaku, Y., Nakagome, K., Kobayashi, T., Hagiwara, K., Kanazawa, M., Nagata, M. (2011). IFN-γ-inducible protein of 10 kDa upregulates the effector functions of eosinophils through β2 integrin and CXCR3. *Respir Res*, 17, 138.

Templeton, D.M. (1994). Titanium, in Handbook on metals in clinical and analytic chemistry. In: Seiler HG, Siegel A, Siegel H, eds. New York: Marcel Dekker, 627-630.

Totlandsdal, A.I., Cassee, F.R., Schwarze, P., Refsnes, M., Låg, M. (2010). Diesel exhaust particles induce CYP1A1 and pro-inflammatory responses via differential pathways in human bronchial epithelial cells. *Part Fibre Toxicol*, 7, 41.

Valko, M., Morris, H., Cronin, M.T. (2005). Metals, toxicity and oxidative stress. *Curr Med Chem*, 12, 1161–1208.

van Der Toorn, M., Smit-de Vries, M.P., Slebos, D.J., de Bruin, H.G., Abello, N., van Oosterhout, A.J., Bischoff, R., Kauffman, H.F. (2007). Cigarette smoke irreversibly modifies glutathione in airway epithelial cells. *Am J Physiol Lung Cell Mol Physiol*, 293, 1156–1162.

Winickoff, J.P., Berkowitz, A.B., Brooks, K., Tanski, S.E., Geller, A., Thomson, C., Lando, H.A., Curry, S., Muramoto, M., Prokhorov, A.V., Best, D., Weitzman, M., Pbert, L.; Tobacco Consortium, Center for Child Health Research of the American Academy of Pediatrics. et al. (2005). State-of-the-art interventions for office-based parental tobacco control. *Pediatrics*, 115, 750–760.

Ward, J.M., Yoon, M., Anver, M.R., Haines, D.C., Kudo, G., Gonzalez, F.J., Kimura, S. *(2001)*. Hyalinosis and Ym1/Ym2 gene expression in the stomach and respiratory tract of 129S4/SvJae and wild-type and CYP1A2-null B6, 129 mice. *Am J Pathol*, 158, 323-332.

Warheit, D.B., Hansen, J.F., Yuen, I.S., Kelly, D.P., Snajdr, S.I., Hartsky, M.A. (1997). Inhalation of high concentrations of low toxicity dusts in rats results in impaired pulmonary clearance mechanisms and persistent inflammation. *Toxicol Appl Pharmacol*, 145, 10-22.

Welch, J.S., Escoubet-Lozach, L., Sykes, D.B., Liddiard, K., Greaves, D.R., Glass, C.K. (2002). TH2 cytokines and allergic challenge induce Ym1 expression in macrophages by a STAT6-dependent mechanism. *J Biol Chem*, 277, 42821-42829.

Wood, L.G., Garg, M.L., Simpson, J.L., Mori, T.A., Croft, K.D., Wark, P.A., Gibson, P.G. (2005). Induced sputum 8-isoprostane concentrations in inflammatory airway diseases. *Am J Respir Crit Care Med*, 171, 426–430.

Wu, W., Samoszuk, M.K., Comhair, S.A., Thomassen, M.J., Farver, C.F., Dweik, R.A., Kavuru, M.S., Erzurum, S.C., Hazen, S.L. (2000). Eosinophils generate brominating oxidants in allergen-induced asthma. *J Clin Invest*, 105, 1455–1463.

Xiao, G.G., Wang, M., Li, N., Loo, J.A., Nel, A.E. (2003). Use of proteomics to demonstrate a hierarchical oxidative stress response to diesel exhaust particle chemicals in a macrophage cell line. *J Biol Chem*, 278, 50781-50790.

Yamada, Y., Nakamura, H., Adachi, T., Sannohe, S., Oyamada, H., Kayaba, H., Yodoi, J., Chihara, J. (2003). Elevated serum levels of thioredoxin in patients with acute exacerbation of asthma. *Immunol Lett*, 86, 199–205.

Yang, W., Omaye, S.T. (2009). Air pollutants, oxidative stress and human health. *Mutat Res*, 674, 45-54.

Yang, Q., Chen, Y., Krewski, D., Burnett, R.T., Shi, Y., McGrail, K.M. (2005). Effect of shortterm exposure to low levels of gaseous pollutants on chronic obstructive pulmonary disease hospitalizations. *Environ Res*, 99, 99–105.

Zelikoff, J.T., Chen, L.C., Cohen, M.D., Fang, K., Gordon, T., Li Y., Nadziejko, C., Schlesinger, R.B. *(2003)*. Effects of inhaled ambient particulate matter on pulmonary antimicrobial immune defense. *Inhal Toxicol*, 15, 131-150.

Zhu, Z., Zheng, T., Homer, R.J., Kim, Y.K., Chen, N.Y., Cohn, L., Hamid, Q., Elias, J.A. *(2004)*. Acidic mammalian chitinase in asthmatic Th2 inflammation and IL-13 pathway activation. *Science*, 304, 1678-1682.

Air Pollution Management and Prediction

Chapter 9

Non-Thermal Plasma Technic for Air Pollution Control

Takao Matsumoto, Douyan Wang, Takao Namihira and Hidenori Akiyama

Additional information is available at the end of the chapter

1. Introduction

The air pollutions from combustion flue gas and industrial gases became worse and cause the environmental problem. It is difficult for the conventional methods such as selective catalytic reduction method and lime-gypsum method to treat exhaust gases energy efficiently and inexpensively. Its energy efficiency and its initial and running costs are still in negative situation for the backward nations. In recent years, the pollution control techniques using non-thermal plasmas have been widely studied because it is one of the promising technologies for pollution control with higher energy efficiency [1]-[7]. The non-thermal plasma could treat multiple toxic molecules simultaneously, and it can be applied to locations where the conventional catalyst methods are difficult to use. In this chapter, a principle of air pollution control by non-thermal plasma, various methods of non-thermal plasma formation and those current situations are introduced.

2. Non-thermal plasma

Plasma, also referred to as "ionized gas" is mixed state of atoms, molecules, ions, electrons and radicals. Plasma has two general states: equilibrium and non-equilibrium. The equilibrium state indicates the temperatures of electrons, ions and neutrals become almost equal, and the background gas is heated from a few thousands to more than ten thousands Kelvin degrees. Because of this, the plasma getting equilibrium state is called as "thermal plasma". On the other hand, the non-equilibrium state means that the temperatures of electrons, ions and neutrals are quite different, and in general the electron temperature is substantially higher than other particles. Therefore, the rise of background gas temperature is quite low in non-equilibrium state and the plasma being non-equilibrium state is called as "non-thermal plasma". Figure 1 shows a typical example of non-thermal plasma [8]. This figure shows the background gas temperature of non-thermal plasma is enough low to touch by a finger. In the non-thermal plasma, the majority of the discharge energy goes into the production of energetic electrons, rather than ion and neutron heating. The energy in the plasma is thus consumed preferentially to the electron impact dissociation and ionization of the background gas for production of radicals that, in turn, decompose the toxic molecules. In short, non-thermal plasma can remove toxic molecules near room temperature without consuming a lot of energy in background gas heating.

For low pressure plasma process such as semiconductor production, the non-equilibrium plasma which is often named "cold plasma" is typically used. Prof. Oda [9] defined that non-thermal plasma is high pressure (typically 1 atmospheric pressure) non-equilibrium plasma. Compared with that cold plasma, the electron temperature and ionization rate are quite lower in non-thermal plasma. Typically, the electron temperature of cold plasma is tens of eV. Meanwhile, in atmospheric pressure, the electron temperature is generally 1 to 10eV and ionization rate is around 0.1%. However, it is important for gas processing in atmospheric pressure because electron and molecular density is overwhelmingly high in comparison to low pressure condition. If the gas processing is done in low pressure condition, the absolute molecule quantity is low. That is, large amount of energetic electron having more than dissociation energy of objective molecules are need in order to generate more radicals and decompose more toxic molecules. Later on, the required value of electron energy for air pollution control is approximately 10eV.

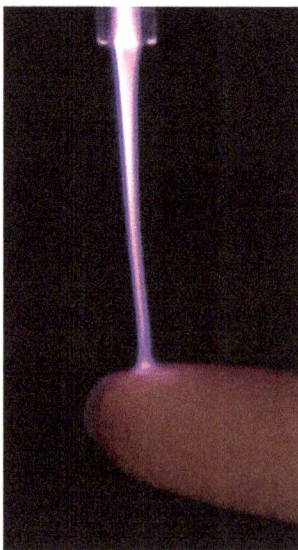

Figure 1. Typical example of non-thermal plasma demonstration [8].

3. Formation methods of non-thermal plasma

Non-thermal plasmas for removal of hazardous pollutants have been produced by an electron beam method and various electrical discharge methods.

3.1. Electron beam

The electron beam irradiation is one of non-thermal plasma formation method. Figure 2 shows schematic representation of electron beam source. In an electron beam method, the electrons are accelerated by high voltage in the vacuum region before being injected into a gas-processing chamber thorough a thin foil window. The energy of electron beam is directly used for dissociation and ionization the background gas. During the ionization by the beam, a shower of ionization electrons is generated, which further produce a large volume of plasma that can be used to initiate the removal of various types of pollutant molecules such as NOx, SOx and VOCs. This exhaust gas treatment technic by an electron beam has a 40 year-old history previously and a lot of pilot plants for air pollution control have been running today [10]-[18].

In particular, an Electron Beam Dry Scrubbing (EBDS) system has been mainly studied at present. Figure 3 shows that the typical principle of EBDS. It is a dry process and does not require an expensive catalyst for NOx removal. In this process, at first, many oxidative radicals such as O, OH and HO_2 were produced by electron beam irradiation into O_2 and H_2O in exhaust gas. Following that, NOx and SOx are oxidized by these radicals to HNO_3 and H_2SO_4. Finally, HNO_3 and H_2SO_4 were converted to ammonium nitrate (NH_4NO_3) and ammonium sulfate ($(NH_4)_2SO_4$ by added ammonia (NH_3) into the treated combustion flue gas. These byproducts are collected by the electrostatic precipitator (ESP) and shipped to outside, because NH_4NO_3 and $(NH_4)_2SO_4$ can be used to make fertilizer. Here it should be noted that a part of NO is reduced to N_2 by N radical which produced by electron beam irradiation. This EBDS system has applicability to a high concentration sulfur-containing coal-fired boiler and a treatment of solid waste. In either case, EBDS could treat NOx and SO_2 in high efficiency. According to the literatures [19], over 95% of SO_2 and over 80% of NOx were removed simultaneously when the flue gas of sulfur-containing (at least 2.5%) coal-fired boiler was used as simulate gas.

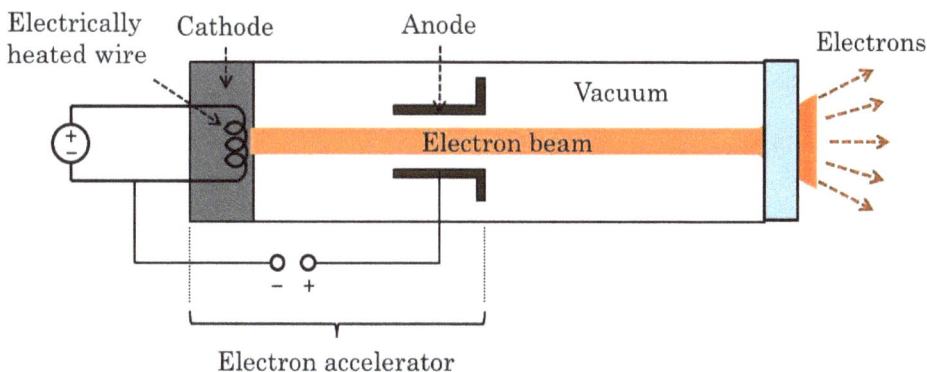

Figure 2. Schematic representation of electron beam source.

The potential of using an electron beam for removal of post combustion toxic gases (NOx, SO₂) was recognized in the early 1970s by the Ebara Corporation (Japan) [20]. Following successful initial batch tests of the Ebara plant, various tests on small pilot plants have been conducted in the Canada [16], Korea [21], Poland [22], and Japan [23], etc. The tests performed in these installations proved that a significant amount of NOx (and SO₂) exhausted from power plants, municipal-waste incinerators, and combustion boilers, etc., could be efficiency removed. In addition, A.G. Ignat'ev [24], B.M. Penentrante [25] and Y. Nakagawa [26] have indicated that using a pulsed electron beam improves the energy efficiency for exhaust gas treatment instead of using a DC electron beam. However, the electron beam methods hasn't put into practical use yet due to the high capital cost of accelerators, X-ray hazard and the unavoidable large energy loss caused by vacuum interface.

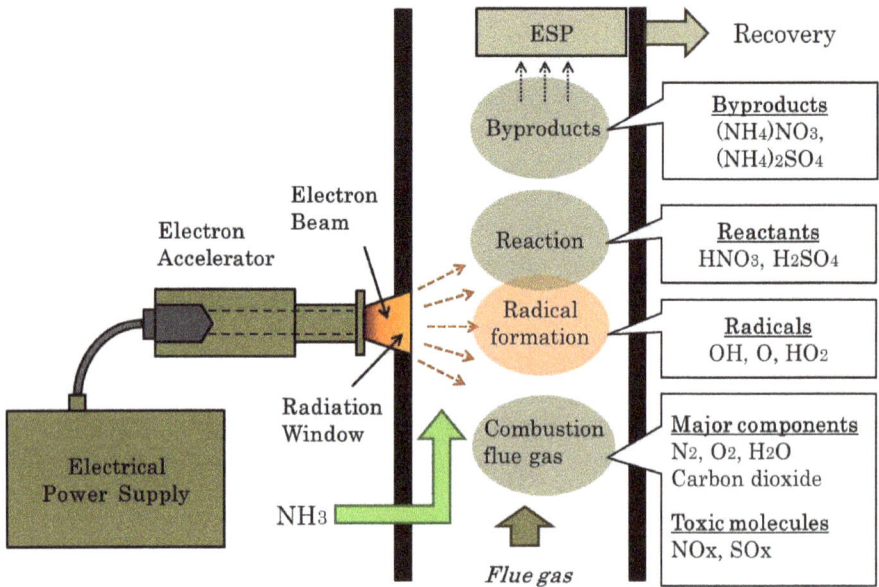

Figure 3. Principle of combustion flue gas treatment by electron beam.

3.2. Electrical discharge

In contrast to the electron beam which produce non-thermal plasma by supplying energetic electrons to objective gas, electrical discharge methods which led objective gas into plasma directly and generate energetic electron and radicals. Electrical discharges could produce non-thermal plasma in atmospheric pressure gases by various power supply such as direct current (DC), alternating current (AC), or pulse power sources. Among them, the dielectric barrier discharge (DBD) method using AC high voltage source and pulsed power discharge method have been developed particularly to this day. In this section, DBD and pulse power discharge are introduced.

3.2.1. Dielectric Barrier Discharge

A schematic representation of typical DBD electrodes is shown in figure 4. The DBD is also called as a silent discharge. In DBD reactor, AC high voltage which is typically 10 to 20 kV and 50 Hz to 2 kHz are applied to electrodes, one or both of which are covered with a thin dielectric layer, such as glass. The gap distance between electrodes is a few hundred of μm to several mm order. The barrier discharge is characterized by millions of small pulsed micro discharge which occur repetitively in gas space. The current density of the micro discharge is approximately 1 kA/cm², the diameter is 0.1 mm and the pulse duration is 3ns. Because of energetic electrons are generated in this micro discharge, various radicals and ions are produced by the electron collision with gas molecules. These radicals defuse into the barrier discharge space and react with background gas. As a result, ozone generation and NOx or VOCs removal are realized.

Dielectric barrier discharge processing is very mature technology, first investigated in 1850's for the production of ozone. Ozone has some effects such as sterilization, deodorization, and decolorization, because of its strong oxidization power, placing it second after fluorine. Furthermore, ozone has no residual toxicity due to its spontaneous decomposition feature. Therefore, ozone has already been put to practical use in water purification instead of conventional sterilization by chlorine. Ozone has been used in Europe for water treatment since early in the 20th century. Initial applications were to disinfect relatively clean spring or well water, but they increasingly evolved to also oxidize contaminants common to surface waters. Since 1950's, ozonation has become the primary method to assure clean water in Switzerland, West Germany and France. More recently, major fresh water and waste water treatment facilities using ozone water treatment methods have been constructed throughout the world. Additionally, new industrial applications of ozone such as wastewater treatment, exhaust gas treatment, odor elimination and semiconductor manufacturing have been studied recently [27-35]. However these new ozone applications demand high concentration of ozone, it was become possible for DBD to produce high concentration (100 to 300 g/m³) of ozone due to the improvement of dielectric materials and electrodes cooling function, and development of ultra-short gap electrodes [27]-[35]. Figure 5 shows a schematic representation of a cylindrical reactor which is modern shape of today's ozonizer. Consequently, the dielectric barrier method is most common way of ozone generation today. In addition, DBD has been studied for flue gas cleaning and toxic gas decomposition. In the literatures, removal of various toxic molecules such as NOx in diesel engine exhaust, greenhouse gas and VOCs such as formaldehyde which causes a sick building syndrome, have been demonstrated.

However, ozone generation by a dielectric barrier discharge is common way at today, it has still some agendas for industrial applications. For example, need of external cooling system for discharge electrodes and its sensitivity narrow gap separation take plenty operation costs. In addition, the narrow gap is sensitive to grit, dust and vibration. Therefore, use of this pollution control by ozone process is limited to a part of well-financed company or state and public institutions. The improvement of energy efficiency for DBD system is strongly demanded in order to spread the ozone processing moreover. In addition, NOx removal and

VOCs treatment using DBD is still in laboratory stage, because DBD could not treat those toxic molecules completely and its energy efficiency is unfavorable at the present stage.

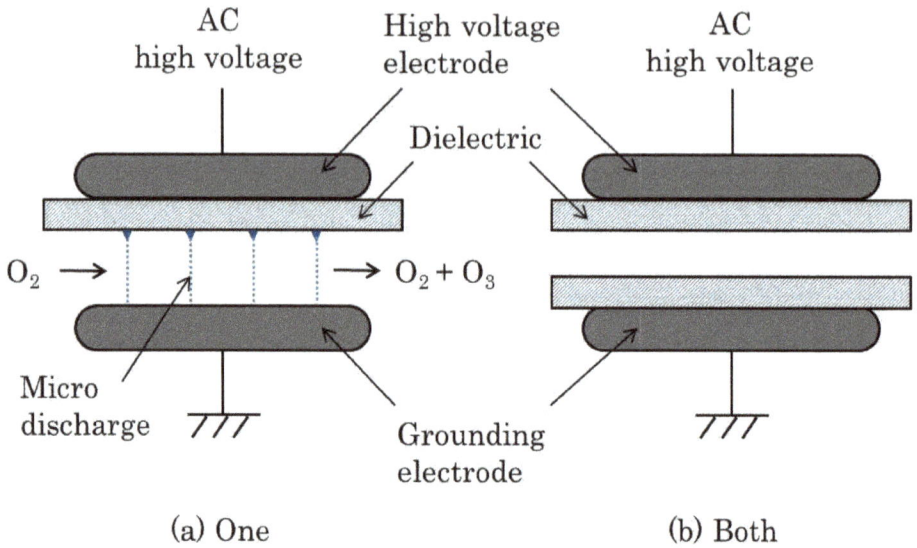

Figure 4. Schematic representation of dielectric barrier discharge electrode.

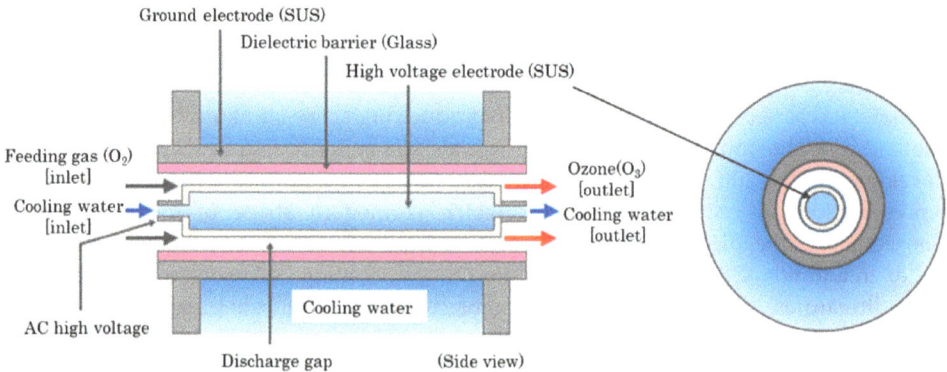

Figure 5. Schematic representation of cylindrical cooled reactor.

3.2.2. Pulse power discharge

Pulsed power discharges have been studied for many years since it is one of the promising technologies for the removal of the hazardous environmental pollutants as well as electron beam and dielectric barrier discharge. Typically, many researchers have reported that the pulsed discharge performed DeNOx process more effectively comparison with direct current (DC) corona discharge [3], [6, 7], [36]-[39]. Because the pulsed discharge is generated by intermittent pulse voltage, it is difficult to transfer to an arc discharge. Therefore, pulsed discharge is possible to apply an overvoltage between electrodes. Moreover, it is noted that the rate of over voltage (applied voltage / DC breakdown voltage) is depend on the voltage rise time and the pulse duration. In this way, since the pulse discharge could apply an overvoltage, it is possible to generate large amount of energetic electrons which have over 10 eV in atmospheric plasma [7]. Generated high-energy electrons could easily dissociate a nitrogen molecule (N_2) which having 9.8 eV of comparatively-high dissociation energy in gas. Therefore, generation of large amount of radicals that contributes to the gas processing become possible, and effectiveness of pulsed discharge could be obtained.

Pulsed power is a technology that concentrates electrical energy and turns it into short pulses of enormous power. Pulsed power technology had been studied for X-ray generation and gaseous discharge from the beginning of twentieth century. At the time, a capacitor discharge which is output from charged capacitor thorough discharge switch has been used for pulsed power generation. Even now, this capacitor discharge method has been adopted widely because this is most simple and low cost method. However, if the operating voltage is critically high, the simple DC charge for the capacitor is impractical. To fix this problem, the Marx circuit system where parallel charged capacitors are connected in series with spark gap switches was invented. From the latter part of the twentieth century, the way of using a pulse forming line (PFL) began to use widely as an intermediate devise of energy storage. This is because, it is recognized that a discharge from a PFL which having constant impedance could obtain more stable output than direct discharge from a capacitor. Additionally, the output pulse duration is shortened by the introduction of PFL. With this, the peak of available power is significantly increased. Furthermore, during the decades, the development of high power semiconductor switch, magnetic core and etc. have allowed us to manufacture the pulse power source having higher energy transfer efficiency. As a result, the pulsed discharge has been recognized as one of the promised non-thermal plasma to practical use in this day.

In addition, recently, it is reported by many researchers that a shorter-duration pulsed power with higher voltage rise time gives significant improvement of energy efficiency of pollutant gas treatment. In the pulsed discharge, the short duration pulse has an effect to prevent the energy loss due to heating by terminating the voltage before the plasma phase shift to thermal plasma. Additionally, it is reported that the faster rise time of applied voltage provides more energetic electrons and a higher energy [7], [40]-[45]. From these factors, it is considered that the development of a short pulse generator is of paramount importance for practical applications. Consequently, pulse power sources for environmental applications had been shifted from a simple condenser source which generates microseconds of pulse power to a pulse forming line (PFL) source which can output sub-microseconds of pulse duration in 1990's. Furthermore, the nanoseconds pulse source has been developed since 2000's and the nanoseconds pulsed streamer discharge has shown remarkable results in energy efficiencies of pollutant control [6, 7], [46]-[49].

Hereinafter, some differences of general pulsed discharge and the nanoseconds pulsed streamer discharge will be introduced. At first, figure 6 shows images of light emissions from conventional pulsed discharges as a function of time after initiation of the discharge current [7]. The peak voltage was +72 kV with 100 ns of pulse duration and 50 ns of voltage rise time. With regard to the coaxial discharge electrode, a rod electrode made of stainless steel, 0.5 mm in diameter and 10 mm in length was placed concentrically in a copper cylinder, 76 mm in diameter. The bright areas of the framing images show the position of the streamer heads during the exposure time of 5 ns. In a rod-to-cylinder coaxial electrode, the positive streamer discharge propagate straight in the radial direction from the coaxial electrode because the interactions between the electric fields near the neighboring streamer heads are the same at somewhere in the coaxial electrode geometry. The streamer heads are associated with a higher density of ionization due to the high electric field therein, and subsequently enhanced recombination, which is followed by increased light emission [7], [40]-[45] (Fig.6). In conventional pulsed discharge, the emission at the vicinity of the rod electrode is observed 10-15ns after pulsed voltage application. The streamer heads were generated in the vicinity of the central electrode and then propagated toward the ground cylinder electrode. After full development of the streamer heads between the electrodes, the discharge phase transformed to a glow-like discharge with a large flow of current in the plasma channel produced by the streamer propagation. Finally, the glow-like discharge finished at the end of the applied pulsed voltage [7], [40]-[45] (Fig.6, Fig.7). Therefore, two stages of the discharge can be clearly defined during the conventional pulsed discharge. The first one is the 'streamer discharge', which means the phase of streamer heads propagation between electrodes. The other is the 'glow-like discharge' that follows the streamer discharge. Here it should be mentioned that in some publications, aforementioned two discharge phases are collectively called as pulse corona discharge. Additionally, the track of the streamer head which propagates from the central rod electrode to the outer cylinder electrode is called as 'primary streamer', and the subsequent streamer head that started from the central electrode at 30 ~ 35 ns (Fig.6, Fig.7) and disappeared at the middle of the electrodes gap is called a 'secondary streamer'.

By the way, as I mentioned in previous section, formation of ultimate non-thermal plasma where only electron has energy is aspired in non-thermal plasma processing. However, energy loss by background gas (ions and neutral molecules) heating starts when the discharge phase shifts to glow-like as shown in figure 6 and 7. In figure 6 and 7, until 50ns, you can see that the discharge is composed with only streamer phase which is a very high level of non-thermal condition. Therefore, if we are able to use this phenomenon which happened until 50 ns, we could become produce radicals in efficiency by using over 10 eV of energetic electrons which are exist in streamer head without ions and neutron molecules heating. Based on this idea, nanoseconds pulse generator has been developed by a lot of researchers recently.

Framing images and streak image of the discharge phenomena caused by a nanoseconds pulsed power generator (NS-PG) having a pulse duration of 5 ns and maximum applied voltage of 100 kV was developed by Prof. Namihira et al. in early 2000s [6, 7], [46]-[49] are

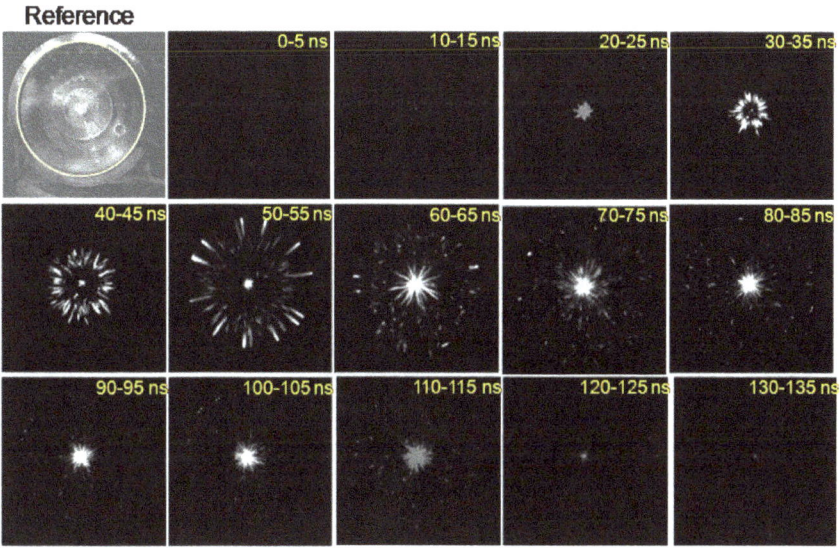

Figure 6. Images of light emissions from positive pulsed streamer discharges as a function of time after initiation of the discharge current.

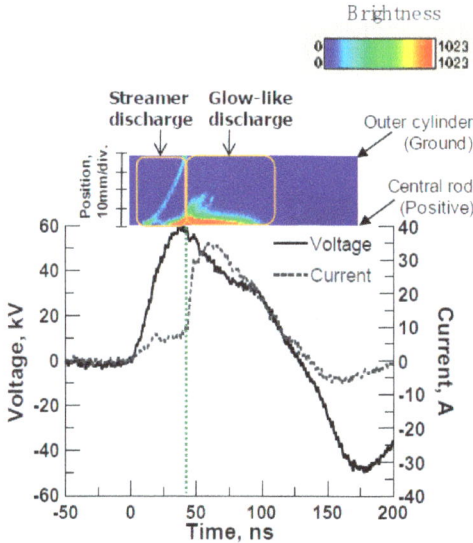

Figure 7. Typical applied voltage and discharge current in the electrode gap, and streak image for the generator with 100 ns of pulse duration. Applied voltage to electrode and discharge current through the electrodes were measured using a voltage divider and a current transformer, respectively. The vertical direction of the streak image corresponds to the position within the electrode gap. The bottom and top ends of the streak image correspond to the central rod and the surface of the grounded cylinder, respectively. The horizontal direction indicates time progression. The sweep time for one frame of exposure was fixed at 200 ns.

shown in Fig. 8, respectively [7]. In case of nanoseconds pulsed streamer discharge, the streamer heads were generated near the central rod electrode and then propagated toward the grounded cylinder electrode in all radial direction of the coaxial electrode as is the case in the conventional pulsed discharge shown in figure 6. The time duration of the streamer discharge was within 6 ns. At around 5 ns, emission from a secondary streamer discharge was observed in the vicinity of the central rod electrode. This is attributed to the strong electric field at the rod. Finally, emission from the pulsed discharge disappeared at around 7ns, and the glow-like discharge phase was not observed. As a result, as can be seen in fig. 8(b), energy loss by background gas heating caused by glow-like phase is suppressed extremely small comparison with 100ns pulsed discharge case (Fig. 7). The average propagation velocity of the streamer heads reported as 6.1 ~ 7.0 mm/ns for a positive peak applied voltage of 67 ~ 93 kV. The average velocity of the streamer heads slightly increased at higher applied voltages. Since the propagation velocity of the streamer heads is 0.1 ~ 1.2mm/ns for a 100 ns pulsed discharge, five times faster velocity is observed with the NS-PG. These characteristics comparison of pulsed discharges is summarized in Table 1. The streamer head always has the largest electric field in the electrode gap, and it is known streamer heads with higher value electric fields have a faster propagation velocity [7]. Therefore, it is understood that the faster propagation velocity of the streamer head means that the streamer head has more energetic electrons and higher energy. Consequently, the electron energy generated by nanoseconds pulsed discharge is higher than that of a general pulsed discharge. Here it should be mentioned that the voltage rise time (defined between 10 to 90%) was 25 ns for a 100 ns general pulsed discharge and 2.5 ns for the 5 ns nanoseconds pulsed discharge. Therefore, the faster propagation velocity of streamer head might be affected by the faster voltage rise time.

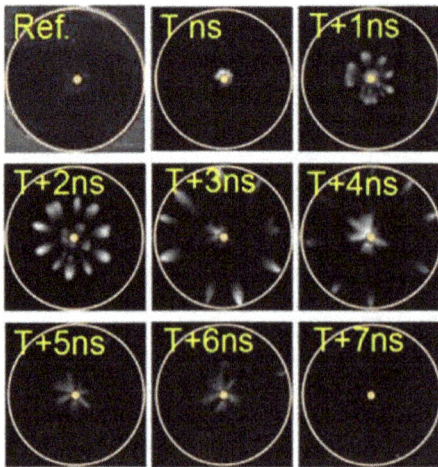

(a) Flaming images   (b) Streak image

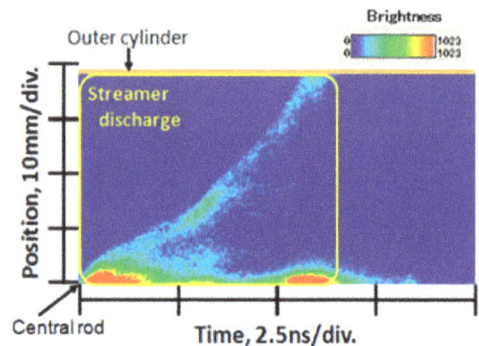

Figure 8. Framing and streak images of nanoseconds pulsed discharge in a coaxial electrode..

This nanoseconds pulsed power has demonstrated extremely high NO removal efficiency and ozone yield [6, 7], [46], [49]. The performances of nanoseconds pulsed discharge is summarized in Table 2. These energy efficiencies of nanoseconds pulsed power are highest value in the recent literatures about pollution control by non-thermal plasma. For the future, nanoseconds pulse power will be expected to verify its practical effectiveness by more practical experiments.

	General pulsed discharge		Nanoseconds pulsed discharge
Voltage rise time	25ns		2.5ns
Voltage fall time	25ns		2.5ns
Pulse duration	100ns		5ns
Discharge phase	Streamer	Glow-like	Streamer
Propagation velocity of streamer heads (Vapplied-peak)	0.1~1.2mm/ns (10~60kV)	-	6.1~7.0mm/ns (67~93kV)
Electrode impedance	5-17kΩ (L=10mm)	2kΩ (L=10mm)	0.3kΩ (L=200mm)

Table 1. Characteristics comparison of pulsed discharges.

	General pulsed discharge	Nanoseconds pulsed discharges	
Pulse duration	50ns	5ns	2ns
Voltage rise time	25ns	2ns	1ns
Voltage fall time	25ns	2ns	1ns
NO removal efficiency Simulated gas: NO (200ppm)/N$_2$ (at 60% of removal ratio)	0.37 mol/kWh	0.52 mol/kWh	0.89 mol/kWh
Ozone yield Feeding gas: Oxygen [at 10g/m^3 of O$_3$ concentration]	30 g/kWh	400 g/kWh	470 g/kWh

Table 2. Gas treatment characteristics comparison of pulsed discharges.

4. Present situation of non-thermal plasma on practical use

As I've discussed, non-thermal plasma has been attracted attention as a new technology of flue gas treatment for the next generation in recent years. Among the many air pollution control of non-thermal plasma, NOx removal and VOCs treatment have been particularly considered as a promising technology. Acid rain is partly produced by emissions of nitrogen oxides such as nitric oxide (NO) and nitrogen dioxide (NO_2) originating from fossil fuels burning in thermal power stations, motor vehicles, and other industrial processes such as steel production and chemical plants. Non-thermal plasmas for removal of NOx have been produced using an electron beam, a dielectric barrier discharge, and a pulsed corona discharge at various energy effectiveness. As explained in previous section, a lot of pilot plant employing electron beam has been running. Also, various electrical discharge methods have been evolved for practical use and some examples of pilot plant using discharge methods is reported at present situation.

However, the energy efficiencies and its performances of air pollution control technique using non-thermal plasma are still unfavorable regrettably. Therefore, a plasma-catalytic hybrid system is currently employed in a practical sense. The complex of a non-thermal plasma and catalyst can be utilized these characteristics of high responsiveness to persistent substance of non-thermal plasma and high reaction selectivity of catalyst. Additionally, there are many merits of the this hybrid system from the point of view of catalyst such as reduction of precious metal catalyst use, regeneration effect of catalyst by plasma irradiation and durability improvement of catalyst by inhibition of reaction temperature etc. [50]-[60]. This hybrid system is commonly combined in one of two ways. The first is the introduction of a catalyst in the plasma discharge (in plasma catalysis, IPC), the second by placing the catalyst after the discharge zone (post plasma catalysis, PPC). Figure 9 shows typical process flow diagrams and description of main functions of IPC and PPC systems. In IPC system, catalyst is activated by plasma exposure. IPC system is a method to improve reaction efficiency and a reaction characteristic by plasma activation of catalyst. In fact, many researchers have reported composite effects such as improvement of decomposition efficiency and reduction of byproduct production by using IPC system. Moreover, it is well known that the catalyst become activated by plasma irradiation in low-temperature region where the catalyst doesn't exhibit catalytic activity. A reactor utilizing these composite effects is named as Plasma-Driven Catalysis (PDC) [58, 59]. For example, however it is considered that the plasma methods have difficulty in treating NOx by reduction, the PDC can run NOx removal by reduction process. In addition, the PDC have a stimulating effect on VOCs decomposition and conversion of VOCs to favorable product of CO_2.

The effect of IPC system differs depending on a combination of the electrical discharge method and the type of catalyst. Therefore, it is considered that the combinatorial optimization is important for IPS system. In addition, it is reported that the influence of reaction field where the catalyst is placed is quite large. Typically, catalyst should be placed on a location where the plasma density is higher in pulse corona discharge reactor or dielectric barrier discharge reactor. Because, more radials and energetic electrons are exist in there. A packed-bed reactor is a typical example of IPC system in common with PDC. A typical schematic diagram and its appearance of packed-bed reactor are shown in figure 10. In this type of reactor, surface discharge and DBD methods is generally adopted as shown in figure 10. Additionally, catalyst or ferroelectric or both are employed as packing material between electrodes. The reason why the ferroelectric is packed is extremely high energetic electrons are produced near the contact points of ferroelectric pellets packed-in the plasma reactor, because of a huge electric field generated near the contact points [53]. As explained in the previous section, the energetic electrons are employed directly to dissociate and ionize the pollutants as well as carrier gas molecules to produce various radicals to react with and convert a part of pollutants. Fundamental characteristics of a dielectric barrier discharge (DBD) in a ferro-electric packed bed reactor have been studied for the Barium Titanate (BaTiO3) based spherical-shaped pellets for the specific dielectric constant from 660 to 104 from the viewpoint of reactor performance improvement [53]. The dielectric constant of pellet packed in the reactor affects discharge characteristics such as power consumption of the reactor, micro discharge onset voltage, number of micro discharge. As the results, the performance of packed bed plasma reactor depends on the dielectric constant and/or material of the pellet packed in the reactor.

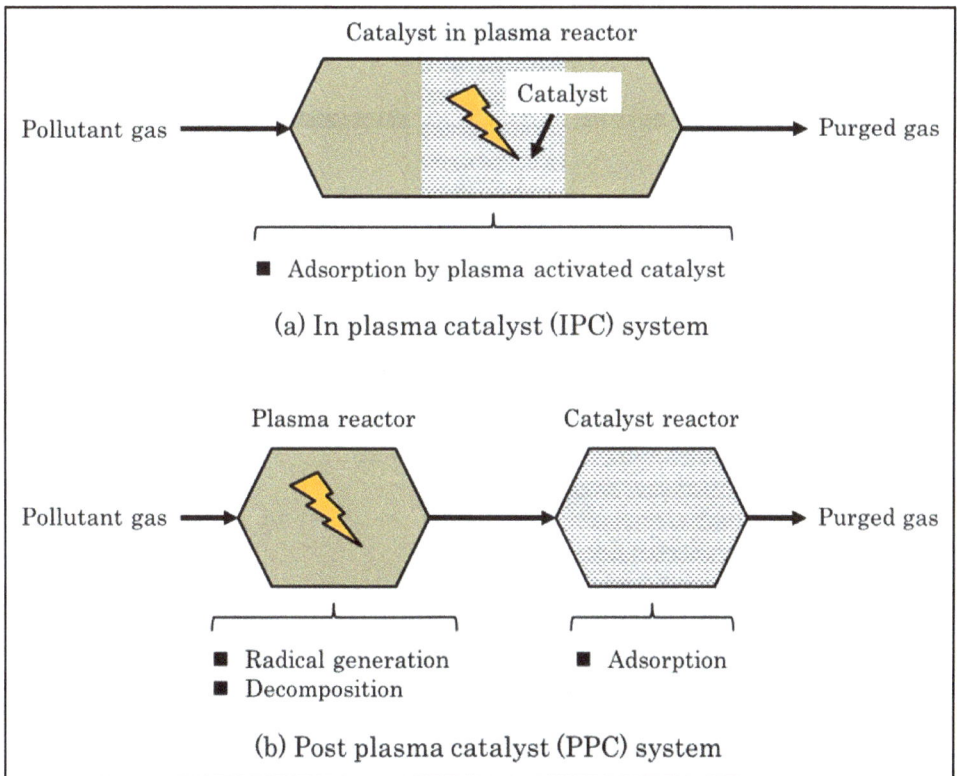

Figure 9. Typical process flow diagrams and description of main functions of Post Plasma Catalyst (PPC) and In Plasma Catalyst (IPC) system [55, 56].

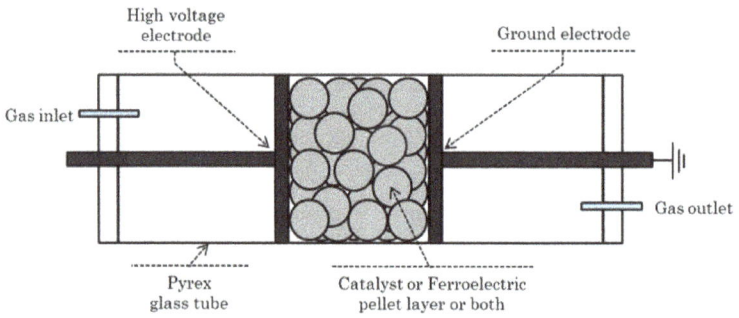

(a) Surface discharge type

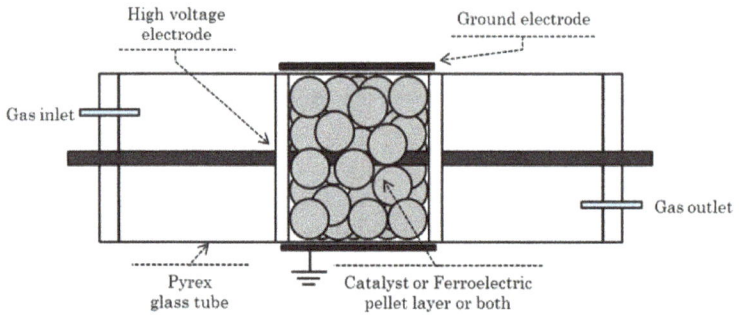

(b) Dielectric barrier discharge type

Figure 10. Schematic diagram and its appearance of packed-bed reactor (Photograph: Prof. Takaki group, Iwate University, Japan)[50], [53].

On the other hand, in the PPC system, the two functions of plasma and catalyst is completely-separated. Therefore, the configuration of reactor and system configuration are nearly independent on each other. As can be seen in figure 9(b), pollutant gas is induced into plasma reactor at first and the toxic molecules are decomposed or oxidized by energetic electrons or radicals which are generated in plasma. After that, residual contaminants that plasma couldn't treat and byproducts are removed by catalyst. Instead, this PPC system is sometimes used so as to extend time for replacement of catalyst. On another front, plasma reactor is sometimes incorporated to generate long-lived radicals such as ozone which work with catalyst as shown on figure 11. In figure 11, ozone was generated in plasma reactor, and then O radical which has stronger oxidative activity than ozone is generated by a reaction of ozone with catalyst. In consequence, VOC is decomposed by O radical to H_2O and CO_2.

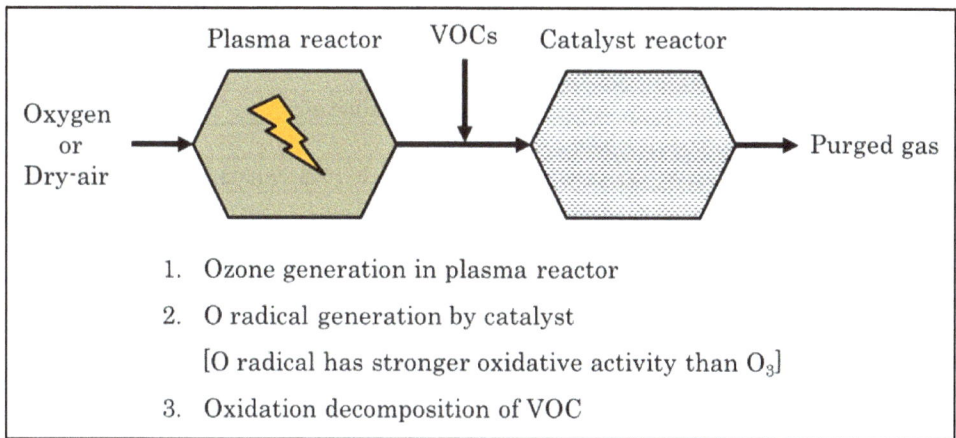

Figure 11. Example of VOC decomposition mechanism using a PPC system [62]-[64].

5. Summary

Previous prodigious studies by esteemed researchers from all ages and cultures have proven that the non-thermal plasma makes pollution control more efficient and effective. In consequence, it is recognized that the non-thermal plasma is one of the promising technologies for pollution control. The advantages of non-thermal plasma process were summarized as follows.

- Unlike conventional processes which need external combustion device or gas enrichment device, non-thermal plasma could treat industrial gas at ambient temperature and atmospheric pressure. Therefore, non-thermal plasma methods have a great advantage in energy efficiency.
- Non-thermal plasma is available for various harmful substances due to its great flexibility for the chemical reaction process (it is mainly depends on ambient gas composition). Therefore, in some case, non-thermal plasma could treat multiple toxic

molecules simultaneously. Table 3 shows that typical harmful substances in various exhaust gases. It has been proved that the non-thermal plasma could treat these toxic molecules in the literatures.

- Catalyst performance is highly improved by the concurrent use of the non-thermal plasma. The typical combined effects are increase of the reaction rate, extension of the catalyst lifetime and decrease of the activating onset temperature. Moreover, the agendas of non-thermal plasma process such as byproducts treatment could be solved by the combined use.

At the present day, a lot of practical trials are being conducted by using pilot plant of plasma reactor and more efficient and effective plasma source is developed by researchers from around the world. Air pollution control using non-thermal plasma has been edge closer to practical use. There are great hopes that air pollution control by non-thermal plasma reduce the environmental cost and make environmental effort accessible to companies and nations.

Type of Exhaust Gas	Containing Harmful Substances
Combustion flue gas	NOx, SOx, CO_2
Diesel gas	NOx, SOx, CO_2 Suspended Particulate Matter (SPM)
Industrial gas	VOCs (Aromatic series [Toluene, Benzene, acetone], Halogenated organics, HCHO), Dioxin, CFC-113, TCE

Table 3. Typical harmful substances in each exhaust gas.

Author details

Takao Matsumoto, Douyan Wang, Takao Namihira and Hidenori Akiyama
Kumamoto University, Japan

6. References

[1] Penetrante, B. M. & Schultheis, S. E., (1993). Nonthermal plasma techniques for pollution control, *Fundamentals and Supporting Technologies*, pt. A, 1-393, Springer-Verlag, New York

[2] Penetrante, B. M. & Schultheis, S. E., (1993). Nonthermal plasma techniques for pollution control, *Electron Beam and Electrical Discharge Processing*, pt. B, 1-397, SpringerVerlag, New York

[3] Penetrante, B. M. (1993). Pollution control applications of pulsed power technology, *Proceedings of 9th IEEE Int. Pulsed Power Conf.*, pp. 1-5, Albuquerque, NM, USA

[4] Müller, S.; Zahn, R. J. (2007). Air pollution control by non-thermal plasma, *Special Issue of 13th Topical Conference on Plasma Technology*, Vol. 47, Issue 7, 520–529

[5] Hackam, R.; Akiyama. H. (2000). Air pollution control by electrical discharge, *IEEE Trans. Dielect. Elect. Insulation*, Vol7, No.5, 654-683.

[6] Namhira, T.; Wang, D.; Akiyama, H. (2009). Pulsed power technology for pollution control", *Acta Physica Polonica A*, Vol. 115, No. 6, 953-955

[7] Wang, D.; Namihira, T.; Akiyama, H. (2010). Pulsed discharge plasma for pollution control, *Air Pollution*, Vanda Villanyi (Ed.), ISBN: 978-953-307-143-5, InTech

[8] Takamura, N., Wang, D., Seki, D., Namihira, T., Yano, K., Saitoh, H., Akiyama, H. (2012). *International Journal of Plasma Environmental Science & Technology*, Vol.6, No.1, 59-62.

[9] Oda, T. (2006). Atmospheric pressure nonthermal plasma decomposition of gaseous air contaminants and that diagnosis, *ICESP X – Australia*, Paper 1A1.

[10] Frank, N. R. (1995). Introduction and historical review of electron beam processing for environmental pollution control, *Radiation Physics and Chemistry*, Vol. 45, 989-1002

[11] Chmielewski, A. G.; Licki. J.; Pawelec, A.; Tyminski, B.; Zimek, Z. (2004). Operational experience of the industrial plant for electron beam flue gas treatment, *Radiation Physics and Chemistry*, Vol. 71, 439–442

[12] Duarte, C. L.; Sampa, M. H. O.; Rela, P. R.; Oikawa, H.; Silveira, C.G.; Azevedo, A.L. (2002). Advanced oxidation process by electron-beam-irradiation-induced decomposition of pollutants in industrial effluents, *Radiation Physics and Chemistry*, Vol. 63, 647–651

[13] Licki, J.; Chmielewski, A. G.; Iller, E.; Zimek, Z.; Mazurek, J.; Sobolewski, L. (2003). Electron-beam flue-gas treatment for multicomponent air-pollution control, *Applied Energy*, Vol. 75, issue 3-4, 145-154

[14] Osuda, Y. (1995). Pilot scale test on electron beam treatment of municipal waste flue gas with spraying slaked-lime slurry, *Radiation Physics and Chemistry*, Vol. 45, 1013-1015

[15] Bhasavanich, D.; Ashby, S.; Deeney, C. & Schlitt, L. (1993). Flue gas irradiation using pulsed corona and pulsed electron beam technology, *Proceedings of 9th IEEE Int. Pulsed Power Conf.*, 441-444, Albuquerque, NM, USA

[16] Chang, J.-S.; Looy, P. C.; Nagai, K.; Yoshioka, T.; Aoki, S. & Maezawa, A. (1996). Preliminary pilot plant tests of a corona discharge-electron beam hybrid combustion flue gas cleaning system, *IEEE Trans. Indust. Appl.*, Vol. 32, 131-137

[17] Penetrante, B. M.; Hsiao, M. C.; Bardsley, J. N.; Merritt, B. T.; Vogtlin, G. E.; Wallman, P. H.; Kuthi, A.; Burkhart, C. P. & Bayless, R. J. (1995). Electron beam and pulsed corona processing of volatile organic compounds and nitrogen oxides, *Proceedings of 10th IEEE Int. Pulsed Power Conf.*, 144-149, Albuquerque, NM, USA

[18] Penetrante, B. M. (1997). Removal of NO from diesel generator exhaust by pulsed electron beam, *11th IEEE International Pulsed Power Confernece*, 91-96

[19] *FEATURES, IAEA BULLETIN*, (1994). Electron beam processing of flue gases: Clearing the air, 7-10

[20] Hirano, S.; Aoki, S.; Izutsu, M.; Yuki, Y. (2000). Ebara electron-beam simultaneous SOx/NOx removal process, *Proc. 25th Int. Tech Conf. Coal Util. Fuel Syst.*, 593-604, Florida, USA

[21] Han, B.; Ko, J.; Kim, J.; Kim, Y.; Chung, W.; Makarov, I.E.; Ponomarev, A.V.; Pikaev, A.K. (2001). Combined electron-beam and biological treatment of dyeing complex wastewater. Pilot plant experiments, *Radiation Physics and Chemistry*, Vol. 64, 53–59

[22] Andrzej G. C.; Anna O. (2010). Electron beam technology for multipollutant emissions control from heavy fuel oil-fired boiler, *Journal of the Air & Waste Management Association*, Vol. 60, 932-938

[23] Doi, T.; Osada, Y.; Morishige, A.; Tokunaga, O.; Miyata, T. (1993). Pilot-plant for NOx, SO2 and HCL removal from flue-gas of municipal waste incinerator by electron beam irradiation, *Radiation Physics and Chemistry*, Vol. 42, 679-682

[24] Ignat'ev, A.V.; Kuznetsov, D. L.; Masyats, G. A.; Novoselov, Y. N. (1992). Cleaning flue gas by pulsed electron beams, *Sov. Tech. Phys. Left.*, Vol. 18, No. 11, 745-746

[25] Penetrante, B. M. (1997). Removal of NOx from diesel generator exhaust by pulsed electron beams, 11th IEEE International Pulsed Power Confernece, 91-96

[26] Nakagawa, Y.; Kawauchi, H. (1998). NOx removal in N2 by pulsed intence electron beam irradiation", Jpa. J. Appl. Phys., Vol.37, L91-L93

[27] Kuzumoto, M. (1998). Extremely narrow discharge gap ozone generator, *Journal of plasma and fusion research/the Japan Society of Plasma Science and Nuclear Fusion Research*, Vol. 74, Issue 10, 1144-1150

[28] Toyofuku, M.; Ohtsu, Y.; Fujita, H. (2004). High ozone generation with a high-dielectric-constant material, *Japanese Journal of Applied Physics*, Vol.43, No.7A, 4368-4372

[29] Kuzumoto, M.; Tabata, Y.; Shiono, S. (1997). Development of a very narrow discharge gap ozone generator -Generation of high concentration ozone up to 20 wt% for pulp bleaching-, *Kami Pa Gikyoushi/Japan Tappi Journal*, Vol. 51(2), 345-350

[30] Zhang, Z.; Bai, X.; Bai, M.; Yang, B.; Zhou, X. (2003). An ozone generator of miniaturization and modularization with the narrow discharge gap, *Plasma Chemistry and Plasma Processing*, Vol. 23(3), 559-568

[31] Takayama, M.; Ebihara, K.; Strycwewska, H.; Ikegami, T.; Gyoutoku, Y.; Kubo, K.; Tachibana, M. (2006). Ozone generation by dielectric barrier discharge for soil sterilization, *Thin Solid Films*, 506-507, 396-399

[32] Park, S. L.; Moon, J. D.; Lee, S. H.; Shin, S. Y. (2006). Effective ozone generation utilizing a meshed-plate electrode in a dielectric-barrier discharge type ozone generator, *Journal of Electrostatics*, Vol. 64, 275-282

[33] Jung, J. S.; Moon, J. D. (2008). Corona discharge and ozone generation charactristics of a wire-plate discharge system with glass-fiber layer, *Journal of Electrostatics*, Vol. 66, 335-341

[34] Sung, Y. M.; Sakoda, T. (2005). Optimum conditions for ozone formation in a micro dielectric barrier discharge, *Surface & Coatings Technology*, Vol. 197, 148-153

[35] Catalog of ozonizer, Mitsubishi Electric Corp., Japan. "http://www.mitsubishielectric.co.jp/service/mizukankyo/catalog/pdf/ozonizer.pdf".

[36] Akiyama, H. (1995). Pollution control by pulsed power, *International Power Electrics Conference*, 1937-1400

[37] Mizuno, A.; Shimizu, K.; Chakrabarti, A.; Dascalescu, L.; Furuta, S. (1995). NOX removal process using pulsed discharge plasma, *IEEE Trans. Indust. Appl.*, Vol. 31, 957-963

[38] Penetrante, B. M.; Hsiao, M. C.; Merritt, B. T.; Vogtlin, G. E.; Wallman, P. H. (1995). Comparison of electrical discharge techniques for nonthermal plasma processing of NO in N_2, *IEEE Transactions on Plasma Science*, Vol. 23, No. 4, 679-687

[39] Yoshida, T.; Tochikubo, F.; Watanabe, T. (1995). Diagnostics of pulsed corona discharge for DeNOX process, *11th International Conference on Gas Discharges and Their Applications*, Vol. 2, 410-413

[40] Wang, D.; Okada, S.; Matsumoto, T.; Namihira, T.; Akiyama, H. (2010). Pulsed discharge induced by nano-seconds pulsed power in atmospheric air", Vol. 38, No. 10, 2746-2751

[41] Namihira, T.; Wang, D.; Katsuki, S.; Hackam, R.; Akiyama, H. (2003). Propagation velocity of pulsed streamer discharges in atmospheric air, *IEEE Transactions on Plasma Science*, Vol. 31, No. 5, 1091-1094

[42] Tochikubo, F.; Teich, T. H. (2000). Optical emission from a pulsed corona discharge and its associated reactions, *Jpn. J. Appl. Phys*, Vol. 39, No. 3A, 1343-1350

[43] Tochikubo, F.; Watanabe, T. (2000). Two - dimensional measurement of emission intensity and NO density in pulsed corona discharge, *International Symposium on High Pressure Low Temperature Plasma Chemistry*, Vol. 1, 219-223

[44] Tochikubo, F.; Miyamoto, A.; Watanabe, T. (1995). Simulation of streamer propagation and chemical reaction in pulsed corona discharge, *11th International Conference on Gas Discharges and Their Applications*, Vol. 1, 168-171

[45] Ono, R.; Oda, T. (2007). Optical diagnosis of pulsed streamer discharge under atmospheric pressure, *International Journal of Plasma Environmental Science & Technology*, Vol. 1, No. 2, 123-129

[46] Namihira, T et al. (2000). Improvement of NOx removal efficiency using short width pulsed power, *IEEE Transactions on Plasma Science*, Vol. 28, No. 2, 434-442

[47] Namihira, T et al. (2007). Characterization of nano-seconds pulsed streamer discharges, *2007 IEEE Pulsed Power and Plasma Science Conference*, Albuquerque, USA, 572-575

[48] Tamaribuchi, H et al. (2007). Effect of pulse width on ozone generation by pulsed streamer discharge, *2007 IEEE Pulsed Power and Plasma Science Conference*, Albuquerque, USA, 407-410

[49] Matsumoto, T.; Wang, D.; Namihira, T.; Akiyama, H. (2011). Performances of 2 nanoseconds pulsed discharge plasma, *Japanese Journal of Applied Physics*, Vol. 50, No. 8, 08JF14-1-5

[50] Takaki, K.; Chang, J.-S.; Kostov, K.G. (2004). Atmospheric pressure of nitrogen plasmas in a ferroelectric packed bed barrier discharge reactor. Part I. Modeling, *IEEE Trans. Diel. Elect. Insul.*, Vol. 11, No. 3, 481-490

[51] Uchida, Y.; Takaki, K.; Urashima, K.; Chang, J.-S. (2004). Atmospheric pressure of nitrogen plasmas in a ferroelectric packed-bed barrier discharge reactor. Part II. Spectroscopic measurements of excited nitrogen molecule density and its vibrational temperature, *IEEE Trans. Diel. Elect. Insul.*, Vol. 11, No. 3, 491-497

[52] Rajanikanth, B. S.; Srinivasan, A. D. (2007). Pulsed plasma promoted adsorption/catalysis for NOx removal from stationary diesel engine exhaust, *IEEE Transactions on Dielectrics and Electrical Insulation*, Vol. 14 (2). 302-311.

[53] Takaki, K.; Takahashi, S.; Mukaigawa, S.; Fujiwara, T.; Sugawara, K.; Sugawara, T. (2009). Influence of pellet shape of ferro-electric packed-bed plasma reactor on ozone generation and NO removal, *International Journal of Plasma Environmental Science and Technology*, Vol. 3, No. 1, 28-34

[54] Yamamoto, T.; Yang, C. L.; Beltran, M. R.; Kravets, Z. (2000). Plasma assisted chemical process for NOx control, *IEEE transactions on industry applications*, Vol. 36, No. 3, 923-927

[55] Chena, L.; Zhanga, X.; Huanga,L.; Le, L. (2010). Application of in-plasma catalysis and post-plasma catalysis for methane partial oxidation to methanol over a Fe2O3-CuO/γ-Al2O3 catalyst, *Journal of Natural Gas Chemistry*, Vol. 19, issue 6, 628-637

[56] Durme, J. V.; Dewulf, J.; Leys, C.; Langenhove, H. V. (2008). Combining non-thermal plasma with heterogeneous catalysis in waste gas treatment, *Applied Catalysis B: Environmental*, Vol. 78, 324–333

[57] Vandenbroucke, A.; Morent, R.; Geyter, N. D.; Dinh, M. T. N.; Giraudon, J. M.; Lamonier, J. F.; Leys, C. (2010). Plasma-catalytic decomposition of TCE, *International Journal of Plasma Environmental Science & Technology*, Vol. 4, No. 2, 135-138

[58] Durme, J. V.; et al. (2008). Combining non-thermal plasma with heterogeneous catalysis in waste gas treatment, *Applied Catalysis B*, Vol. 78, 324-333

[59] Fan, H. Y.; et al. (2009). High-efficiency plasma catalytic removal of dilute benzene from air, *J. Phys. D: Appl. Phys.*, Vol.42, 225105

[60] Kim, H. H.; Oh, S. M.; Lee, Y. H.; Ogata, A.; Futamura, S. (2005). Decomposition of gas-phase benzene using plasma-driven catalyst (PDC) reactor packed with Ag/TiO2 catalyst, *Applied Catalysis B: Environmental*, Vol. 56, 213-220

[61] Kim, H. H.; Lee, Y. H.; Ogata, A.; Futamura, S. (2006). Effect of different catalysts on the decomposition of VOCs using flow-type plasma-driven catalysis, *IEEE Transactions on Plasma Science*, Issue 3, Vol. 3, 984-995

[62] Ogata, A.; Saito, K.; Kim, H. H.; Sugasawa, M.; Aritani, H.; Einaga, H. (2010). Performance of an ozone decomposition catalyst in hybrid plasma reactors for VOC removal, *Plasma Chem. Plasma Process*, Vol. 30, 33-42

[63] Kim, H. H.; Ogata, A.; Futamura, S. (2007). Complete oxidation of volatile organic compounds (VOCs) using plasma-driven catalysis and oxygen plasma, *International Journal of Plasma Environmental Science & Technology*, Vol.1, No.1, 46-51

[64] Demidiouk, V.; Chae, J. O. (2005). Decomposition of volatile organic compounds in plasma-catalytic system, *IEEE Transactions on Plasma Science*, Vol. 33, No. 1, 157-161

Chapter 10

Mitigating Urban Heat Island Effects in Tehran Metropolitan Area

Parisa Shahmohamadi, Ulrich Cubasch, Sahar Sodoudi and A.I. Che-Ani

Additional information is available at the end of the chapter

1. Introduction

The majority of cities are sources of heat, pollution and the thermal structure of the atmosphere above them is affected by the so-called "heat island" effect. In fact, an UHI is best visualized as a dome of stagnant warm air over the heavily built-up areas of cities [1]. The heat that is absorbed during the day by the buildings, roads and other constructions in an urban area is re-emitted after sunset, creating high temperature differences between urban and rural areas [2]. The exact form and size of this phenomenon varies in time and space as a result of meteorological, location and urban characteristics [3]. Therefore, UHI morphology is strongly controlled by the unique character of each city. Oke [3] stated that a larger city with a cloudless sky and light winds just after sunset, the boundary between the rural and the urban areas exhibits a steep temperature gradient to the UHI, and then the rest of the urban area appears as a "plateau" warm air with a steady but weaker horizontal gradient of increasing temperature towards the city centre. The uniformity of the "plateau" is interrupted by the influence of distinct intra-urban land-uses such as parks, lakes and open areas (cool), and commercial, industrial or dense building areas (warm). In metropolitan areas especially in Tehran, Iran, the urban core shows a final "peak" to the UHI where the urban maximum temperature is found. The difference between this value and the background rural temperature defines the "UHI intensity" (ΔT_{u-r}). The intensity of the UHI is mainly determined by the thermal balance of the urban region and can result in a temperature difference of up to 10 degrees [2]. The UHI intensity varies in a recognizable way through the day under ideal weather conditions. At night, stored heat is released slowly from the urban surface, contrary to the rapid heat escape from rural surfaces. Thus, the UHI intensity peaks several hours after sunset when rural surfaces (and consequently surface air temperatures) have cooled yet urban surfaces remain warm. After sunrise rural areas warm more quickly than urban areas. If the difference in heating rates is great enough, rural air temperatures may equal or exceed urban temperatures. This reduces the UHI intensity to a daytime minimum, and may even generate an urban cool island.

2. Problem statement

These questions might strike the mind that why is UHI crucial problem in urban areas? And why should it be considered? In order to answer these questions, it is imperative to study the negative impacts of UHIs. Their negative impacts affect so many people in so many ways. Wong and Chen [4] summarized major negative impacts of UHI as below:

1. Air quality (environmental factor): UHI effect increases the possibility of the formation of smog created by photochemical reactions of pollutants in the air. The formation of smog that is highly sensitive to temperatures since photochemical reactions are more likely to occur and intensify at higher temperatures. Atmospheric pollution can be aggravated due to the accumulation of smog. In addition, the increased emissions of ozone precursors from vehicles is also associated with the high ambient temperature;
2. Human mortality and disease (social factor): the UHI effect also involves the hazard of heat stress related injuries which can threaten the health of urban dwellers; and
3. Waste of natural resources (economical factor): higher temperatures in cities also increase cooling energy consumption and water demand for landscape irrigation. The peak electric demand will be increased as well. As a result, more electrical energy production is needed and this will trigger the release of more greenhouse gas due to the combustion of fossil fuel. The side effects also include the increased pollution level and energy costs. A feedback loop occurs when greenhouse gases eventually contribute to global warming.

Growing concern for the future of cities and for the well-being of city dwellers, stimulated by trends in world urbanization, the increasing number and size of cities, and the deterioration of many urban environments, has focused attention on the problems of living in the city. Citizens in cities around the world want clean air, clean water, reduced noise, more vegetation and protection of habitat areas, and safety. These are all seen as contributing not only to their health but also to their quality of life. Cities have been blamed for causing environmental catastrophes, diminishing the quality of life. Cities are also at risk from industrial hazards, natural disasters, and the specter of global warming. The likely negative impacts of global warming include increasing storms, flooding, droughts and the probable destruction of some ecosystems. In urban areas, there is an "urban heat island" effect resulting from the production and accumulation of heat in the urban mass. So, how will UHI affect the cities of the future?

The majority of people in the world live in metropolitan areas subject to new and potentially traumatic climatic conditions. A better conceptual approach is needed to understand the role and effects of UHI in cities and to consider in urban design guidelines and implementation measures. As Glantz [5] declared;

> It must be understood that cities are under real, not imagined, threats from global climate change and must be redesigned to deal with this reality.

Increasing urbanization and industrialization in Tehran metropolitan area in recent decades caused the urban environment to deteriorate. Tehran suffers from raised temperatures in the city core, generally known as the heat island effect. Raised temperatures, especially in summer, turn Tehran city centers into unwelcome hot areas, with direct effects on energy consumption for cooling buildings and morbidity and mortality risks for the population. These raised temperatures in Tehran city centre derive from the altered thermal balances in urban spaces, mainly due to the materials and activities taking place in cities, by far different to those in rural areas. The increasing numbers of buildings and construction in Tehran caused that vegetation and trees replaced by buildings. Thus, air temperature increases especially in high-density areas. The general lack of vegetation and the low albedo of urban surfaces are strong characteristics of the formation of UHI effect in Tehran metropolitan area. The geometry between a vegetated area and the density-morphology of an urban area are completely different, which has a direct effect on wind and shade distributions. Human activities taking place in Tehran urban areas are responsible for anthropogenic heat release (transport, space and water heating, cooling and the like) and air pollution, the latter affecting clouds cover. The combination of these factors determines the way in which heat is absorbed, stored, released and dispersed in the urban environment, expressed as a temperature increase in the urban area.

Therefore the majority of citizens are suffering from outdoor environment discomfort and this issue has a deeper and problematic dimension in the case of Iran especially in the city of Tehran. This research is an effort to recognize the radical cause of UHI in the city and will suggest some appropriate recommendations to solve this matter.

This research addresses the following objectives:

1. To identify the possible causes of UHI in Tehran metropolitan area;
2. To investigate the severity and impact of UHI on the environmental conditions of Tehran metropolitan area; and
3. To explore, develop and verify the various potential measures/models that could be implemented to mitigate the UHI effects in Tehran.

3. Conceptual framework

According to the Oke [6] model, different climatic events happen in different scales in cities and affect each other. These scales can be divided into two categories:

1. Horizontal scales include: micro-scale, local scale, and meso-scale.
2. Vertical scales (or different types of UHI) include: Air UHI (UCL UHI and UBL UHI), Surface UHI, and Sub-surface UHI.

Since it requires more spaces to explain all the scales, this research concentrates on UCL UHI. Figure 1 shows the conceptual framework of this research. UCL UHI forms via interaction between meteorological and urban structure factors. Since there are many factors which contribute to form UHI, this paper only picks the significant factors, vegetation covers and albedo materials, because other factors such as location, the size of the population and city, density and the like require long term planning. Then, in next stage, it develops a model with the title of "Natural Ventilator of the City" (NVC).

3.1. Interaction between meteorological and urban structure factors

Urban climate is concerned with interactions between the atmosphere and human settlements. As Oke [7] stated urban climate includes the impact of the atmosphere upon the people, infrastructure and activities in villages, towns and cities as well as the effects of those places upon the atmosphere. Therefore, climate has major impact on urbanization. Different climatic parameters affect the design of the city in terms of its general structure, orientation, building forms, materials and the like. Wong and Chen [4] stated that climate has impacts on buildings in terms of their thermal and visual performances, indoor air quality and building integrity. For example, a properly oriented building receives less solar heat gain and result in better thermal performance. In addition climate can influence the pattern of energy consumption.

It is not always one-way influence from climate toward urbanization. Urbanization also has more influence on climate. Buildings in cities influence the climate in five major ways [8]:

1. By replacing grass, soil and trees with asphalt, concrete and glass;
2. By replacing the rounded, soft shapes of trees and bushes with blocky, angular buildings and towers;
3. By releasing artificial heat from buildings, air conditioners, industry and automobiles;
4. By efficiently disposing of precipitation in drains, sewers and gutters, preventing surface infiltration; and
5. By emitting contaminants from a wide range of sources, which with resultant chemical reactions can create an unpleasant urban atmosphere.

Urban areas are the sources of anthropogenic carbon dioxide emissions from the burning of fossil fuels for heating and cooling; from industrial processes; transportation of people and goods, and the like [9, 10, 11]. Increased in pollutant sources both stationary (industrial) and non-stationary (vehicles) result in worsening atmospheric conditions [12]. The urban environment affects many climatological parameters. Global solar radiation is seriously reduced because of increased scattering and absorption [11]. Many cities in the tropics experience weak winds and limited circulation of air

In parallel, the urban environment affects precipitation and cloud cover. The exact effect of urbanization depends on the relative place of a specific city with respect to the general atmospheric circulation [11].

The city affects both physical and chemical processes in the atmospheric boundary layer (the lowest 1000m of the atmosphere) [13, 14] including: 1. Flow obstacles; 2. The area of an irregular elevated aerodynamic surface roughness; 3. Heat islands; and 4. Sources of emissions, such as sulphate aerosols that affect cloud formation and albedo.

One of the well-known phenomena of the urban climate is the UHI The term UHI denotes the increased temperature of a city compared with the temperature of the surrounding rural.

The temperature difference is raised with an increase in population and building density, which it is caused higher temperature than rural areas, as well as lower humidity due to low albedo and non-reflective materials, lower wind speed due to high density, and create various types of UHIs in different layers of urban climate. Hence, according to these data, Figure 2 illustrates the interaction between urban structure and climatic factors. The effect of urban structure factors on climate and vice versa is caused the formation of UHIs in different layers.

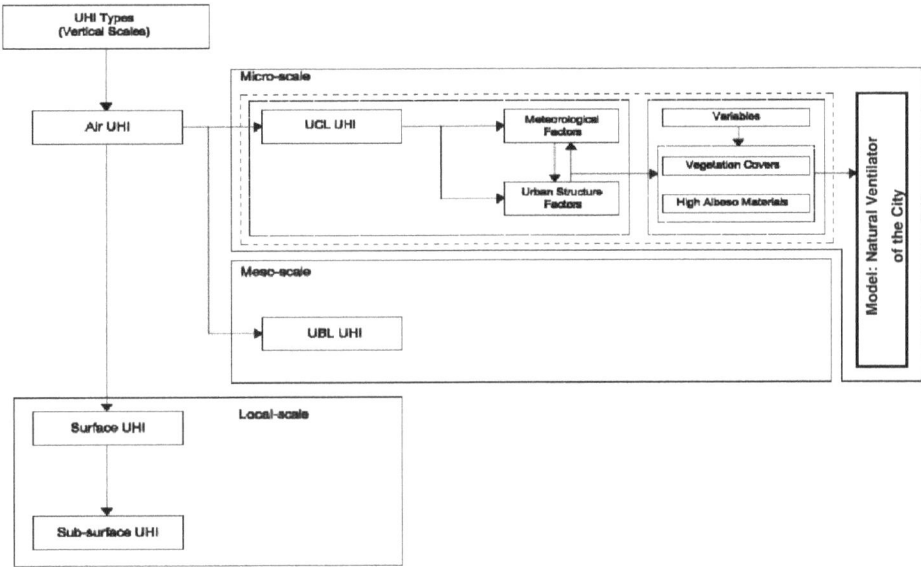

Figure 1. Conceptual frameworkwhich helps the accumulation of pollutants [12]. The wind speed in the canopy layer is seriously decreased compared to the undisturbed wind speed and its direction may be altered. This is mainly due to the specific roughness of a city, to channelling effects through canyons and also to UHI effects [11]. In addition, higher temperatures increase the production of secondary, photochemical pollutants and the high humidity contributes to a hazy atmosphere.

4.2. The effect of vegetation covers and high albedo materials over meteorological Factors

4.2.1. The effect of vegetation covers over meteorological factors: benefits of greenery in built environment

Green spaces contribute significantly to cool our cities and reduce UHI effects. Vegetation covers have extreme impacts on various aspects of life include filter pollutions, reduce air temperature, energy savings, help to mitigate greenhouse effect, provide an appropriate and pleasant environment for people. In fact, vegetation covers with their related benefits, play an important role in preventing the urban ecosystem from facing its ecological downfall. The ability of urban trees to improve the thermal comfort conditions in the surroundings is a function of the seasons, background climate, size of green area, type of surface over which trees are planted, and the amount of leaf cover [1]. Akbari et al. [15] discussed that the effectiveness of vegetation depends on its intensity, shape, dimensions and placement. But in general, any tree, even one bereft of leaves, can have a noticeable impact on energy use. In fact, trees in paved urban areas intercept both the sensible heat and the long wave radiation from high temperature paved materials such as asphalt [16, 17].

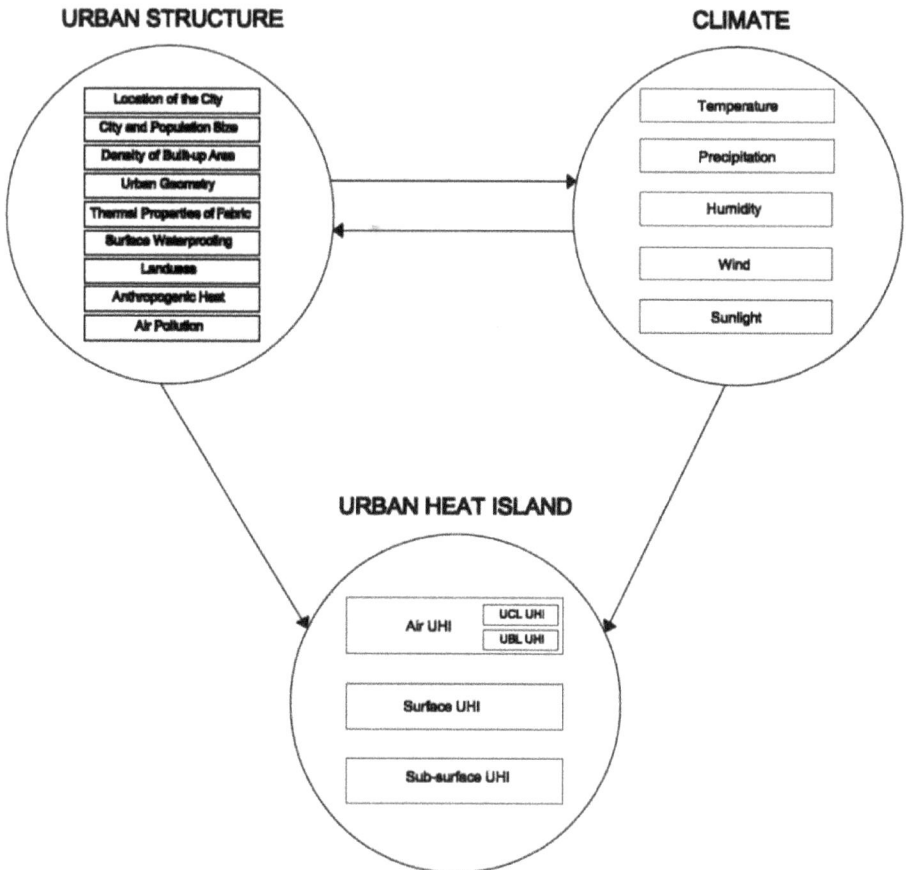

URBAN STRUCTURE

CLIMATE

- Location of the City
- City and Population Size
- Density of Built-up Area
- Urban Geometry
- Thermal Properties of Fabric
- Surface Waterproofing
- Landuses
- Anthropogenic Heat
- Air Pollution

- Temperature
- Precipitation
- Humidity
- Wind
- Sunlight

URBAN HEAT ISLAND

Air UHI | UCL UHI
       | UBL UHI

Surface UHI

Sub-surface UHI

Figure 2. Interaction between urban structure and climate factors

Wong and Chen [4] declared that greenery in a built environment has benefits in all aspects of life such as environment, economic, aesthetic and social.

1. Environmental benefits

Plants can offer cooling benefits in a city through two mechanisms, direct shading and evapotraspiration, which lead to alleviate UHI effects and provide pleasant environment. These benefits are:

- Reduce urban air temperature;
- Reduce air pollution and improve air quality; and
- Provide best ventilation condition.

2. Economic benefits

Economical benefits are associated with the environmental benefits brought by plants in a built environment. These benefits are:

- Energy saving;
- More usable space; and
- Reduce cooling resources through better insulation.

3. Aesthetic benefits

The aesthetic benefits are:

- Improve aesthetic appeal;
- Hide ugly roof tops services; and
- Integrate well with the building aesthetically.

4. Social benefits

The social benefits are:

- Foster community interaction;
- Facilitate recreational and leisure activities; and
- Therapeutic effects and improve health of its users.

4.2.2. The effect of high albedo materials over meteorological factors

The role of building materials, which is mainly determined by two characteristics including technical and optical characteristics [11, 1], is critical in mitigation of UHI effect. The technical characteristics of the materials used determined to high degree of energy consumption and comfort conditions of individual house, as well as open spaces. The optical characteristics of the materials used in the urban fabric largely define its thermal balance [11]. Two significant factors, albedo (reflectivity) which is the ratio of the amount of light reflected from a material to the amount of light shining on the material and emissivity which is the ratio of heat radiated by a substance to the heat radiated by a blackbody at the same temperature, are the most important parameters of optical characteristic [4]. Generally, urban surfaces tend to have lower albedo than surfaces in the rural environment (e.g. vegetation), thus absorb more solar radiation. This causes higher surface temperatures than air temperature; they can become 30-40°C higher than ambient air temperature [18]. Use of high albedo materials reduces the amount of solar radiation absorbed through building envelopes and urban structures and thus keeps their surfaces cooler. Emissivity controls the release of long-wave radiation to the surrounding. The albedo and emissivity, aspects related to the durability, cost, appearance and pollution emitted by the materials have to be considered. Using scale models, Simpson and McPherson [19] reported slightly better energy consumption performance under a white roof than a silver-colored roof, indicating the importance of emissivity in addition to albedo. Santamouris [11] reported asphalt temperatures close to 63°C and white pavements close to 45°C. Higher surface temperatures contribute to increasing the temperature of the ambient air and the UHI intensity.

Porosity is another characteristic of material that can affect urban temperature and UHI intensity. Porous surfaces absorb water (e.g. soil) account for quite significant latent heat flux in the atmosphere. Lack of porosity materials in urban surface, a high percentage of non-reflective, water-resistant surfaces and a low percentage of vegetated and moisture trapping surface create an evaporation deficit in the city caused UHI intensity. Vegetation, especially in the presence of high moisture levels, plays a key role in the regulation of surface temperatures even more than may non-reflective or low-albedo surfaces [20] and a lack of vegetation reduces heat lost due to evapotranspiration [21].

According to above description these three characteristics of materials are responsible for formation of UHI. Figure 3 shows that low quality of materials such as low albedo and low emissivity and the lack of porosity increase temperature, energy consumption, pollution and finally UHI, while Figure 4 shows that by increasing the quality of materials, UHI intensity can be decreased. The existence of these characteristics of materials properly helps to balance temperature, energy consumption and pollution in so far as reduce UHI effects and achieve ideal condition (UHI=0) (Figure 5). This process can be described in the following way:

$$Al.\downarrow +Em.\downarrow +Por.\downarrow = Temp.\uparrow +En.\uparrow +Pol.\uparrow = UHI \uparrow$$
$$Al.\uparrow +Em.\uparrow +Por.\uparrow = Temp.\downarrow +En.\downarrow +Pol.\downarrow = UHI \downarrow$$

Therefore, the existence of all characteristics of materials and integration between them can extremely contribute to reduce UHI effects rather than existence of one characteristic.

4.2.3. Model: Natural ventilator of the city

The most important part of any models is to pick the significant variables. It is realistic to present a range of key components involved and discuss how much interaction and impacts affect the basic system.

By deducing to researches, vegetation covers and high albedo materials have direct impact on mitigating of urban temperature and UHI effects. In this way, the increasing concern for the UHI impacts and the air quality is believed to be the motivation focusing on these key components and making a conceptual model. According to pervious discussion, many researches declared that greenery and high albedo materials could extremely affect the UHI. In facts, these researches experimented these two variables and observed their impacts on the UHI intensity separately. It is obvious that there are many conceptual models that can control the UHI effects. They all share the same goal that is to reduce UHI effects caused by interaction between two factors including meteorological and urban structure factors. Aside from all this, the study tried to develop the model and observe the impacts of vegetation covers and high albedo materials on UHI in parallel.

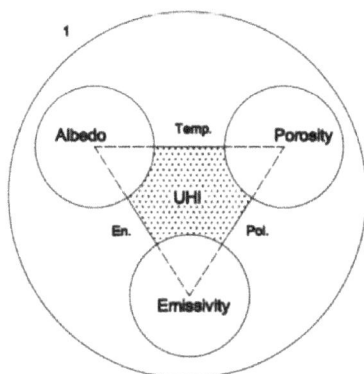

Figure 3. Low quality of material and UHI intensity

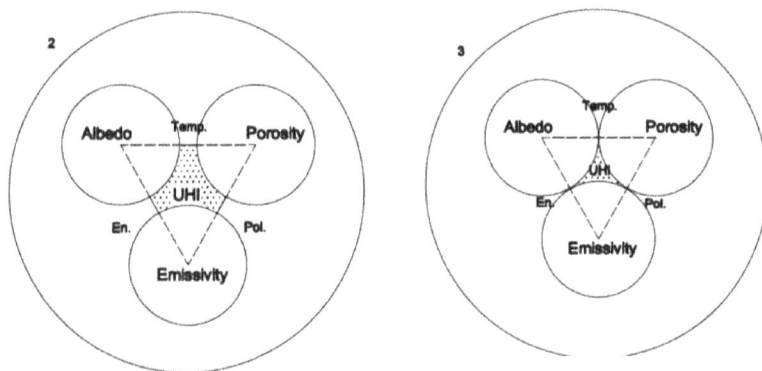

Figure 4. Mitigation of UHI intensity by increasing the quality of the materials

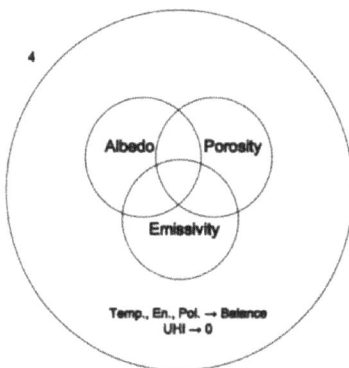

Figure 5. Integration of characteristics of material, creation of balance and reduction of UHI intensity

The key components of the study can be divided into two categories, first meteorological and urban structure factors which with their interactions form UHI over the urban areas; second vegetation covers and high albedo materials which contribute to mitigate the produced UHI. In addition, since focusing on other factors which classified in urban structure factors such as location, population, city size, density of built-up area and the like require long term planning, an optimal and realistic solution is to focus on thermal properties of fabric and surface waterproofing, which can be manipulated and achieved the good results quickly for mitigating UHI effects.

Compiling the four key components into a specific model is meaningful in promoting the passive climate control brought by vegetation covers and high albedo materials in an urban area. The interactions among the four key components and how variables can contribute to mitigate the UHI effects are presented in the model shown in Figure 6. The constituent parts of the model are the impacts of vegetation covers and high albedo materials over meteorological factors. Components with the solid circles indicate relatively stable conditions, while the dashed circles imply their potential variation in an urban area which by changing the amount of them can adjust the urban temperature and mitigate UHI effects. VU is the amount of vegetation covers introduced into an urban area. This can be enforced when more greenery is introduced into the urban area, such as vertical and horizontal green spaces, roof gardens and the like. VM is the ability of greenery to control meteorological factors. HAU is the amount of reflectively, emissivity and porosity of materials introduced into an urban area. This can be enforced when more high albedo materials is introduced into the urban area. HAM is the ability of high albedo materials to control meteorological factors.

In Figure 7, the shaded area represents the UHI intensity which created by interaction between meteorological and urban structure factors. A greater interaction leads to higher UHI intensity that encounters an urban area with imbalance condition.

For achieving balance condition and mitigating the UHI intensity, two variables, vegetation covers and high albedo materials contribute to approach the lower UHI intensity (Figure 8) and achieve ideal condition (UHI=0) (Figure 9).

Based on the model, three hypotheses can be generated:

$$VU = \frac{1}{8}U$$

$$HAU = \frac{1}{8}U$$

$$VU + HAU = \frac{1}{4}U \Rightarrow UHI \downarrow$$

Hypothesis1: if the amount of vegetation cover and high albedo material all together cover approximately a fourth of urban area, the effect of UHI can be reduced (Figure 7):

Hypothesis 2: if the amount of vegetation cover and high albedo material all together cover approximately a third of urban area, the effect of UHI can be extremely reduced (Figure 8):

$$VU \approx \frac{1}{6}U$$

$$HAU = \frac{1}{6}U$$

$$VU + HAU = \frac{1}{3}U \Rightarrow UHI \approx Min$$

Hypothesis 3: if the amount of vegetation cover and high albedo material increase and cover half of urban area, the effect of UHI can be achieved zero which is the ideal condition (Figure 9):

$$VU \approx \frac{1}{4}U$$

$$HAU \approx \frac{1}{4}U$$

$$VU + HAU = \frac{1}{2}U \Rightarrow UHI = 0$$

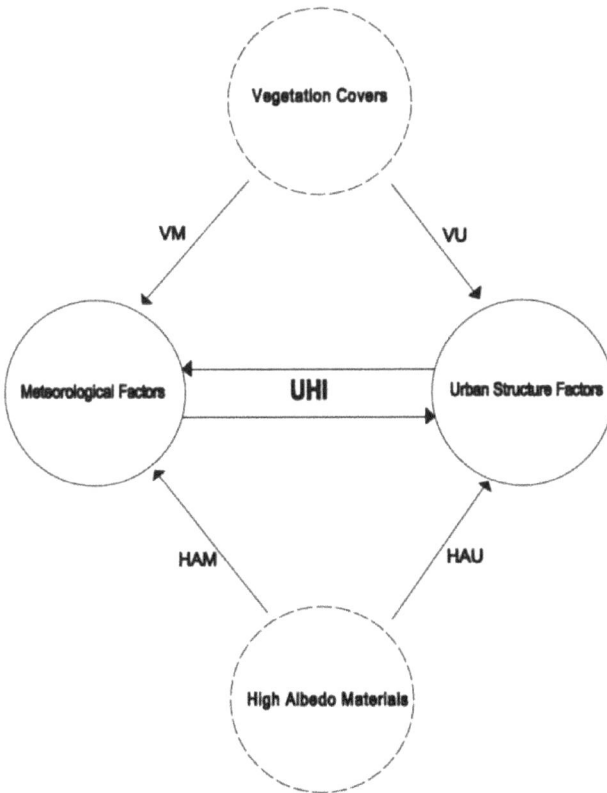

Figure 6. Conceptual model, vegetation covers and high albedo materials are considered to be the major components of UHI mitigation

$$VU \downarrow + HAU \downarrow = VM \downarrow + HAM \downarrow = UHI \uparrow$$

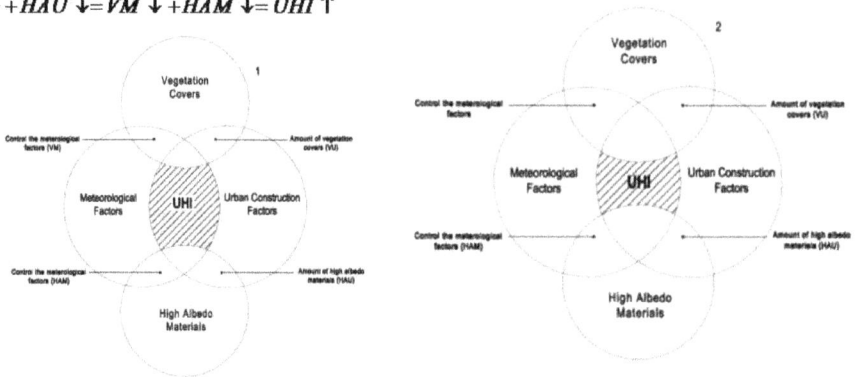

Figure 7. Increasing the amount of greenery and high albedo materials mitigate the UHI effects

$$VU \uparrow + HAU \uparrow = VM \uparrow + HAM \uparrow = UHI$$

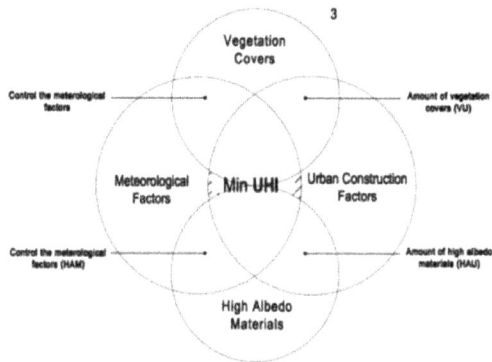

Figure 8. Approach to lower UHI intensity with increasing greenery and albedo of materials

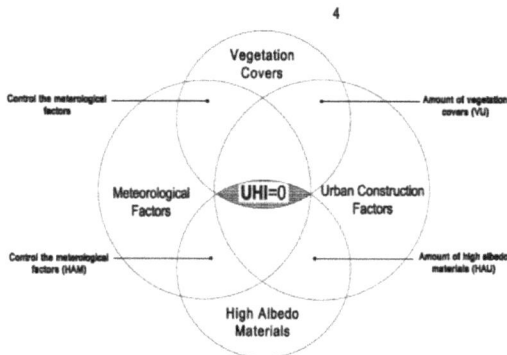

Figure 9. Achieve ideal condition by covering urban areas with approximately 50% greenery and high albedo materials

5. Methodology

In order to test the model, UHI measuring, modeling and simulation have been carried out which described in following way:

5.1. Methodologies used for Urban Heat Island measurements

The methodologies employed for measuring UHI are:

1. Satellite images: broad and visible instantaneous observed;
5. Historical weather data: long-term observation; and
6. Mobile survey: observation of given area within a designated period.

5.1.1. Urban Heat Island measurement through satellite image

A Landsat ETM7+ satellite image obtained on 18 July 2000 was selected (Figure 10). Satellite image with a thermal band was processed to obtain an instantaneous impression of the UHI. In order to map out the UHI, mapping of land surface temperature (LST) and normalized difference vegetation index (NDVI) were necessary. It aimed to overlay two images (NDVI and LST images) and extract maximum temperature value for both urban and rural area as well as identify the possible hot spots in the metropolitan area. Figure 11 shows the process of UHI mapping.

Figure 10. Landsat-7 ETM+ image of Tehran acquired on 18 July 2000 (band combination RGB 7 5 3)

5.1.2. Urban Heat Island measurement through historical weather data

In order to measure UHI intensity during a 25 years period, this study has chosen two stations (Mehrabad station in urban area and Karaj station in rural area). The stations selected to be used from weather station network sources, which under the governmental organization named as Iran Meteorological Organization.

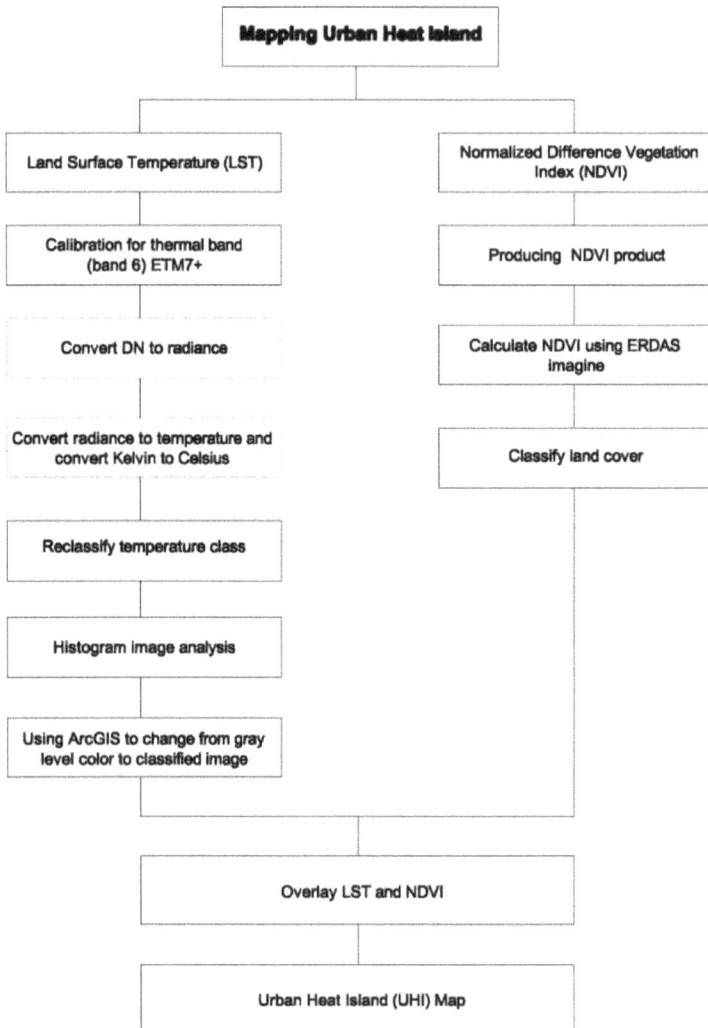

Figure 11. The process of UHI mapping

5.1.3. Urban Heat Island measurement through mobile survey

This survey entailed travelling on a predetermined path throughout a district, stopping at representative locations to take reading using just a single set of weather instrumentation. Using professional measuring instrument: Anemometer, Hygrometer, Thermometer and Light meter called Lutron LM-8000 (4 IN 1).

Method of transport taken in this measurement is to cycle between measurement locations. Since the measurement must be taken in specified period, using car or public transportation was not logical because of traffic jam.

5.2. Methodology used for modeling

This research used modeling based on GIS analysis, which is divided into two analyses (Figure 12); 1. 3D analysis; and 2. Spatial analysis in order to have the classification and the area of vegetation cover, albedo material and both together in current situation of 6 urban district of Tehran.

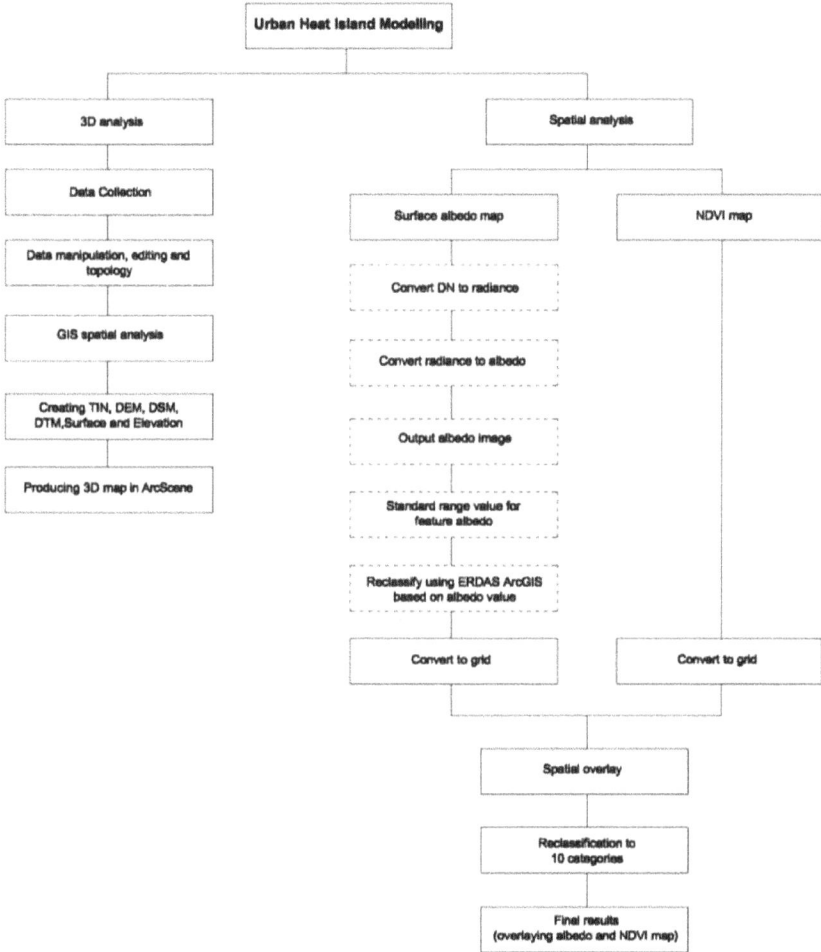

Figure 12. The process of UHI modeling based on GIS

5.3. Methodology used for simulation

This chapter used ENVI-met, three dimensional non-hydrostatic microclimate model, for simulating 'natural ventilator of the city' model with three scenarios. The Figure 13 shows the process of simulation with ENVI-met.

```
┌─────────────────────────────────────────────────┐
│     Creating model area: using ENVI-metEddi4      │
└─────────────────────────────────────────────────┘
                        │
                        ▼
┌─────────────────────────────────────────────────┐
│  Giving input data to model: using ENVI-metcedit  │
└─────────────────────────────────────────────────┘
                        │
                        ▼
┌─────────────────────────────────────────────────┐
│   Run the model: using ENVI-met40-270*200*36      │
└─────────────────────────────────────────────────┘
                        │
                        ▼
┌─────────────────────────────────────────────────┐
│     See the output data: using leonardo40         │
└─────────────────────────────────────────────────┘
```

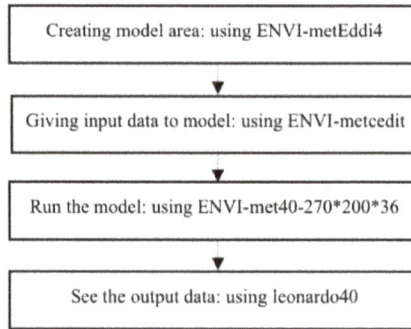

Figure 13. The process of simulation with ENVI-met

6. Model area: 6 urban district of Tehran

This paper put the model and its hypotheses in the context of 6 urban district of Tehran (Figure 14) for following reason:

- with high density of built-up area and low albedo and non-reflective materials, higher production of anthropogenic heat due to the transportation, cooling and heating system and cooking plays an important role on formation of UHI;
- Located near the centre of gravity of Tehran;
- Located on main axes of the city (Enghelab and Vali-e-Asr streets) (Figure 15);
- Surrounded the district by main urban axes (highways)(Figure 15);
- Concentration of superior activities and urban central functions; and
- Concentration of pollutions over central part of Tehran brought by west prevailing wind and increase inversion in Tehran.

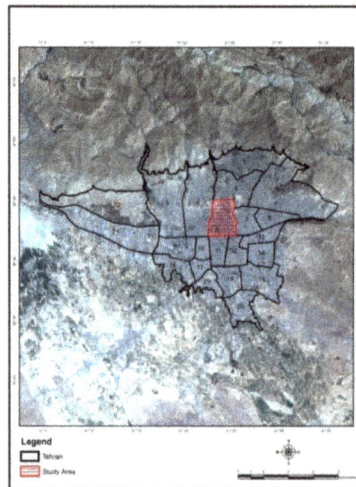

Figure 14. The location of 6 urban district of Tehran (in the central part of Tehran)

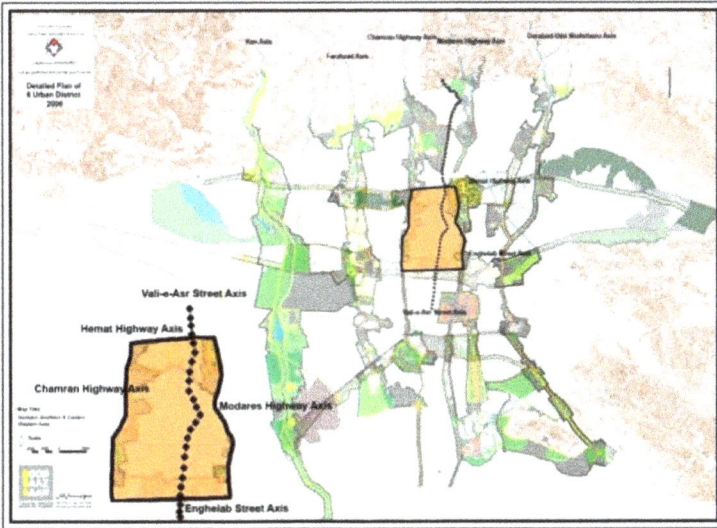

Figure 15. Main northern-southern and eastern-western axes surrounded 6 urban district

The area of the simulation has been shown in Figure 16 where has the higher intensity of UHI. The model area has a size of 230*234 m, resulting in 94*92*25 cells with a resolution of 2*2*2 meters. Within the area only residential buildings with average height of 16 m are located. The geographic coordinates of the model area were set to 35.73° latitude and 51.50° longitude.

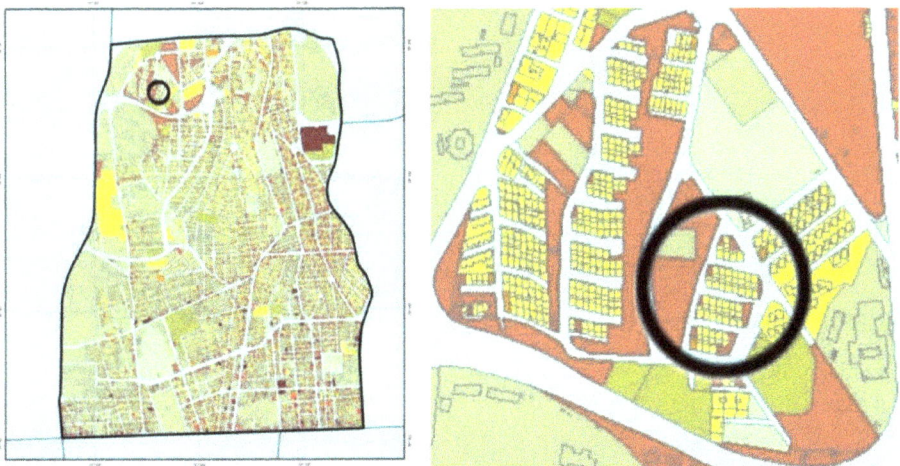

Figure 16. The certain area of simulation in 6 urban district of Tehran with higher intensity of UHI

7. Discussion of results

7.1. Urban heat island measurement

7.1.1. Satellite image

The clear observation is the surface temperature of Tehran in 18 July 2000 (Figure 17). The warm region where is represented by red and yellow colour, is mostly located in the central, western and southern part of Tehran where CBD, industrial area and airport are located respectively. On the other hand, northern part of Tehran is relatively cool with green colour. This is due to the concentration of greenery and water bodies as well as less impact from the densely placed urban developments. The contrast between urban and rural areas hints at the prevalence of the UHI effect in Tehran, although the satellite image only provides the instantaneous observation during the daytime.

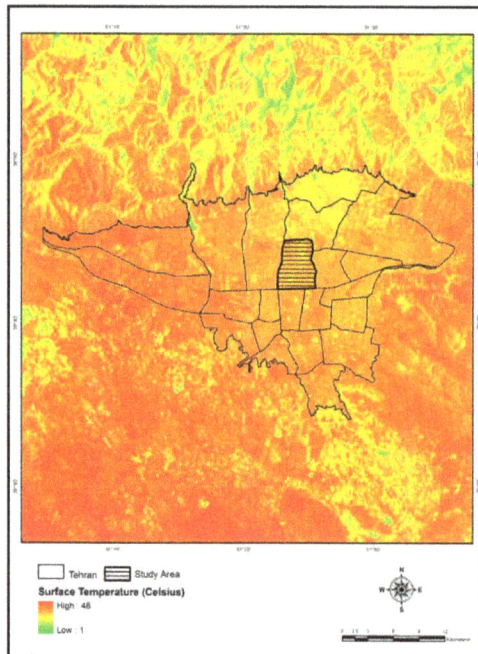

Figure 17. Figure 17. Tehran surface temperature map

The UHI intensity of Tehran is:

Urban max = 39 °C
Rural max = 27°C
UHI = (39-27) °C = 12°C

Therefore, daytime Tehran surface UHI shows 12°C of differences between urban and rural areas.

The UHI intensity of 6 urban district is:

Urban max = 40 °C
Rural max = 27°C
UHI = (40-27) °C = 13°C

Daytime UHI shows a 13°C of difference between urban and rural area. In fact, UHI intensity of 6 urban district is 1°C higher than Tehran.

Study area (6 urban district) image reveals some spots with either high or low surface temperature. As shown in Figure 18, Region 1 represents some of governmental organization, such as energy organization and some commercial and residential land uses. They experienced the highest temperature during daytime especially in the north and west parts of the region mainly because of lack of extensive landscape and being close to the two main highways (Hemat in north and Chamran in west) as well. Similarly, higher temperature was observed in eastern north and east parts of the Region 2. This is also reasonable since the exposed runway absorbs a lot of incident solar radiation during the daytime and incurs high surface temperature. It is due to the bus terminal station located in the Abassabad lands as well. Region 3 represents the most crowded area with higher traffic congestion, which the majority of commercial land uses are located in this region. Region 5 and 6 are close to the CBD of Tehran and neighbour with the Enghelab street, the most crowded street, but the existing of one of the biggest parks of Tehran, Laleh park, in Region 5 was able to reduce the higher temperature partly. Region 4 also represents some of the commercial and residential land uses. Furthermore, as shown in Figure 18, the areas around the main axis of the district which separates the regions of 1, 3 and 5 from regions of 2, 4 and 6, have lower urban temperature which it can be mainly due to the trees axis in the Vali-e-Asr street. This axis is not the worst scenarios, however, some red spots can be seen in this area. Therefore, the worst scenarios have been occurred in eastern north, western north and also some areas in west of the district. These all can be due to the lack of vegetation covers and low albedo materials and higher density of population and production of anthropogenic heat.

7.1.2. Historical weather data

The investigation of a 25 years period of urban (Mehrabad station) and rural (Karaj station) temperature in Tehran makes clear the temperature difference between these two areas. The selection of these two stations is because of the long records and validity. Mehrabad station is located within the west region of Tehran which airport situated there. In Karaj station, the major land use is agriculture. At Mehrabad and Karaj stations, the annual maximum, mean and minimum dry-bulb temperatures indicate a slow upward trend towards warming or cooling during the period 1985-2009 (Figure 19 and 20).

An exploration of the mean temperature trends of Mehrabad (urban) and Karaj (rural) as well as temperature differences between the two locations was made. The results show that (Figure 21) for the first five years temperature difference increased from 1.9°C to 3.6°C. This was an increase of 1.7°C for the period. It dropped by 1.7°C in 1991. Over the next 12 years there were fluctuations. It started increasing from 1.7°C in 2003 to 3.2°C in 2004 and it reached to peak on 2005 before the temperature difference decreased by about 3.6°C to around 2.6°C in 2009. The highest temperature differences is 3.6°C occurred in 1989, 1990 and 2005, while the lowest one with 1.5°C occurred in 1993, 1995 and 1999. It means that from 2003 forward there is higher intensity of UHI.

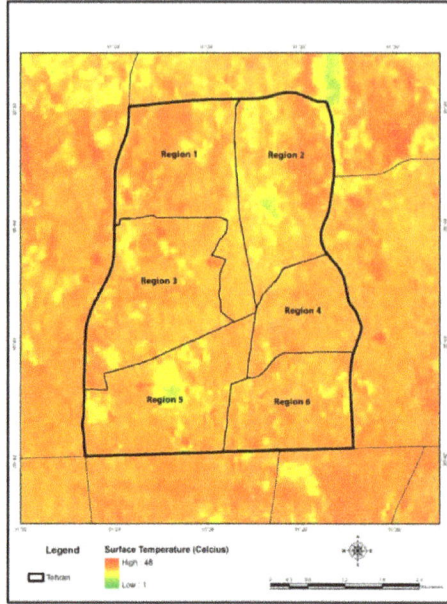

Figure 18. 6 urban district's surface temperature map

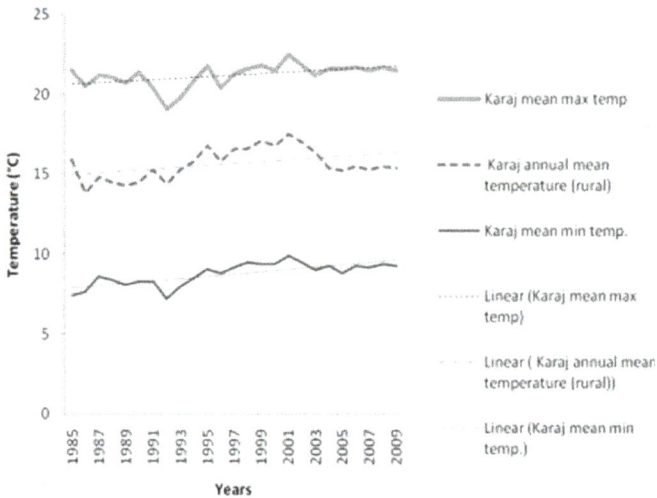

Figure 19. Analysis of the past 25 years' weather data at Mehrabad station

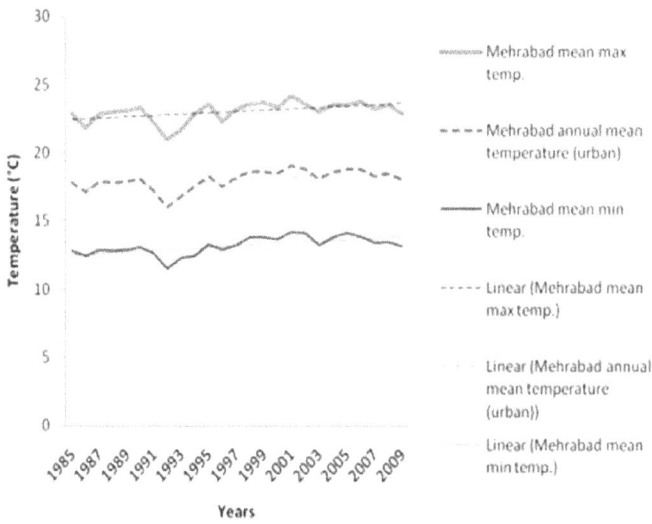

Figure 20. Analysis of the past 25 years' weather data at Karaj station

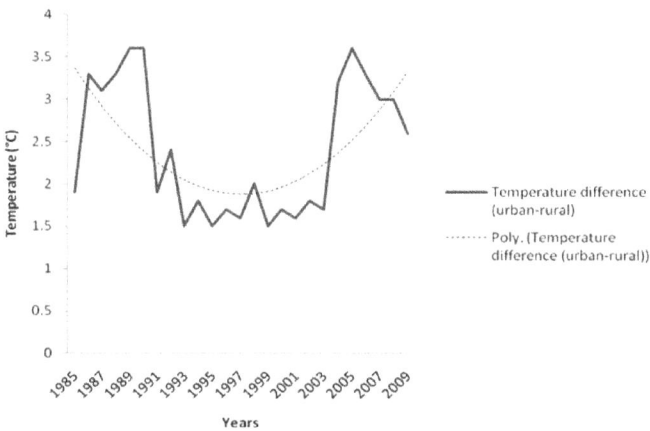

Figure 21. Mehrabad-Karaj mean temperature differences

7.1.3. Mobile survey

Field measurements are used for measuring the air temperature of current situation of the area. In this way, it was chosen 31 points in three parts of the district. In fact, three ways were traversed, two narrow strips around (from point number 1 to 11 and 12 to 21) and the central part of the district (from point number 22 to 31), to cover whole areas of the district (Figure 22). Since there were only two hours with higher radiation intensity to measure points, this study has selected three consecutive days to measure them exactly in these two hours. It investigated the correlation between temperatures and different land uses in current condition of 6 urban district.

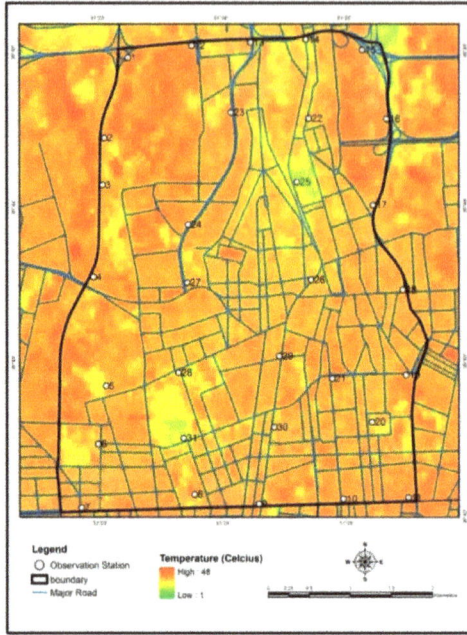

Figure 22. Selection of 31 points in 6 urban district with 1 Km distance for mobile survey

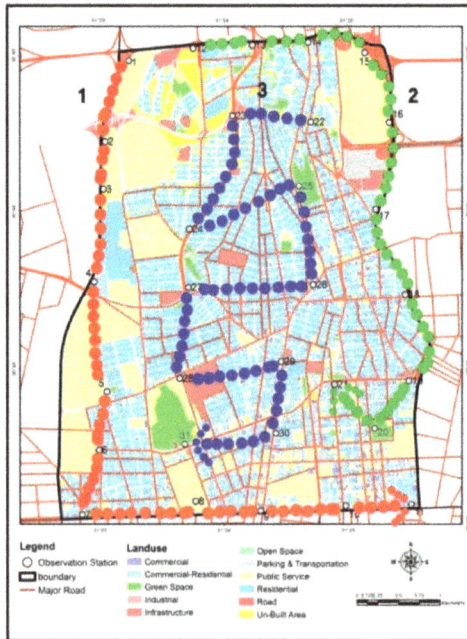

Figure 23. Three routes selected for measuring air temperature in 6 urban district

First route running from north to south and then to east (1 to 11), second route from west to east and then to south (12 to 21) and third route which cover central part of the district (22 to 31) running like zigzag movement to cover whole area of the central part passed through quite a number of different land uses (Figure 23). In order to save time and measure the defined points in defined time, bicycle was selected. According to high traffic jam in Tehran reaching to all points in the exact time by car was impossible. Therefore using vehicle equipped with observation tube which can automatically record ambient temperature was difficult.

The maximum air temperature, 42 °C, was observed in the route number 1 in industrial, commercial and public services. The lowest temperature, 38.5 °C, was observed in residential area (Figure 24). In route number 2, the maximum temperature was also 42°C in industrial and public services and the lowest one was 34°C in park (Figure 25). In route number 3, the highest temperature was 40.5 °C in industrial area and the lowest one was 30 °C in park (Figure 26).

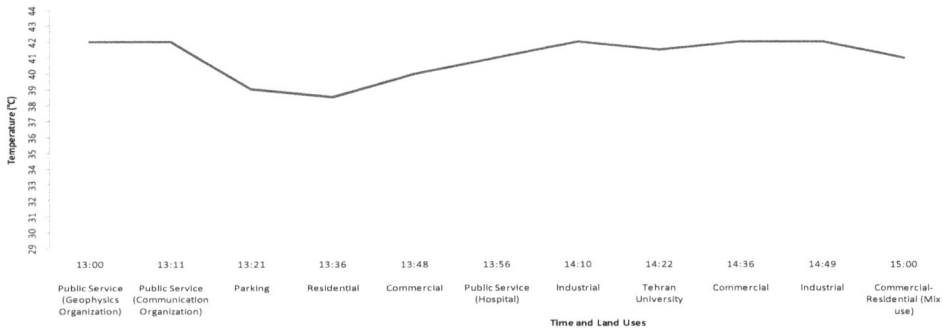

Figure 24. Maximum air temperature in different land uses in route number 1

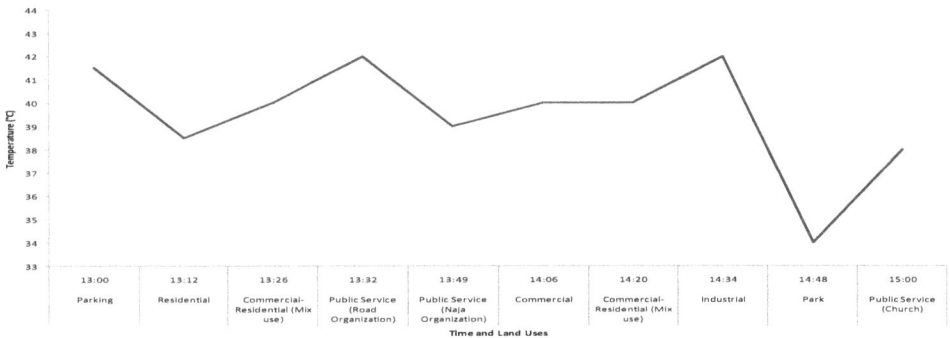

Figure 25. Maximum air temperature in different land uses in route number 2

As seen routes number 1 and 2 have the same highest temperature in industrial area and public services but the route number 3 has 1.5 °C less than the other routes. It is due to the location of routes number 1 and 2 and they are located next to the main highways (Hemat, Chamran, Modares and Enghelab highways) which have the highest traffic jam, air pollution, production of anthropogenic heat and last but not least the lack of vegetation covers which can ventilate air and using low albedo and non-reflective materials. The industrial areas generally have low-rise buildings and the high temperature recorded in these areas is related to the extensive usage of metal roofing in the buildings. The high temperature of commercial and public services buildings is related to use concrete and dark stones (Figure 27) that absorb a huge part of the solar radiation incident on it and later release it to the atmosphere. In addition, calculation of averaging the temperature in every route shows the highest mean temperature in route number 1 with 41 °C, route number 2 with 39.6 °C and route number 3 with 38.25°C respectively (Table 1). From the results it is observed that the daytime temperature seemed to be dominated more by the solar radiation component rather than by the reradiated temperature, which is the main cause of daytime UHI. The average of observations obtained during daytime in three days in 2009 shows that the temperature is 2.2 °C higher than the average of temperature derived from satellite image in 2000. It means that there were more constructions in these 9 years and made the condition much more worse.

Routes	Mean max temperature (°C)	Mean min temperature (°C)
1	41	30.06
2	39.60	29.05
3	38.25	28.90

Table 1. Mean min and max temperatures in three different routes

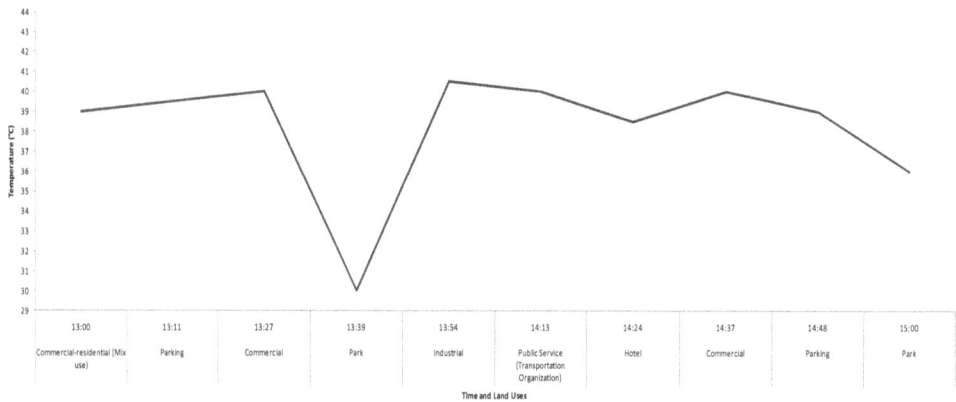

Figure 26. Maximum air temperature in different land uses in route number 3

In urban areas, the night time temperatures varied between 25°C and 35.5°C and it was found that the CBD area was around 7 °C hotter than the locations with greenery (Figure 28, 29 and 30). This also indicates the center of the night time UHI effect which has shifted from the industrial areas during the daytime to the CBD area. The average temperature in every route in night time also shows the highest mean temperature in route number 1 with 30.06 °C, while route number 2 has 29.05°C and route number 3 with 28.90 °C (Table 1). It means that at night time also route number 1 has the highest temperature. Comparing these results with satellite image results shows that west, north-west and east parts of the district in satellite image (2000) have higher intensity of UHI, while mobile survey (2009) shows the condition much worse.

Figure 31 shows the correlation between GIS and mobile survey. P-value, which shows 0.05, is significant difference between GIS and mobile survey. The Pearson correlation which is 0.67 shows that the difference is reliable. This means the reading between these two sources is complement each other.

Figure 27. Using dark stone, concrete and metal materials

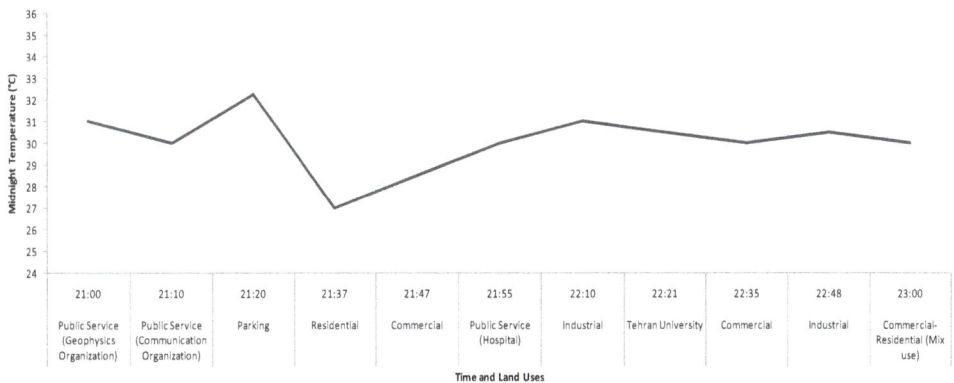

Figure 28. Minimum air temperature in different land uses in route number 1

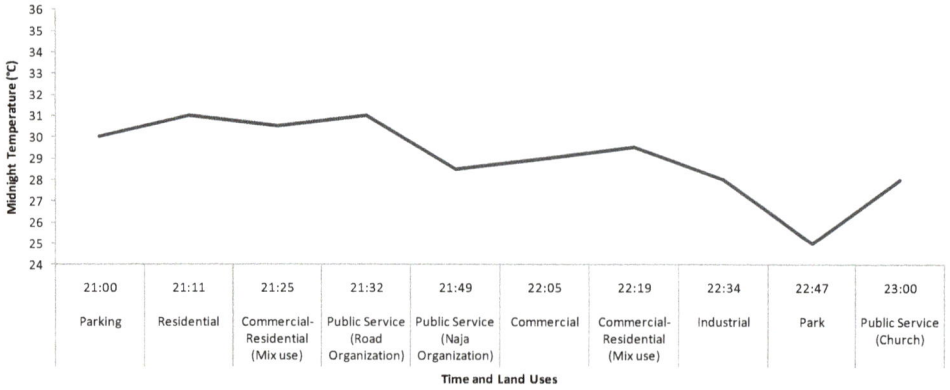

Figure 29. Minimum air temperature in different land uses in route number 2

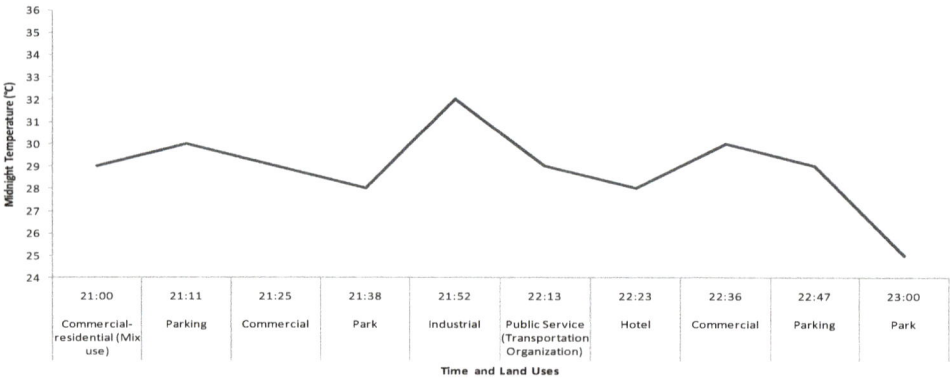

Figure 30. Minimum air temperature in different land uses in route number 3

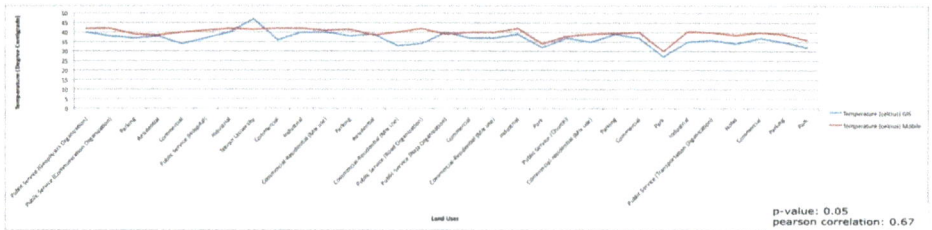

Figure 31. Figure 31. Correlation between GIS and mobile survey

7.2. Modelling based on GIS analysis

The results obtained from modelling based on GIS, 3D and spatial analysis, are described in following way:

7.2.1. 3D analysis

This analysis gives visual views of the district that can help to better understand of the site. The 3D of district (Figure 32) shows that the northern part of the district has some ups and downs that contribute to cause an unequal distribution of pollution and provide warm air canopy over this area. As shown in Figure 18, the northern part of the district has the hottest surface temperature that is due to the concentration of the pollutions and cause to form the UHI.

7.2.2. Spatial analysis

To prove the hypotheses spatial analysis based on ArcGIS has been done in order to estimate the area of albedo (lower, medium and higher), vegetation covers and then overlay them. Results obtained from creating albedo (Figure 33) and NDVI grid maps (Figure 34) in vector format.

1. Albedo Grid Map

As shown in Table 2, albedo has been classified into three groups including (Figure 33):

1. Higher albedo with value of 0.170-0.310 occupied 37% of whole area of 6 urban district. According to the Table 3, this range of albedo belongs to concrete (0.10-0.35);
7. Medium albedo with value of 0.140-0.160 occupied 50% of whole area of 6 urban district. According to Table 3, this range of value belongs to asphalt (0.05-0.2) or corrugated iron (0.10-0.16); and
8. Lower albedo with value of 0.064-0.130 occupied 13% of whole area of 6 urban district. According to Table 3, this range of value belongs to gravel (0.08-0.18), smooth-surface asphalt (0.07) or black coloured materials.

2. NDVI Grid Map

As shown in Table 4, the land cover types have been divided into 4 categories including (Figure 34):

1. Vegetation with value of 0.0-0.7 from very poor to very high density. This type covers 51.57 hectare, 2.4% of whole area of 6 urban district;
2. Non-vegetation with value of -0.0- -0.4 including urban area, desert, mountain area and cloud. This type covers 2087.52 hectare, 97.37% of whole area of 6 urban district;
3. Water with value of -0.4- -0.7 constituted 1.7 hectare, 0.08% of whole area; and
4. There are also some land covers that their types were not recognizable which cover 3.21 hectare, 0.15% of whole area of 6 urban district.

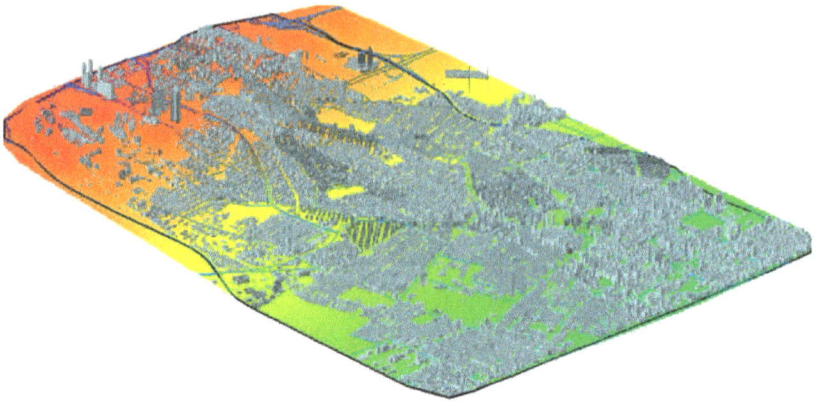

Figure 32. 3D of 6 urban district

Albedo Classification	Albedo Value	Area (Hectare)	Percent of 6 Urban District Area
Higher albedo	0.170-0.310	792	37%
Medium albedo	0.140-0.160	1071	50%
Lower albedo	0.064-0.130	280	13%
		2144	100%

Table 2. Albedo classification with related values and area in 6 urban district

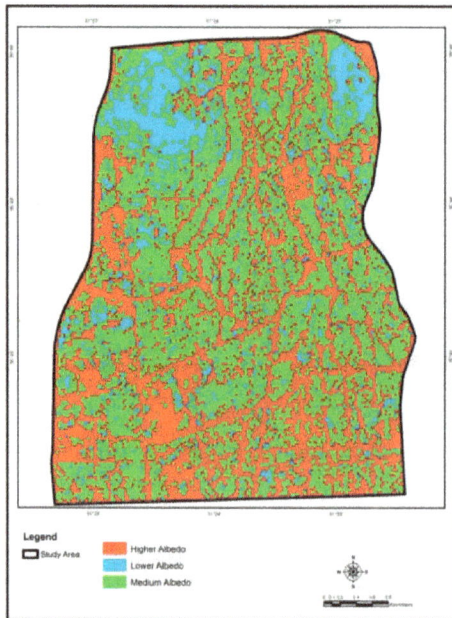

Figure 33. Albedo grid map of 6 urban district

Surface	Albedo
Streets	
Asphalt (fresh 0.05, aged 0.2)	0.05-0.2
Walls	
Concrete	0.10-0.35
Brick/Stone	0.20-0.40
Whitewashed stone	0.80
White marble chips	0.55
Light-coloured brick	0.30-0.50
Red brick	0.20-0.30
Dark brick and slate	0.20
Limestone	0.30-0.45
Roofs	
Smooth-surface asphalt (weathered)	0.07
Asphalt	0.10-0.15
Tar and gravel	0.08-0.18
Tile	0.10-0.35
Slate	0.10
Thatch	0.15-0.20
Corrugated iron	0.10-0.16
Highly reflective roof after weathering	0.6-0.7
Paints	
White, whitewash	0.50-0.90
Red, brown, green	0.20-0.35
Black	0.20-0.15
Urban areas	
Range	0.10-0.27
Average	0.15
Other	
Light-coloured sand	0.40-0.60
Dry grass	0.30
Average soil	0.30
Dry sand	0.20-0.30
Deciduous plants	0.20-0.30
Deciduous forests	0.15-0.20
Cultivated soil	0.20
Wet sand	0.10-0.20
Coniferous forests	0.10-0.15
Wood (oak)	0.10
Dark cultivated soils	0.07-0.10
Artificial turf	0.50-0.10
Grass and leaf mulch	0.05

Table 3. Albedo of typical urban materials and areas [22,3]

NDVI value	Vegetation density	Land cover type	Area (Hectare)	Percent of 6 Urban District Area
0	Unknown	Unknown	3.21	0.15%
-0.4 - -0.7	Non-Vegetation	Water	1.7	0.08%
-0.0 - -0.4	Non-Vegetation	Urban area, desert, mountain area and cloud	2087.52	97.37%
0.0 - 0.1	Very Poor	Vegetation	51.57	2.4%
0.0 - 0.2	Poor			
0.2 - 0.3	Moderate			
0.3 - 0.5	High			
0.5 - 0.7	Very High			
			2144	100%

Table 4. Land cover types with related value and areas in 6 urban district

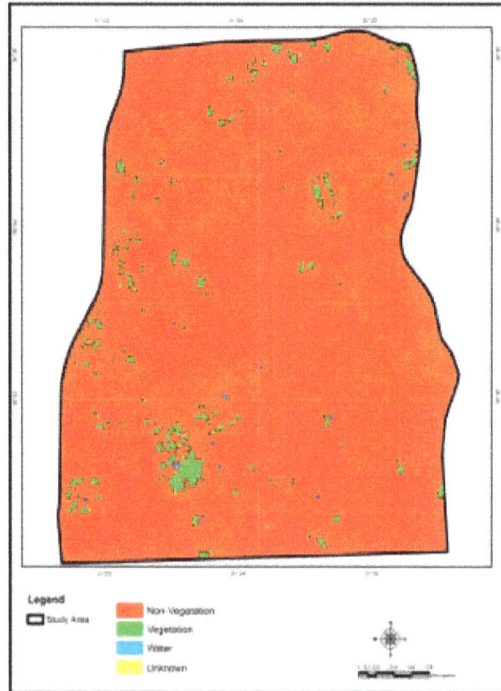

Figure 34. NDVI grid map of 6 urban district

3. Overlaying the albedo and NDVI grid maps

Results obtained from overlaying the albedo and NDVI grid maps have been shown in Figure 35 and Table 5.

No.	Albedo Classification	Land Cover Type	Area (Hectare)
1	Higher Albedo	Unknown	3.05
2	Higher Albedo	Non-Vegetation	739.16
3	Higher Albedo	Vegetation	48.80
4	Higher Albedo	Water	1.06
5	Medium Albedo	Unknown	0.16
6	Medium Albedo	Non-Vegetation	1068.08
7	Medium Albedo	Vegetation	2.76
8	Medium Albedo	Water	0.48
9	Lower Albedo	Non-Vegetation	280.28
10	Lower Albedo	Water	0.16
			2144

Table 5. Results of overlaying albedo and NDVI grid map of 6 urban district

Figure 35. Overlaying albedo and NDVI map of 6 urban district

Each number shows the combination of albedo and land cover types. It can be divided into following groups:

1. Higher albedo with non-vegetation and vegetation covers;
2. Medium albedo with non-vegetation and vegetation covers; and
3. Lower albedo with non-vegetation cover (as seen in Table 5 the combination of lower albedo and vegetation cover does not exist).

Number 2 (Orange colour) shows the combination of higher albedo and non-vegetation with 739.16 hectare area, number 6 (yellow colour) shows the combination of medium albedo and non-vegetation with 1068.08 hectare area, number 9 (green colour) shows the combination of lower albedo and non-vegetation with 280.28 hectare area, which occupied 34.5%, 50% and 13% of 6 urban district respectively. Number 3 (red colour) shows the combination of higher albedo and vegetation with 48.80 hectare area, number 7 (brown colour) shows the combination of medium albedo and vegetation with 2.76 hectare area which occupied 2.27% and 0.12% of 6 urban district respectively. Other numbers are negligible which are not observed in Figure 35. Although number 3 includes high albedo and vegetation cover, it encompass the very low percentage of area which not only it has not impact on mitigation of UHI, but also this value of albedo (with value of 0.17-0.310) with lower reflectivity can increase the UHI intensity. The area of vegetation cover is negligible in comparison with whole area of the district. Number 6 with 50% of whole area of 6 urban district including non-vegetation cover and medium albedo materials (with value of 0.14-0.160) has been widely distributed in the district and provided worse condition for this district. After number 6, number 2 is in the worse condition with 34.5% of whole area of 6 urban district including non-vegetation cover and high albedo materials (with value of 0.17-0.310). Then number 9 with 13% of whole area of 6 urban district and the combination of lower albedo and non-vegetation cover stands in the next rank. It has been widely distributed in region 1 and 2 that the topography of these regions also provided higher UHI intensity. It is also observed in Figure 16 that these regions have higher UHI impacts.

7.3. Analyze the model of natural ventilator of the city

The area of 6 urban district is 2144.33 hectare with population of 232583. According to Table 4, vegetation covers constitute 51.57 hectare of 6 urban district area. Therefore, per capita of vegetation cover in this district is $2.21 m^2$.

Population of 6 urban district = 232583
Vegetation cover area = 51.57 hectare = 515700 m^2
Therefore, per capita of vegetation cover = 2.21 m^2

It means that only 2.4% of whole area of 6 urban district is composed of vegetation cover with 2.21 m^2 per capita.

Based on studies and investigations of United Nation (UN) Environment, acceptable per capita of green spaces in cities is of between 20 and 25 m^2 for each person [23].

In fact, 6 urban district with 2.21 m^2 per capita of vegetation cover is around 18 m^2 less than indicator of UN that makes the situation worse and increases UHI intensity in given area.

According to hypotheses, if a fourth area of 6 urban district is covered with vegetation covers and high albedo materials, therefore:

Area of 6 urban district = 21443300 m^2

21443300 ÷ 4 = 5360825 m^2

$$5360825 \div 2 = 2680412.5 \text{ m}^2 \longrightarrow \text{Vegetation covers}$$
$$2680412.5 \text{ m}^2 \longrightarrow \text{Higher albedo materials}$$

Therefore, per capita of vegetation cover is:

$2680412.5 \div 232583 = 11.5 \text{ m}^2$

In comparison with the UN indicator (20-25 m²), it is 8.5 m² less.

Therefore, this hypothesis is not applicable

If a third area of 6 urban district is covered with vegetation covers and high albedo materials, therefore:

Area of 6 urban district = 21443300 m²

$21443300 \div 3 = 7147766.67 \text{ m}^2$

$$7147766.67 \div 2 = 3573883.33 \text{ m}^2 \longrightarrow \text{Vegetation covers}$$
$$3573883.33 \text{m}^2 \longrightarrow \text{Higher albedo materials}$$

Therefore, per capita of vegetation cover is:

$3573883.33 \div 232583 = 15.3 \text{ m}^2$

In comparison with the UN indicator (20-25 m²), it is still 5 m² less.

Therefore, this hypothesis also is not applicable.

If the amount of vegetation cover and high albedo material increase and cover half of urban area, the effect of UHI can be achieved zero which is the ideal condition, therefore:

Area of 6 urban district = 21443300 m²

$21443300 \div 2 = 10721650 \text{ m}^2$

$$10721650 \div 2 = 5360825 \text{m}^2 \longrightarrow \text{Vegetation covers}$$
$$5360825 \text{ m}^2 \longrightarrow \text{Higher albedo materials}$$

Therefore, per capita of vegetation cover is:

$5360825 \div 232583 = 23 \text{ m}^2$

In comparison with the UN indicator (20-25 m²), it is acceptable.

Therefore, this hypothesis is applicable.

In addition, as Reagan and Acklam [24] calculated, when the reflectivity of the rest of area (5360825 m²) with poorly insulated building is increased from 0.25 to 0.65, the heat gains through the roof are reduced by half. It means that the albedo values mentioned in Table 2 is higher albedo in current classification of 6 urban district and it does not have higher reflectivity.

Figure 36 shows the areas with higher intensity of UHI chosen to implement natural ventilator model as shown in Figure 18. These areas cover approximately half area of 6 urban district with vegetation cover along with high albedo material and they act as ventilation holes and mitigate UHI effects.

Figure 36. Covering the half area of 6 urban district with vegetation cover along with high albedo material

7.4. Simulation through ENVI-met

ENVI-met was employed to simulate "natural ventilator of the city" (NVC) model. This simulation compares the current situation of 6 urban district of Tehran with three scenarios according to the variable of the NVC model. These three scenarios were created as follows:

- Scenario 1: change current low albedo material to high albedo materials;
- Scenario 2: cover the model area with vegetation cover; and
- Scenario 3: cover the model area with vegetation cover along with high albedo material.

The boundary condition was set according to the current situation of model area based on weather data obtained from the mobile survey. One typical time scenario, 1200 hr, was selected for analysis.

Figure 37 illustrates the current situation of the model area and three scenarios in ENVI-met. The material used for buildings, in current situation of 6 urban district, is concrete with albedo of 0.30, for roofs and roads is asphalt with albedo of 0.14-0.16 and some parts of the area have covered by loamy soil with albedo of 0.17-0.23. There is the lack of vegetation cover in this area. In scenario 1, low albedo materials were changed to high albedo one, asphalt to bright asphalt with albedo of 0.55, concrete was covered with white coating with albedo of 0.85, and loamy soil to light colored soil with albedo of 0.6. In scenario 2, horizontal and vertical surfaces were covered by vegetation cover. In fact, these two scenarios show that how vegetation and high albedo material can contribute to UHI mitigation separately. In scenario 3, the model area was covered with high albedo material along with vegetation cover in order to see that how these two variables can contribute together to mitigate the effect of UHI.

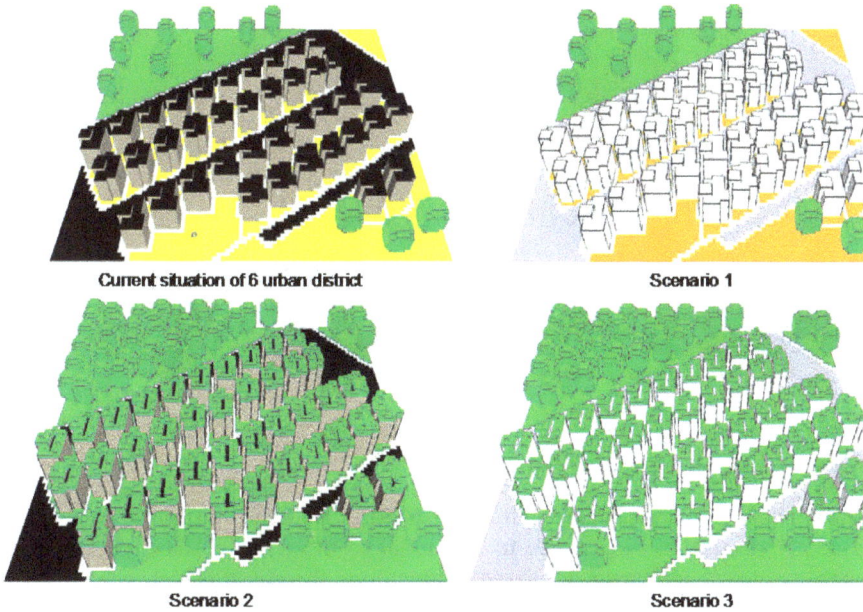

Current situation of 6 urban district

Scenario 1

Scenario 2

Scenario 3

Figure 37. Current situation of the model area and three scenarios in ENVI-met

Figure 38 shows the daytime (at 1200 hr) simulation of the current situation of 6 urban district and three scenarios. As seen in this Figure, in current situation of 6 urban district, higher temperature (above 295.80 K) occurs in roads and roof of buildings with low albedo material such as asphalt and the areas with less vegetation cover. The simulation results show that when the points are closer to the green area (east north), lower temperatures (294.60 K) were observed.

In scenario 1, the cooling effect of high albedo materials can be seen in the simulation (Figure 37). In the east north part, it is not seen higher temperature while in current situation there is higher temperature and it decreased to around between 294.20 and 295.20 K and in building area to around less than 294K. It means that high albedo materials have extreme effects on decreasing the ambient temperature.

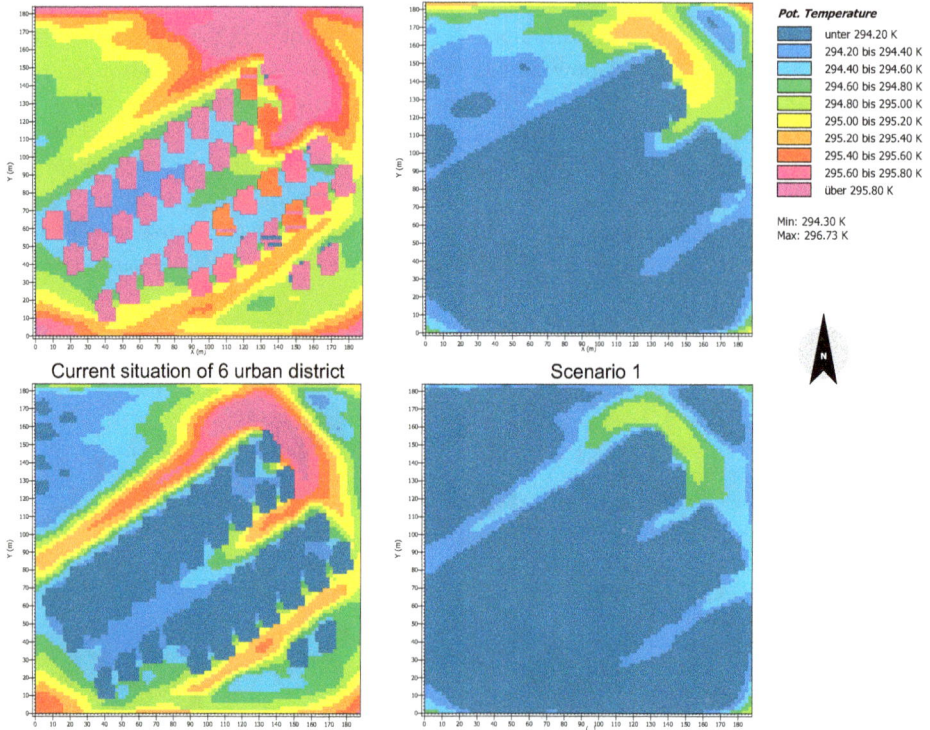

Figure 38. Daytime simulation of 6 urban district and three scenarios in ENVI-met

When the vegetation is added in scenario 2, the temperature in the areas with trees and roof gardens has been reduced from around 295 K to 294.20 K. In fact, the moisture levels in the soil dose not cause the temperature to be similar to those on the hard pavement areas. Although the vegetation cover decreased temperature, there is still higher temperature in roads. Green roofs contributed to decreased the temperature in housing area. While the vegetation is replaced with hard pavement (current situation), it can be seen that the whole area now has a higher temperature at about above 295.80 K. The qualitative analysis of the temperature data showed that the coolest areas were in the Saee and Laleh parks located in route number 3. It means that field measurement has also shown the same results that greenery can decrease temperature. The reduction of the air temperature in the areas with more vegetation cover can reach 0.8 °C. In the comparison of the scenarios' 1 and 2 and current situation for temperature, scenario 1 has more effect on the surrounding built-up area than vegetation cover.

In scenario 3, the combination of vegetation and high albedo material has been examined in order to test that how these two variables can affect the surrounding built-up area in parallel. As seen in Figure 38, in scenario 3, it is obvious that the combination of these two variables can affect to reduce the temperature around 2.43°C. Scenarios' 1 and 2 also contribute to reduce the temperature singly, while in scenario 3 which is the combination of scenario 1 and 2 has extreme contribution to mitigate the air temperature.

Therefore, in the cross-comparison of the three scenarios for temperature, the best cooling effect on the surrounding built-up area is observed in the third scenario and cooling effect of greenery along with high albedo material can be confirmed by the simulation.

Author details

Parisa Shahmohamadi, Ulrich Cubasch and Sahar Sodoudi
Institut für Meteorologie, Freie Universität Berlin, Germany

A.I. Che-Ani
Department of Architecture, Faculty of Engineering and Built Environment,
Universiti Kebangsaan Malaysia, Selangor, Malaysia

8. References

[1] Emmanuel M.R 2005. An Urban Approach to Climate-Sensitive Design; Strategies for the Tropics. London. Spon Press.

[2] Asimakopoulos D.N, Assimakopoulos V.D, Chrisomallidou N, Klitsikas N, Mangold D, Michel P, Santamouris M, Tsangrassoulis A (2001) Energy and Climate in the Urban Built Environment. M. Santamouris (Ed.). London. James & James Publication.

[3] Oke T.R (1987) Boundary Layer Climates (2an edn.). New York. Methuen and Co. Ltd.

[4] Wong N.H, Chen Y (2009) Tropical Urban Heat Islands: Climate, Buildings and Greenery. London and New York. Taylor & Francis Press.

[5] Glantz M.H (2005) What Makes Good Climates Go Bad?. Geotimes 50(4):18-24.

[6] Oke T.R (2006) Initial Guidance to Obtain Representative Meteorological Observations at Urban Sites.Instruments and Observing Methods. Canada. WHO. No. 81.

[7] Oke T.R (2006) Towards Better Scientific Communication in Urban Climate. Theoretical and Applied Climatology 84: 179-190.

[8] Bridgman H, Warner R, Dodson J (1995) Urban Biophysical Environments. Melbourne and New York: Oxford University Press.

[9] Grimmond S (2007) Urbanization and Global Environmental Change: Local Effects of Urban Warming. Cities and Global Environmental Change 83-88.

[10] Oke T.R (1981) Canyon Geometry and the Nocturnal Urban Heat Island: Comparison of Scale Model and Field Observations. Journal of Climatology 1: 237-254.

[11] Santamouris M (2001) Energy and Climate in the Urban Built Environment. London. James & James Publication.

[12] Roth M (2002) Effects of Cities on Local Climates, Proceedings of Workshop of IGES/APN Mega-City Project, 23-25 January 2002, Kitakyushu Japan.

[13] Mayer H (1992) PlannungsfaktorStadtklima. Münchner Forum, Berichte und Protokolle, 107:167–205.

[14] Fezer F(1995) Das KlimaderStädt. Gotha, Justus PerthesVerlag.

[15] Akbari H, Davis S, Dosano S, Huang J,Winnett S (Editors) (1992) Cooling our Communities: a Guidebook on Tree Planting and Light-Colored Surfacing. United States Environmental Protection Agency, Washington, D.C.

[16] Halvorson H, Potts D (1981) Water Requirements of Honeylocust (Gleditsiatriacanthos f. inermis) in the Urban Forest. USDA Forest Service Research Paper. NE-487.

[17] Heilman J, Brittin C, Zajicek J (1989) Water Use by Shrubs as Affected by Energy Exchange with Building Walls. Agricultural and Forest Meteorology 48: 345-357.

[18] Akbari H, Pomerantz M, Taha H (2001) Cool Surfaces and ShadeTrees to Reduce Energy Use and Improve Air Quality in Urban Areas.Solar Energy 70(3): 295–310.

[19] Simpson J.R, McPhersonE.G (1997) The Effect of Roof Albedo Modification on Cooling Loads of Scale Model Residences in Tucson, Arizona. Energy and Buildings 25: 127-137.

[20] Goward S.N, Cruickshanks G.D, Hope A.S (1985) Observed Relation between Thermal Emission and Reflected Spectral Radiance of a Complex Vegetated Landscape. Remote Sensing of Environment 18: 137-146.

[21] Lougeay R, Brazel A, Hubble M (1996) Monitoring Intra-Urban Temperature Patterns and Associated Land Cover in Phoenix; Arizona Using Landsat Thermal Data.Geocarto International 11: 79-89.

[22] Bretz S, Akbari H, Rosenfield A, Taha H (1992) Implementation of Solar Reflective Surface: Materials and Utility Programs. LBL Report 32467. University of California.

[23] United Nation Environment (2010)http://www.unep.org/.

[24] Reagan J.A, Acklam D.M (1979) Solar Reflectivity of Common Building Materials and its Influence on the Roof Heat Gain of Typical Southwestern USA Residences. Energy and Buildings 2: 237-248.

Chapter 11

Managing Emergency Response of Air Pollution by the Expert System

Wang-Kun Chen

Additional information is available at the end of the chapter

1. Introduction

Recently, the emergency preparedness of environmental disaster has been grown because of the climate change and growth of new technology in industry. The need to reduce the risk of major event of air pollution is of great concern. To ensure the quality of response management and reduce the loss in the air pollution event, it is necessary to design a reliable emergency response system. However, the phenomenon of air pollution is very complicated so it is very difficult to consider all possible factors in one system.

A well prepared response management plan should include the prediction and recommendation for the policy makers so as to reduce the possible damage of the disaster. This chapter sets out the method to improve both planning for emergency response of air pollution and recommendations to improve the effectiveness of this system.

The effect of air pollution includes the long-term and short term. Long-term effect of air pollution was controlled by the abatement program of source reduction. However, the short-term episode is more difficult to control because the emergency response is usually very complicated and related to many people in the neighboring area.

The environmental disaster, both from the natural and man-made release, has to be controlled by the well-designed management program. However, the consequence of the disaster was related to so many actions and regulations, therefore it is very difficult to make a quick and correct response measure only by the human. The supplementary system with the aid of computer system become more important in the decision making process.

The decision making system for air pollution management has to consider the appropriate method to avoid the damage from the pollution. Therefore, a complete database includes all the possible reason and consequence results should be included in this system. Beside, the system should be able to deduce the possible consequence and suggest the best choice for the decision makers.

A case study was presents in this chapter. This study uses the experience in Taiwan as an example. Since Taiwan is a very small island with highest population density, the air pollution also causes severe problem for the public. Because the population density is as high as the on the top of the world, the air pollution response management system have also received more attention in environmental management.

The chapter was written in the following structure. The concept of air pollution management was described in the second section. Then the structure of knowledge bank was proposed. The data base and inference system was proposed and written in the following. Finally a conclusion of this system and the suggestion for the future research was followed.

2. Concept of air pollution management system

2.1. Definition of air pollution disaster and risk management pattern

Before going into the detail, we have to know the concept of air pollution management. An air pollution management system for emergency response could be described by figure 1. In this system, there includes a knowledge database, an inference mechanism, and the interface with the users and another resource. Because the system will influence many people and interest groups, so it has to be designed more carefully in order to get the optimum decision. This data base and inference mechanism is just used to ensure the reliability of the effectiveness of this system.

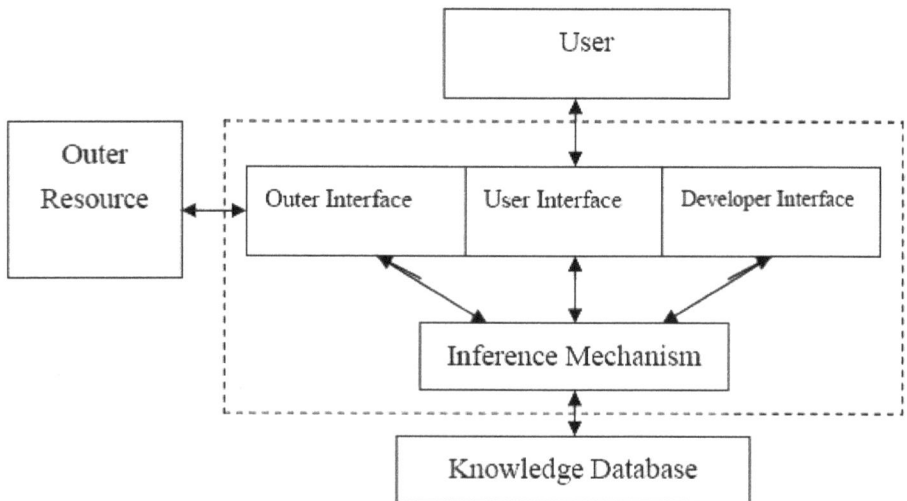

Figure 1. Expert system for air pollution emergency response

The design of this system includes the following steps as: (1) Identify problem characteristics; (2) Find concept to represent knowledge; (3) Design structure to organize knowledge; (4) Formulate rules to embody knowledge; (5) Validate rules that organize knowledge.

Every air pollution event has its characteristics; it can be represents by an appropriate knowledge. Then we design the structure to organize this knowledge, and formulate the mathematical rules to embody the knowledge so that it can be inference in the system. Finally, we have to validate the rules and organize all of this knowledge for the future forecasting.

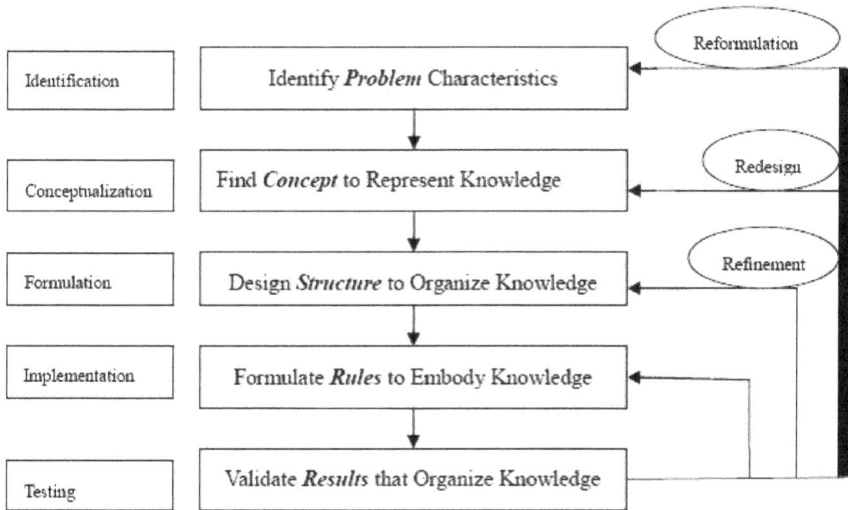

Figure 2. Flow chart for constructing an expert system

2.2. Identify the pattern system of risk management

The air pollution disaster is an accidental phenomenon in the environment; it can be represented by a mathematical pattern. The pattern structure of air pollution management can be categorized as the following four types as shown in figure 3. The first pattern is the characteristic of the air pollution episode itself. The second pattern is the change pattern in the ecosystem. The third pattern is the loss pattern in the economic system. And the final pattern is the response pattern for the disaster management.

Different episode has different characteristics, such as the dust storm, forest fire, explosion, and toxic gas leakage, etc. it will cause various types of damages. These will cause the damage in ecosystems and loss in economic system. Therefore, the change in these two systems has to be the well prepared management systems. A good management system should be able to concern all of these factors together. And suggest an optimum decision for the decision maker. However, it will include many criteria in the thinking, so the expert system has to be applied in the solution.

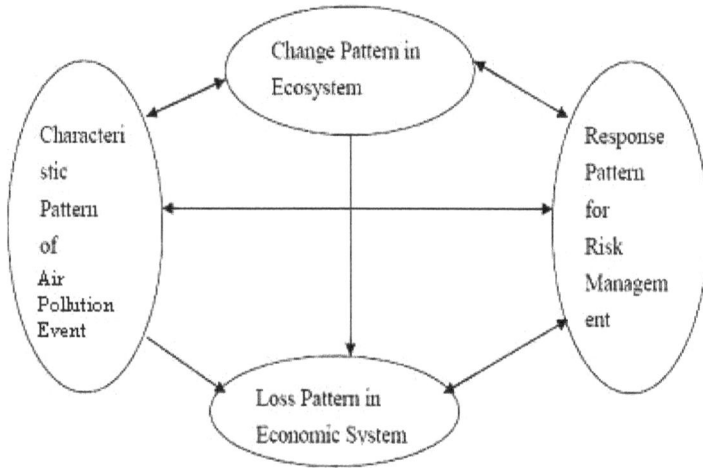

Figure 3. Pattern structures of environmental risk and its management

2.3. Define the space-time information system of air pollution risk management

In designing the expert system, the risk of air pollution event should be discussed first. Since the air pollutant or hazardous materials can be released into the atmosphere by accidents at plants, chemical processing, and other facilities. They may also release by transportation accidents. All of these events can cause the risk for the residents in the neighboring communities. Thus a precise way to estimate the possible damage to the community of the environment is indeed very important. The formulation of a disaster event has to be derived first.

The process of air pollution disaster management is a space-time information problem. In general, the space geographical information can be represented by the following equation:

$$I = \sum_{i=1}^{m} \sum_{j=1}^{n} \left[\begin{array}{c} S_{ij}\left(T_{ij}\right) \\ A_{ij}\left(T_{ij}\right) \\ T_{ij} \end{array} \right]$$

$$T_{ij}(t_b, t_e)_{ij}$$

(1)

Where I is the collection of space geographical information; it is the individual vector for the ith item, j is the state of this item; $S_{ij}(T_{ij})$ and $A_{ij}(T_{ij})$ represent the characteristics of this item in time t_b to t_e. The inferences of the above equation are the following:

[Inference 1]If i is constant, then this equation represents the time series data of the same characteristic.

[Inference 2] If j is constant, then this equation represents the characteristic distribution in the same time.

From the above equation, all the events can be described and all the influence of this event in different space can be explained. The remaining parts are to transform the actual events into the mathematical forms for further inference.

2.4. Define the emergency response measure for risk management

For an air pollution disaster management system, there exists a domain of emergency response measures, defined by the following equation:

$$
\begin{aligned}
M &= \left(Action_1, Action_2, \cdots\cdots, Action_n \right) \\
&= \left(m_1, m_2, \cdots\cdots, m_n \right) \\
&= \sum_{i=1}^{n} m_i
\end{aligned}
\tag{2}
$$

Where M is the collection of emergency response measure; Action i is the individual vector for the ith measure. Each action is represented by a symbol m; and there are m measures in the action domain. If the risk management system is good enough, there should have enough measure to solve the problem encountered by the air pollution. Therefore, we have the following inferences:

[Inference 3] If i is constant, then for each $S_{ij}(T_{ij})$ and $A_{ij}(T_{ij})$, there exists a measure m_i in the emergency response measures domain.

[Inference 4] If j is constant, then for each $S_{ij}(T_{ij})$ and $A_{ij}(T_{ij})$, there exists a measure m_i in the emergency response measures domain.

Different actions have different effectiveness. For example, the authority can stop the emission of air pollutants from the plant in case of necessary. Or they may restrict the activity of community people when there is a need to reduce the emission from the sources. Most of the action related to the benefit of the community people, so different divisions of the government has to be involved, like the local government, environmental protection agency, chamber of commerce, and regional development agency etc.

Not every action has a significant effect on the reduction of disaster, and they costs different budget. Thus we have to be very careful in selecting the actions. All the assembly of these actions is the policy of the government. It is recommended by the expert system.

3. Knowledge-bank analysis

3.1. Construct the modeling base for the expert system

The knowledge bank of an expert system is shown as figure 4. There are different types of knowledge storaged in this system, like the events pattern, the change pattern of economic system, the change pattern of ecological system, and others.

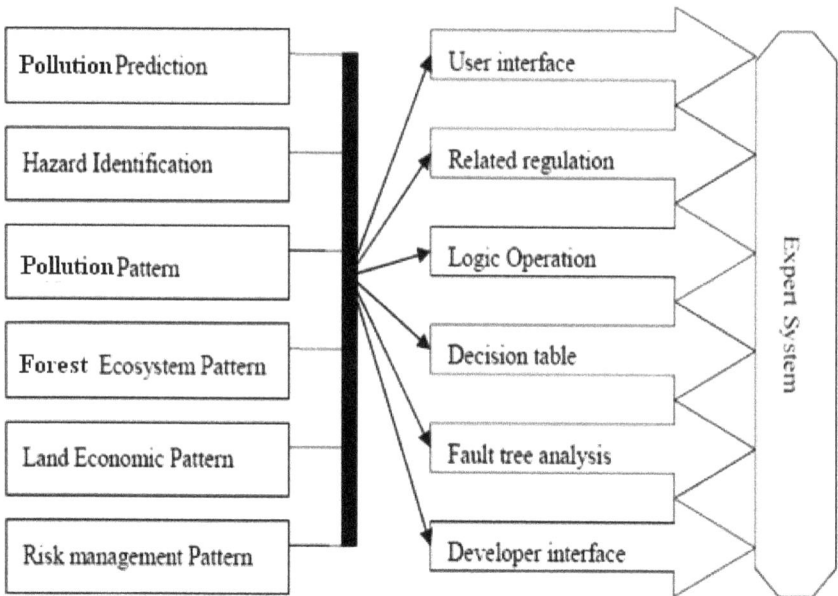

Figure 4. Intelligent knowledge-based expert system

The entire possible pattern, such as the pollution prediction, hazard identification, pollution distribution, forest ecosystem, land economic, and risk management are included in the knowledge database. With this information, the system is able to forecast the possible outcome of the pollution disaster so that the residents can determine the best prevention strategy. Other tools like the logic operation, decision table, and fault tree analysis technique should be included. Finally, a developer interface and the user interface have to be designed very carefully.

The knowledge bank contains the model bank, pattern bank, and regulation bank as below. [5] There are three main modeling activities which included in the expert system: (1) contingency modeling, (2) short term modeling, and (3) accidental, or release modeling [6].

Contingency modeling is to present concentrations for specific chemicals and emission, which may be encountered at a possible release place. Short term modeling is to calculate concentrations occurred in a short periods. The third modeling, accidental release modeling, is perhaps the most critical to emergency managers, which includes natural or accidental release. This type of release modeling is performed soon after a release occurs and is proposed to give immediate responses.

3.2. The model bank for expert system

In designing the expert system for air pollution management, we have to analysis the necessary model as the tool for choosing the correct response measures. Three major model banks should be contained in the system, which are meteorological model, air quality model, and economic model.

1. Meteorological model:

The meteorological model includes the following:

1. Wind field model
2. Temperature variation model
3. Pollutant path prediction model
4. Terrain model
5. Cloud model
6. Vertical wind distribution models
7. Remote sensing generated meteorological parameter model

The wind field model helps us to know the possible damage of the episode. Temperature variation model provide us the diffusion capacity of the atmosphere. The pollutant path prediction model helps us to identify the duty of the polluter. A terrain model provides us the information for the safety management of this event. The cloud model is benefit for the estimation of precipitation of pollutant. And the vertical wind distribution model provides us the understanding of vertical diffusion capability of the atmosphere.

2. Air quality model:

There are several approaches to calculate air pollution diffusion. The most famous are the following three types.

1. Gaussian diffusion model
2. Trajectory model
3. Grid model

The above three model has different capabilities. Gaussian diffusion model is suitable for the near field forecasting. Trajectory model are often used to know the source-receptor relationship and suggest the possible decision for pollution abatement. The grid model can treat the photochemical reaction and often used in the implementation management program of air pollution.

Recently, the improvements in computer technology have significantly improved the speed and accuracy of air quality models. These models have been found in many different areas from ensuring regulatory compliance to assessing human exposure to natural disaster, accidental release, and intentional air pollutant transport.

3. Economic models:

The economic models include the following:

1. Housing damage model
2. Personal injury and death model
3. Agricultural loss models
4. Indirect economic loss models

5.   Post-disaster reconstruction costs model

3.3. The pattern bank for expert system

Three mathematical methods could be applied in the treatment of pattern in this research, which are: (1) Statistical pattern ;(2) Fuzzy pattern ;(3) Neural-network pattern. [2] [3] [4] [7]

In the system, we define the following systems: (1) Wind pattern; (2) Weather pattern; (3) Source pattern ;(4) Population pattern.

3.4. The regulation bank for expert system

The risk management should follow the present regulations; therefore, a regulation bank for response measure is necessary in the management system.

3.5. The action bank for expert system

The action includes different economic models as the following.

1.   Reinsurance compensation model
2.   Super fund models
3.   Major disaster securities market model
4.   Social public disclosure models
5.   Education and training model
6.   Emergency response models
7.   Human resource models

4. Data base analysis and inference system

4.1. Geographical information systems

The tool capable to handle the figure and characters simultaneously is necessary for the research of air pollution emergency response system. The concept of geographical information system (GIS) could be the answer. GIS are tools that allow for the handing of spatial data into information. A lot OF GIS has been developed and applied in diverse field. For example, the GPS satellite system, the web-digital map, and the remote sensing technique, etc. They are all built with GIS as the core technology.

Geographical information system has several advantages over the traditional database. The major advantages include different treatment of characters, the ability to treat the map data, and desirable property to pose the data on internet. The tool contains report generating a summary to analyze the area affected by the air shed.

The GIS has the following subsystems:

1.   A data input subsystem that collects and preprocesses spatial data from various sources. This subsystem is also largely responsible for the transformation of different

types of spatial data(i.e. from isoclines symbols on a topographic map to point elevations inside the GIS) [1]

2. A data storage and retrial subsystem that organizes the spatial data in a manner that allows retrieval, updating, and editing.
3. A data manipulation and analysis subsystem that performs tasks on the data, aggregates and disaggregates, estimates parameters and constraints, and performs modeling functions.
4. A reporting subsystem that display all or part of the database in tabular, graphic, or map form.

4.2. Inference mechanism for the expert system

Logical formula operators allow us to compare values and evaluate the results. When two values are compared using logical operators, the result is either true or false. Logical operators are available in the Compute Wizard if/then/else formula menu as the following:

The four quadrants	
Conditions	Condition alternatives
Actions	Action entries

Table 1. Example for conditions and actions

The inference mechanism in the decision supporting system includes the decision table, logic gate, decision tree, and fault tree etc. Decision tree analysis will be applied in the system for decision support. A decision tree is a decision support tool that uses a tree-like graph or model of decisions and their possible consequences, including chance event outcomes, resource costs, and utility.

In this study, we use the IF/THEN in the emergency response system. Table 2 is the example of decision for an episode by the IF (information) / THEN (action) operator.

4.3. Models for air pollution emergency management

Different kinds of pattern model for air pollution management are listed in table 3.

5. Case study results and discussion

5.1. Expert system for air pollution management

In this study, we use the dust storm as an example for the emergency response system. A framework for the expert system was described in this section. This system provides a easy-to-use, real-time access to pollution concentration predictions and consequence analysis. The system enables us to rapidly determine hazard areas, affected population, meteorological conditions, and relevant geographical information

IF	Logical operator	example (information)	THEN(action)
>	greater than	concentration	activities prohibited
<	less than	distance to hot spot	warning
>=	greater than or equal to	emission amount	warning
<=	less than or equal to	distance to hot spot	activities prohibited
==	equal to (comparison)	path	broadcast
!=	not equal to	trajectory	dust storm
&&	AND	location and path	broadcast
and	AND	wind direction and wind speed	broadcast
|	OR	rainfall or runoff flow	alert
or	OR	radius or distance	warning
!	NOT	particle concentration	warning

Table 2. IF (information)/THEN (action) operator

	Disaster Pattern	Air model pattern	Economic Pattern	Risk management Pattern
1	Wind field model	Gaussian diffusion model	Housing damage model	Reinsurance compensation model
2	Temperature variation model	Trajectory model	Personal injury and death model	Super fund model
3	Pollutant path prediction model	Grid model	Agricultural Loss Model	Major disaster securities market model
4	Cloud model	Hybrid model	Roads and bridges damage model	Social public disclosure model
5	Vertical wind distribution model		Indirect economic loss model	Education and training model
6	Terrain model		Post-disaster reconstruction costs model	Emergency Response
7	Remote sensing generated meteorological parameter model -		-	Human resource model

Table 3. Four types of pattern in the emergency response system

Dust storm is a meteorological disaster which comes from the Mongolia area of northern China. The main reason for the formation of dust storm is the overdeveloping and the global warming which induce the soil become desert. The strong wind blow also increase the number of dust storm event year by year. This phenomenon also affects the neighboring area such as Korea, Japan, and Taiwan etc. [6]

In order to realize the dust storm disaster, many researches has been made recently and the database was build. Most of the information about the dust storm was monitored by the meteorological and environmental monitoring system followed by the data processing procedure.

The information generated in the process was largely in the form of figure or character. Although it is convenient to understand, the time consuming in processing these information is too long. The new architecture in this study is capable to offer a function that enables us to search the map data directory from the dust storm event. The main advantage for the new architecture is to simplify the search work and save the time for searching dust storm event.

5.2. Design and capability of the emergency management systems

The case study described here referring to the design and algorithm of a dust storm event response system. The system was combined with the geographical information system and was called "DSGIS (Dust Storm Geographical Information System". DSGIS is an interactive geographical information system for dust storm research and has been developed to enhance the understanding of dust storm phenomenon and to offer a more convenient environment for the researchers and public.

The concept of geographical information system and supporting database system was applied in DSGIS for planning of the figureizational operating system of dust storm event. It enables use of digitization information to search and treat the dust storm event information.

There are three major concerns in implementing a dust storm geographical information system as the following: programming language, database, GIS tool, and interface.

Visual Basic was chosen to be the programming language of this system. The database applied in the preliminary system is "ACCESS" developed by Microsoft. GIS Design tool"ArcGIS ENGINE" supplied by ESRI was used in this system. And there are three interfaces in the system: (1) User Interface (2) Developer Interface (3) Outer Interface.

5.3. Design results of the expert system for managing air pollution

The dust storm events were gathered in the DSGIS database system and combined with the air quality monitoring station data.

In planning the database structure, the monitored data were collected first. A standard form was suggested to be the format of this database. There are five index of air pollutants in each monitoring station, they are total suspended particulate (TSP), sulfur dioxide (SO_2), nitrogen oxide (NOₓ), carbon monoxide (CO), and ozone (O₃). The database was designed based on the above monitored information. However, the content of this system will be revised and expanded in case there is any change of demand for this system.

The dust storm determined the air quality, mainly in particulate. However, the concentration change during the dust storm period was also increased by the researchers. The air quality index PSI is automatically calculated by the above five air pollution concentration value and categorized as suggested by EPA. The core system and algorithm for DSGIS is described in the system. Following the database structure discussed in previous section, the infrasture of DSGIS is shown in figure 5.

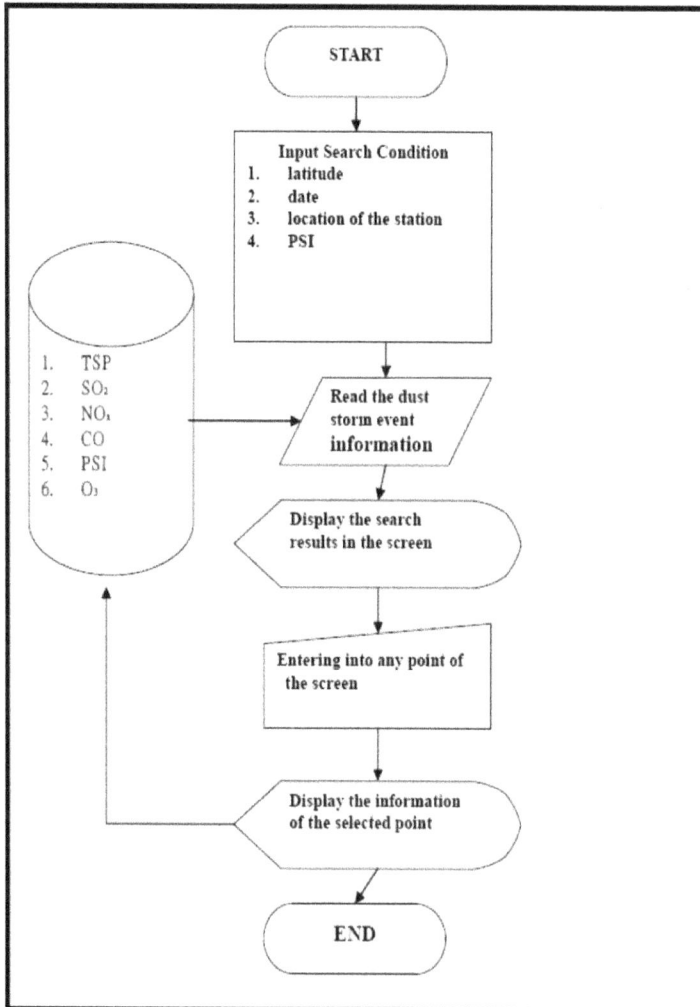

Figure 5. The Dust Storm GIS system (DSGIS)

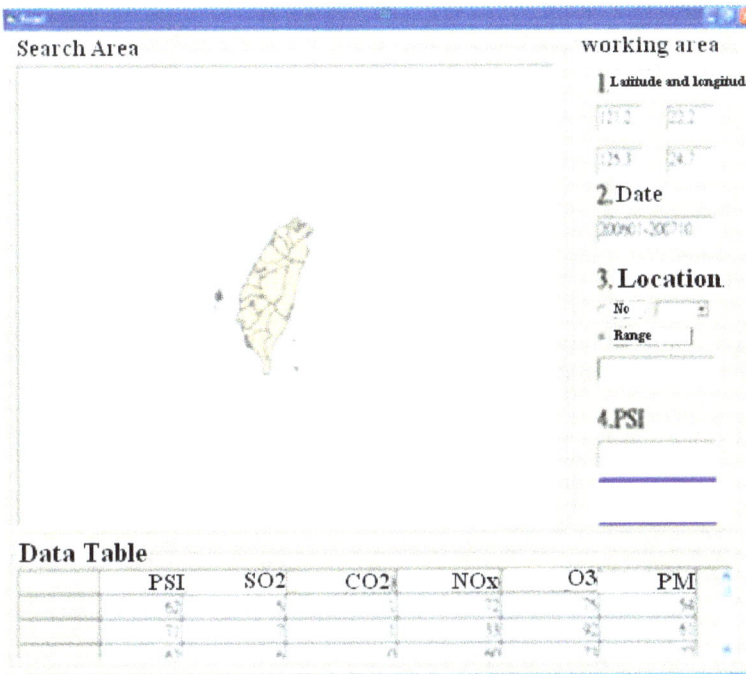

Figure 6. Operation environment

The order of the system structure can be explained as in figure 6. The selected place or dust storm event can be input from the interface of the screen. Also, a "map searching" method was developed in this study. This method enables us to search a station directly from and display the information the users want to know. The detail of each step is described as the following.

Entering into the system

The DSGIS will download the data automatically from the database of the selected zone where the user entering into the interface of the operating environment.

Input the search condition

Four search conditions can be used as the search condition, they are:

i. Latitude: input two sets of number of longitude and latitude such as (123, 23) and (121, 25). The number sets represent all the geographical information within the four points of the four numbers.

ii. Date: input year, month, date, these data can be input simultaneously or separately, e.g. the data 20070409 represents all the data in April, 2007.

iii. Location of the station: Input the station's number or the name of the station. It is also permissible to input two longitude and latitude to include all the stations within this area.

iv. Select the number of PSI as it is defined and all the information within this range can be retrieved.

Display the search results (Fig. 6)

The information consistent with the search condition will be displayed The information consistent with the search condition will be displayed in the screen on its location with the following sign as◎, ●, ★, ☆, ▲, Δ, ♂, etc. .

View the search results (Fig.7)

When the search results were displayed, the DSGIS also allows the users the select the station directly through the mouse acted on the screen. More information about this station will be displayed consequently.

The version one of DSGIS has already completed. The structure of each component remains very flexible for the future application and adjustment of this system.

Figure 7. Search results

5.4. Numerical weather prediction bank of the expert system

In order to have precise results in the expert system, the numerical weather prediction model has to be applied. The prognostic data from numerical weather prediction models is suggested. The weather models predict future three dimensional atmosphere states by solving the conservation equations for mass, momentum, and thermodynamic energy. These models represent the relevant physical processes for moisture, cumulus convection, and radiation, and sub grid-scale turbulence.

5.5. Atmospheric transport and diffusion models of the expert system

Some models were suggested in this system, as listed in Table 4.

Air model pattern	Name of the models
Gaussian diffusion model	1. ISCST
	2. AERMOD
Trajectory model	1. CALPUF
	2. GTx
Grid model	1. TAQM
	2. CAMx
	3. WRF

Table 4. Atmospheric transport model used in the system

5.6. System validation and supporting databases of the expert system

The supporting database is important because the changes in metrological conditions and in emission strengths may affect the air quality.

A supporting database for the monitoring of the air quality data is important as explained in figure 8. the source distribution example of a county is shown in figure 9. the calculated pollution concentration in northern is shown in figure 10. Finally, the estimated social cost cause by air pollution in each district was shown in figure 11 as a for the decision maker.

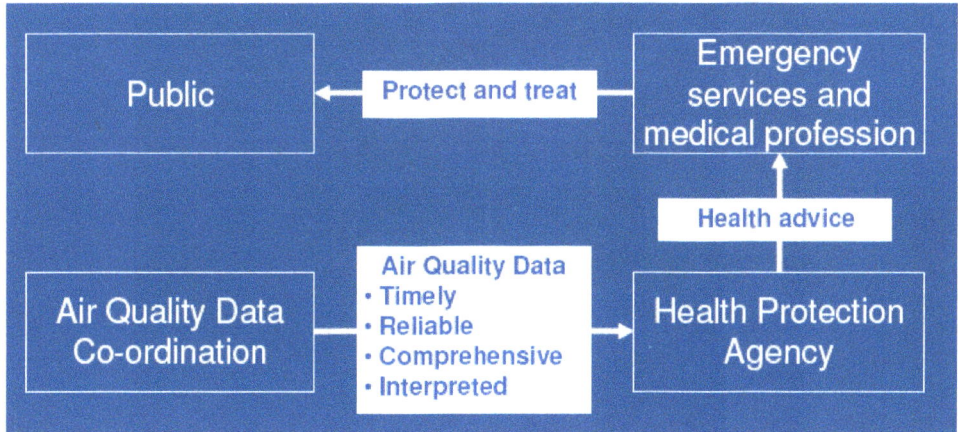

Figure 8. The use of air quality monitoring data in the emergency response management system

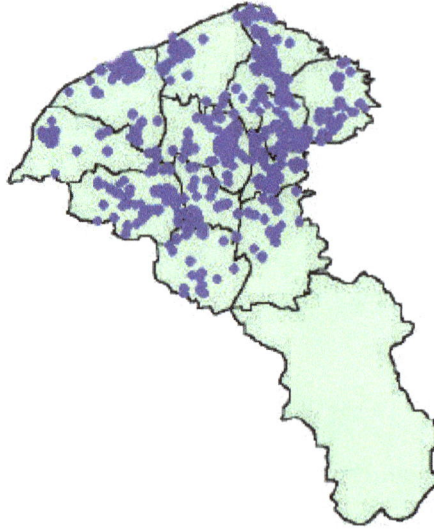

Figure 9. Example of the source distribution diagram in a county located in northern Taiwan

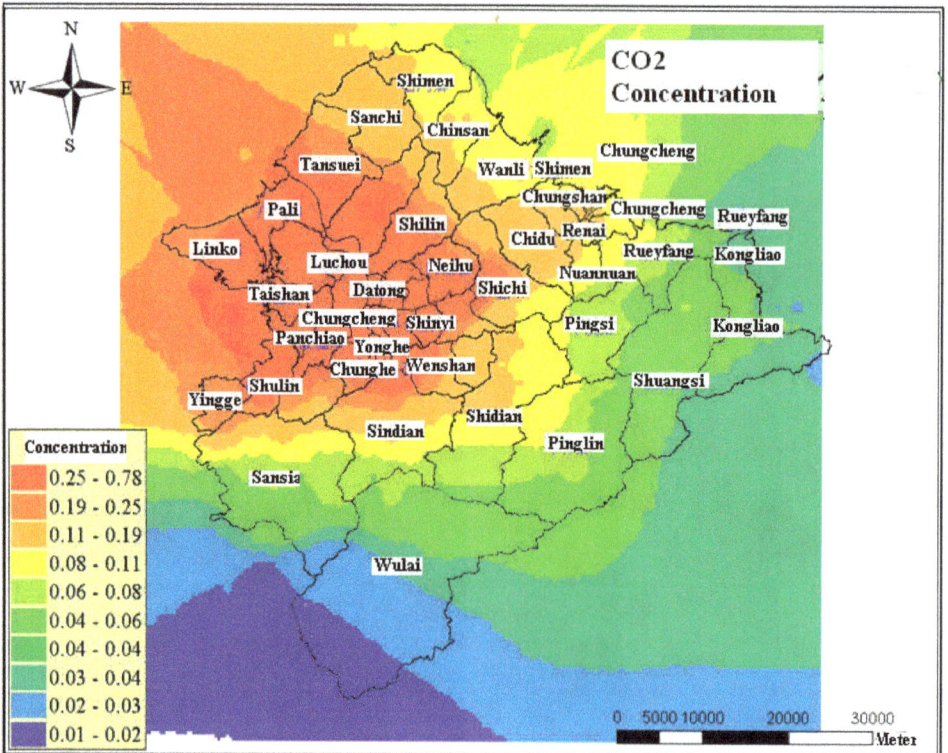

Figure 10. Predicted distribution of the pollutant concentration in northern Taiwan

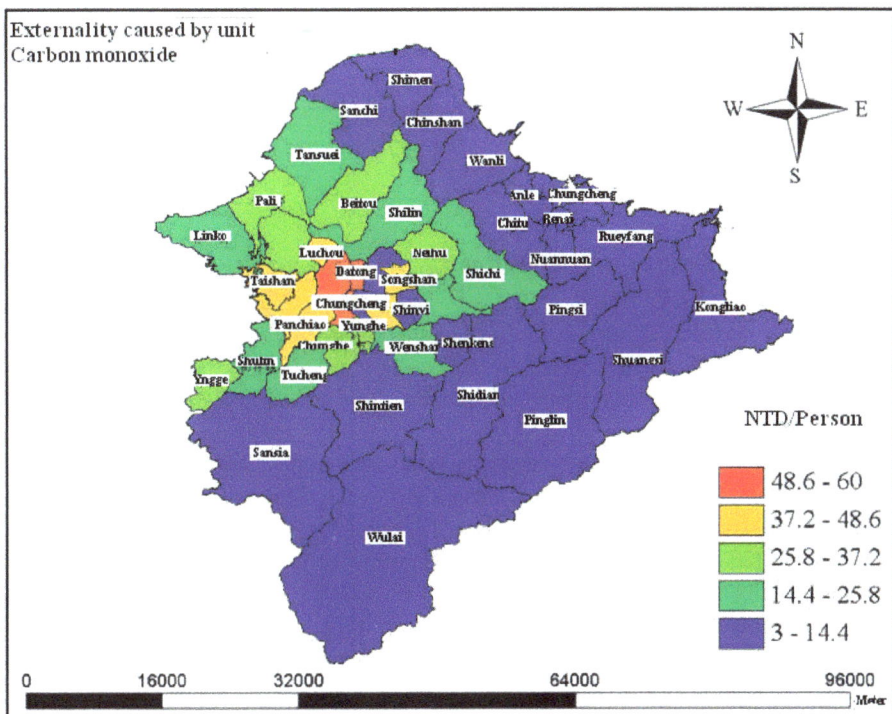

Figure 11. Predicted social cost of pollution in northern Taiwan

6. Conclusion and suggestion

The intelligent expert system for air pollution emergency response was established in this study. The dust storm event geographical information system was studied and a knowledge-based decision support system for emergency response and risk management was established. The mathematical pattern relationship of air pollution effects on neighboring area and the corresponding response measures were included in this system. The decision maker can specify the procedure and minimize their human error in the decision process.

The performance results indicate that the function of DSGIS is acceptable. Generally speaking, DSGIS is a useful tool for taking the necessary knowledge and information about the dust storm. In addition, it also provide more convenient operating interface for the users. The concept of "map searching" is more convenient than the traditional searching methods. The performance results also show that the effectiveness of the DSGIS in searching the event is reliable and acceptable.

Since this system is designed on a module-based feature, it is easy to expand the application to more cases. Future work includes development of other independent module for individual event and gathers more information about the event into the database.

Author details

Wang-Kun Chen
Jinwen University of Science and Technology, Department of Environment and Property Management, Taiwan

7. References

[1] DeMers, M. N., 2000, *"Fundamentals of Geographical Information System"* 2nd ed. John Wiley and Sons, Inc. USA.

[2] Duda P. O., Hart P. E., Stork D. G. (2001) *Pattern Classification* (2nd), Wiley, New York, ISBN 0-471-05669-3..

[3] Schalkoff R., *Pattern Recognition: Statistical, Structural, and Neural Approaches.* John Wiley & Sons, 1992.

[4] Schuermann, J. (1996): *Pattern Recognition,* Wiley&Sons, 1996, ISBN 0-471-13534-8.

[5] Taiwan EPA (1996) *"Air Pollution Emergency Response System"* Taiwan Environmental Protection Administration, Taipei, Taiwan.

[6] Turpin, R. (2004) Air Plume Modeling, Planning or Diagnostic Tool. *Environmental Protection Agency.* Retrieved February 17, 2007 From http://www.ofcm.gov/ atdworkshop/proceedings/ session1/campagna.pdf

[7] Yie C.R., 2002,*"The effect of Mainland dust storm to the acid air pollutants in central Taiwan"*, National Chung-Hsin University.

[8] Zadeh L. A. (1965) "Fuzzy sets". *Information and Control* 8 (3) 338–353.

Chapter 12

Environmental Control and Emission Reduction for Coking Plants

Michael Hein and Manfred Kaiser

Additional information is available at the end of the chapter

1. Introduction

Coke is a necessary component for the production of iron and steel. Nearly 65 % of the worldwide steel production takes place via so-called pig iron (hot-metal route), which is produced in the blast furnace from iron ore by use of coke.

The importance of coke as raw material for the steel production has been approved during the last years while the worldwide need for steel has strongly increased. Since 1990 the steel production has nearly doubled and reached 1.417 mio. t in 2010 (Worldsteel, 2012). Coke production from hard coals was increased by 70 % in the same period resulting in approx. 593 mio. t in 2010 (Re-Net, 2011)(Fig. 1).

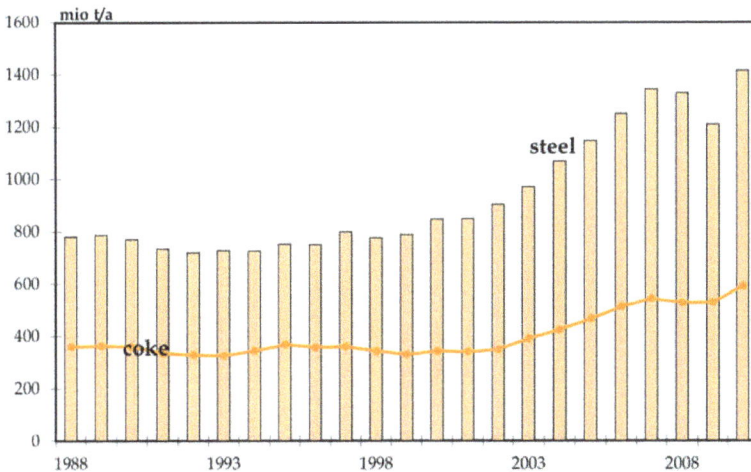

Figure 1. Worldwide crude steel and coke production (Re-Net, 2011, Worldsteel, 2012)

One can assume that this trend will continue in the next future, too. That means, that similar than in the recent years, new coke making capacities will be built and older and smaller plants will be replaced by high performance coke plants, in the future. This will be the case in China, India, Southeast-Asia and South America in particular. Already today approx. 65 % of the coke worldwide is produced in China.

There is a lack of an official statistic from which one can derive the total number of coking plants worldwide. However, it is to assume that this will be in the range of 500 plants, not including so-called primitive ovens, that means smaller coking plants without any technical equipment for operation.

Three principles will still characterize prospective projects for new coking plants: improvement of economics of coke production as well as optimization of the coke quality. A third principle has prevailed during the last four decades because of more stringent becoming legislation: reduction of the impact of the coking process on the environmental, and on the ambient air in particular. Due to the legal demands, coke plant operators were obliged to improve techniques for emissions control, to revamp batteries, or, in some cases, to shut down a battery and built a new one if the new standards could not be fulfilled under economic and technical reasons.

Progress made in emission control at coking plants can be read from an improvement of air quality in the Rhine-Ruhr area in Germany, which is the center of the German cokemaking industry till today (LANUV, 2012). Besides the shrinking importance of coal use in homefiring the reduction of coke plants´emissions is the reason for the continuous decline of Benzo(a)pyrene (BaP) as a highly carcinogenic aromatic hydrocarbon in the ambient air of this area during the last 20 years (Fig. 2).

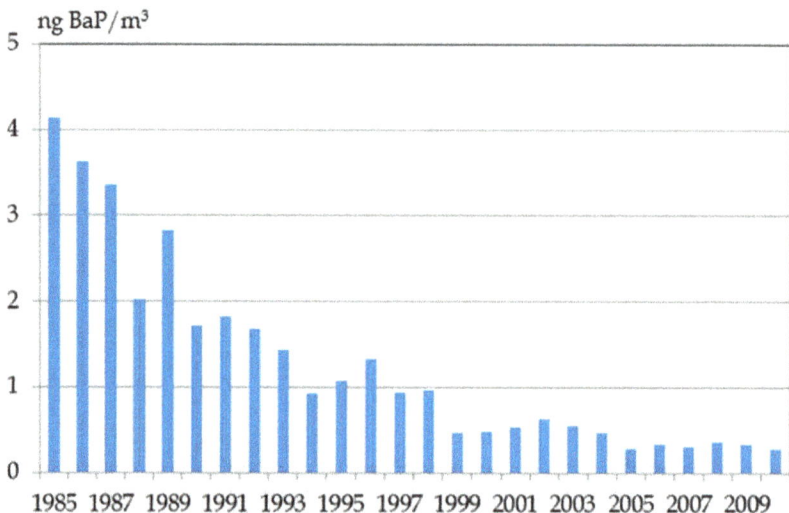

Figure 2. Benzo(a)pyrene (BaP) in ambient air of the Rhine-Ruhr area (LANUV, 2012)

Benzo(a)pyrene plays an important role with regard to the environmental assessment of the coking process. Very often it is used as a guide substance for polycyclic aromatic hydrocarbons (PAH) which can be emitted from leaks at the coking chambers. In order to reduce these fugitive emissions, measuring methods are necessary by which the made progress can be quantified. Reliable statements on the amount of emitted BaP are indispensable, too, for making a forecast on the BaP burden in ambient air of the surrounding.

2. Modern cokemaking technology

2.1. Generals

The bulk of the worldwide coke production in 2011 was effected in conventional coking plants including a recovery of gas and coal chemicals. These plants are very often called by-product coking plants, too. Approx. 5 % of the total coke production originate from the non-recovery technology, which does not recover gas and coal chemicals. Both technologies display a quasi continuous process with charge-wise coke production in several ovens connected in a battery.

A scheme of the total process of conventional coking is shown on Fig. 3. The process can be devided in the two steps: battery operation (left side of Fig. 3), and coke oven gas (COG) cleaning and by-product plant, respectively (right side of Fig. 3).

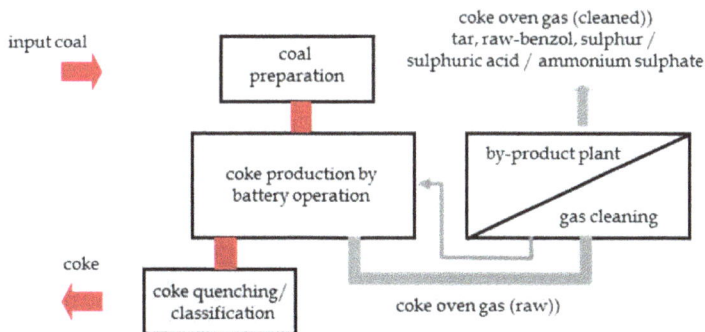

Figure 3. Scheme of conventional cokemaking

2.2. Conventional coking plant – by-product plant

By-product coking plants are comprised of single oven chambers, being 12 to 20 m long, 3 to 8 m tall, and 0.4 to 0.6 m wide, in which the input coal is heated up indirectly. Several chambers are grouped to form one battery (multi-chamber-system; Fig. 4). A single battery may consist of up to 85 ovens. The front-end sides of the individual ovens are sealed with doors. The ovens are charged through charging holes in the oven top. As an alternative, the oven can also be charged from the side via one opened door after the input coal was stamped before in order to build a formed cake (stamp charging). Subsequently to a 15 to 25 hours coking time the doors are opened and the built coke is pushed by the coke pusher machine out of the oven into a coke quench car. Then the coke is quenched in a dry or wet quenching facility. The oven chamber is sealed again, initiating a new carbonization cycle. The gas evolving on coal carbonization leaves the oven chamber through a standpipe (offtake) and is passed on via a common gas collecting main to the gas treatment facilities and to the by-product recovery plant. The ovens are run at a slightly positive pressure of 10 to 15 mm water column.

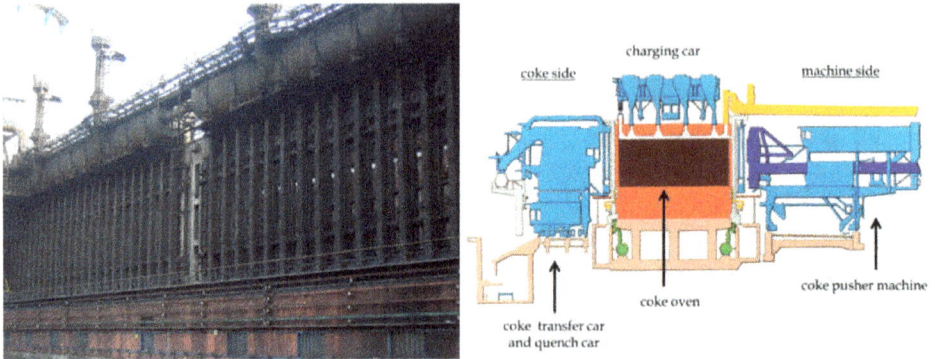

Figure 4. View on the doors of a coke oven battery of the coking plant Zdzieszowice, Poland (left side); schematic drawing of the machines for battery operation (right side)

As outlined in Fig. 7, the oven chambers are heated through heating flues, located between the chambers, in which cleaned coke oven gas or blast furnace gas is combusted. The temperature in the heating flues lies between 1150 and 1350 °C usually.

Battery operation, i.e. charging and pushing is carried out by large machines (Fig. 5) which very often are running automatically.

Figure 5. Pusher machine of the coke plant Huckingen (left) and charging car of the former coking plant Kaiserstuhl III (right)

Coke oven gas (COG) as built during the coking process is unsuited for use as underfireing gas for the coke oven batteries and for other applications, because of technical, and of environmental related reasons in particular. The necessary cleaning is made in the so-called by-product plant which comprises a complex chemical plant. For a coking plant with an annual coke production of 1 mio. t, the design capacity for the by-product plant is about 61,000 Nm³ COG/h.

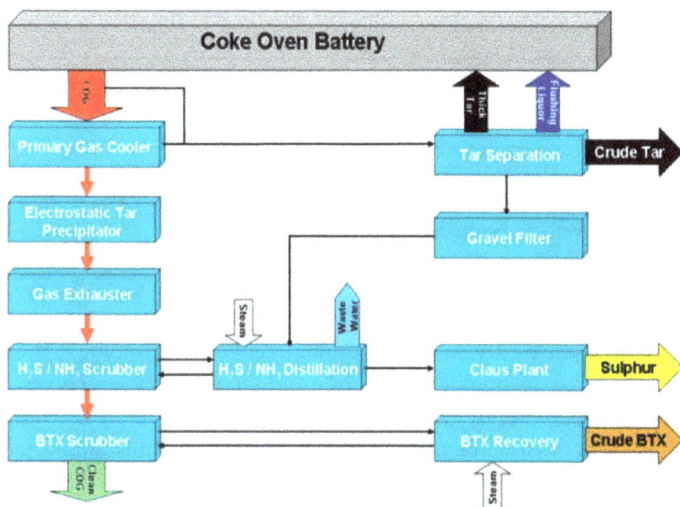

Figure 6. Scheme for a modern by-product plant

A general simplified process diagram is shown in Fig. 6. Coke oven gas leaving the battery ovens has a temperature of 800 to 1000 °C, and just before entering the collecting main it is sprayed with flushing liquor (ammonia water) coming from tar separation. After spraying the gas comes down to temperatures in the range of 80 °C. At this temperature most of the raw tar is condensed, therefore a separation into gas and liquid phase is possible in a downcomer. The liquid phase flows from here to the tar separation unit to separate water and crude tar; crude tar is one by-product.

The raw gas is directed to the primary gas cooler were it is cooled down to 21 °C by indirect cooling. The next step is the electrostatic tar precipitators, where the residual amounts of tar fog are almost completely removed, down to maximum 20 mg/Nm³. After this step COG is sucked off by exhausters keeping the necessary pressure for exhausting the gas from battery and is led to the subsequent gas treatment. There exist two techniques for H₂S removal from COG, in principle (see section 5.2). In Fig. 6 only the ASK process (Ammonium-Sulphur cycle process, ASK), combined with a subsequent Claus plant for sulphur production, as a high value by-product, is shown as the most common desulphurization process in Europe. In section 5.2 this technique is described more in detail.

The last optional gas treatment step is BTX and naphthalene removal in a scrubber using washing oil. The crude BTX is a further by-product.

Most of the water used in the by-product plant is recycled in the process. Only a small amount of waste water, which mainly represents the water content of the input coal, is produced as effluent of the ammonia still and has to be treated in biological waster water treatment plant.

Typical figures for the quality of coke oven gas befor and after gas cleaning are shown on Table 1. The Figures can be varied due to the coal quality and the coking process itself.

	crude coke oven gas	cleaned coke oven gas	unit
Tar	60-110	0.1	g/m^3
BTX	28 – 35	< 5	g/m^3
NH_3	7-9	< 0.1	g/m^3
H_2S	4 – 8	< 0.5	g/m^3

Table 1. Quality of coke oven gas before and after cleaning

2.3. Non-recovery plant – heat-recovery plant

The most essential features by which the non-recovery technology differs from the conventional cokemaking technology with by-product recovery are given in Fig. 7. In contrast to conventional coking by which the coke is heated indirectly by combustion of gas within the heating flues outside the oven chamber, exclusively, during non-recovery coking the necessary heat is transferred both directly and indirectly into the oven chamber as described in the following.

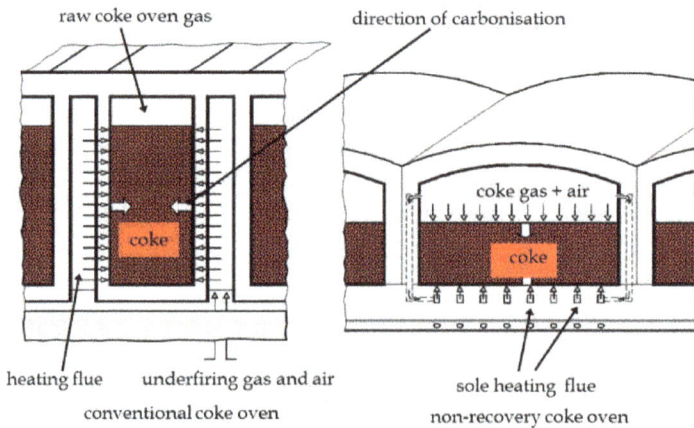

Figure 7. Principle drawings of conventional and non-recovery cokemaking (Hein, 2002).

The basis for modern non-recovery plants is the so-called Jewell-Thomson oven, several ovens of which are grouped together to form one battery (Fig. 8). The ovens are characterized by a tunnel-like shape with a rectangular ground area and an arched top. The dimensions of the chambers of modern plants run up to 14 x 3.6 x 2.8 m (L x W x H). Coal charging (up to 50 t) of the ovens is accomplished through the open pusher side door. Very often the coal is stamped before, and then the coal is charged into the hot oven chamber. Typical charging levels lie at 1000 mm. The carbonization process is started by the heat still existing from the preceding carbonization cycle. The released coke oven gas is partly burnt by addition of ambient air through the doors and passed through so-called down comers into the heating flues situated in the oven sole. By way of a further supply of air, the complete combustion of raw gas is effected here at temperatures between 1200 and 1400 °C. With plants according the state of the art, the hot waste gas is utilized to generate energy, and subsequently is subjected to desulphurization before exited into the atmosphere.The coking time in Jewell-Thomson ovens amounts to approx. 48 hours. After that time, the coke is pushed out and quenched in wet mode, normally.

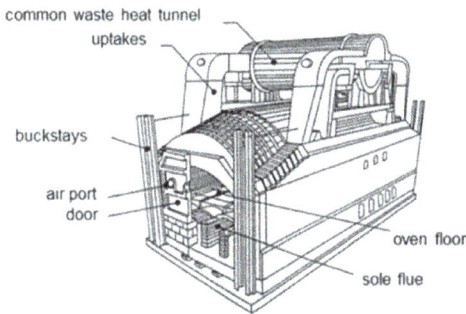

Figure 8. Schematic drawing of the Jewell-Thomson oven (Hein, 2002) (left) and view on the ovens of the heat-recovery cokung plant of the Shanxi Xishan Coal Gasification Co. Ltd., Gujiao, China (right)

Due to the negative pressure, under which the coking process is running, emissions from leaks at the doors are avoided in principle. Dust emissions occurring during coke pushing are exhausted via a coke side shed. Very often suction devices are installed at the pusher side, too, in order to capture emissions caused during charging.

As the techniques for emission control during charging, pushing and quenching are similar to those applied at conventional coking, and fugitive emissions at the ovens are excluded by principle reasons, it is resigned to address emission related issues regarding non-recovery cokemaking in a separate section.

3. Emission sources on conventional coking plants

Typical emission sources with regard to battery operation are shown on Fig. 9. These are directed and fugitive emission sources. Fugitive emissions mainly occur from leaks at the closed openings of the coke oven batteries (doors, charging hole lids and offtakes) or are caused by non-captured emissions during coke pushing and coal charging. These emissions can not be avoided completely, also when considering closure facilies according state of the art in technology and being under best state of maintenance, and contain dust, polycyclic aromatic hydrocarbon compounds (PAH) and Benzene as most relevant components. Carcinogenic Benzo(a)pyrene is very often used as guide substance for the group of PAHs.

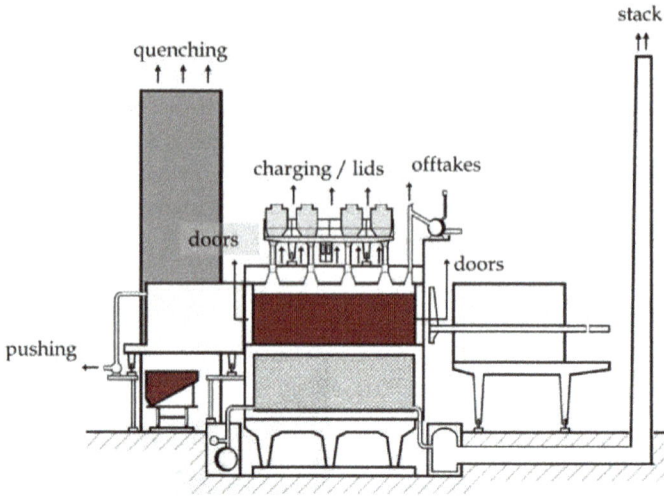

Figure 9. Schematic drawing of typical emission sources at a conventional coking plant

Emissions from directed sources are created at the stack for the off-gas from battery underfiring. The most important compounds which are emitted here are dust, NOx, SOx and CO_2. Dust is emitted also by the offgas of the pushing emission control as well as during coke quenching. Emissions caused at preparation of charging coals, and at classification of coke, respectively, are not addressed here because well-proven dust removal systems are available to cope with them.

Emissions from the by-product plant are bearing secondary importance in contrast to emission from battery operation. This is valid for emissions from open tanks, leaks in the piping system and at flanges, pressure valves, pumps, etc., as well as for the off-gas from the technical facilities for sulphur-removal (sulphuric acid plant, Claus plant). On the other hand, more relevance is to be attached to the efficiency of the devices for H_2S removal from the coke oven gas (see section 5.2). Remaining H_2S will influence the amount of SO_2 in the off-gas at the stack of the battery in case of using cleaned coke oven gas for battery heating.

4. Legislation on emission control

4.1. Germany

4.1.1. Generals

Starting, it should be emphasized that legal rules given by the European Union (EU) have a significant impact on the national legislations of the member states. While regulations of the EU becomes immediately enforceable as law in all member states, directives are only binding for member states with regard to the achievable target, while they leave it up to the member states to decide on the form and means needed to realize the commonly set targets within the framework of their national legal system.

In Germany, the most important legal rule with reagard to industrial emission control represents the Technical Instruction for Air Quality Control – Technische Anleitung zur Reinhaltung der Luft – the so-called TA Luft. The first issue of TA Luft was enacted in 1964 and was amended for several times in the following years. The TA Luft is the most essential guide for implementation the demands of the German Federal Immission Control Act - Bundes-Immissionsschutzgesetzes (BImSchG) – which was released in 1974.

The Federal Immission Control Act, amongst others, is based upon the two fundamental principles of "risk defense" and "precaution". The precautionary principle is expressed in the approval of new plants and flows into the demand for compliance with what is called the state of the art in technology in the construction and operation of industrial plants with special regard to environmental control.

The state of the art is basically stipulated in the TA Luft which at the same time generally prescribes ambient air quality standards that must not be exceeded in the vicinity of a new plant after its commissioning. To this effect it is required to calculate the additional burden of the pollutants, which are to be expected upon commissioning of the planned plant, by dispersion calculations (see also section 8.2). Furthermore for precaution, the TA Luft prescribes emission limit standards, especially for directed sources, which shall be examined for compliance within regular intervals.

In view of the "risk defense" principle of the Federal Imission Control Act its 22nd Decree stipulates air quality standards for various hazardous substances, the compliance of which shall be achieved, for example, by implementing so-called air pollution control plans. This area-related rule concerns all plants, that means also those for which a permission has already been granted, and may necessitate an obligation for retrofitting the plant.

The TA Luft amendments which came into force in 1986 gained special importance for the coking plants which were built in the 1980th in Germany. Although the permits for the new constructions of the coke plants Prosper, Huckingen, Salzgitter and Dillingen are dated before the enactment of TA Luft 1986, its demands have to be fulfilled by the new plants to the greatest possible extent.

Compliance with the TA Luft 1986 without any extension, that means including the demand for operation of a coke dry quenching unit, was necessary for the new construction of the coke plant Kaiserstuhl III which was operated in Dortmund between 1992 and 2000.

Due to the progresses reached in emission control in Germany since 1986, an emendment of the TA Luft came into force in the year 2002 (TA Luft, 2002). The permits of the coke plants Schwelgern and of battery no.3 of the Saar central coking plant (Dillingen) were affected from this amendment, which disclaims on dry quenching as the only mode for coke cooling. More informations on the coking plants mentioned before will be given in section 6. The most important features of the current TA Luft with regard to emission control on coking plants will be described in the following sections.

4.1.2. Techniques to apply on coking plants with regard to emission control

As a measure for precaution the TA Luft sets standards for the technical equipment for emission control on industrial plants, and specifies how to operate the plant in a most environment-friendly way. Table 2 contains the most important techniques and work practice standards to apply on the coke oven batteries with regard to the TA Luft-amendments of the year 2002 (TA-Luft, 2002). Most of the standards of the German TA Luft were adopted by the BREF-document of the European Union (EU, 2012) nearly complete. Most of them are described in section 5 more in detail.

techniques
- gravity charging: emission free charging by transfer of charging gases to the main and into the neighbour oven, as an option
- stamp charging: combustion of not transferred gases
- doors with technical gas-proof sealings
- water-sealed lids at offtakes
- single chamber pressure control should be applied
- coke side emission control including a mobile hood and a stationary control device
- coke quenching by dry or wet quenching mode
work practice standards
- additional sealing of lids of charging holes
- regulary, and preferential automatic, cleaning of closure facilities

Table 2. Techniques for emission control and work practice standards as demanded by (TA Luft, 2002)

4.1.3. Limit values for emissions at directed sources

In order to reduce atmospheric emissions from industrial plants as far as possible TA Luft sets limit values which have to be checked regularly. Table 3 contains limit values for emissions at the outlets of directed sources of coking plants. In contrast to the US Clean Air Act (section 4.3) TA Luft contains no legal demands for fugitive emissions by setting standards for the allowed number of visible emissions.

process	emission	limit value	
stamp charging	dust:	10	mg/Nm³
battery underfiring	dust	10	mg/Nm³
	NOx	0.50	g/Nm³
	sulfur*	0.8	g/Nm³
pushing	dust	5	mg/Nm³
or	dust	5	g/t$_{coke}$
quenching			
dry	dust	15	mg/Nm³
wet (new plants)	dust	10	g/t$_{coke}$
wet (existing plants)	dust	25	g/t$_{coke}$

Table 3. Emission limit values for battery operation according (TA Luft, 2002); *: sulfur content of the heating gas before combustion

Special emission limits are set for the off-gas of a sulfuricacid-plant and of a Claus-plant for sulfur recovery, if exist as part of the by-product plant.

4.2. European union

In the European Union, there are in principle two directives that influence coke plant operation:

- „IED Directive" (EU, 2010) on industrial emissions (integrated pollution prevention and control)
- „Air Quality Directive" (EU, 2008)

As mentioned in section 4.1.1. Directives of the EU are only binding for member states with regard to the target to be achieved; they have to be transformed to the national legislation of the member state.

The IED-Directive addresses the conditions for plant operation and sets standards for emission control. This directive stipulates that the "best available technique BAT" which has to be applied is to be described in a so-called BREF document („Best available technique Reference" document) for certain industrial plants. For coking plants, the set-up of such a BREF document was finalized in the year 2000. An amendment was promulgated in 2012 (EU, 2012), and it assigns "Associated Emission Lewels AEL" to the BATs. BAT-AELs give ranges for emission lewels which can be achieved by application of emission control techniques according BAT. AELs which are relevant for cokemaking operation are described on Table 4. A more detailed description of the BATs is given in section 5.

The Air Quality Directive (EU, 2008) and its so-called 4. Daughter Directive (EU, 2004) describe the targets and principles of the air quality policy pursued by the European Union. Ambient air standards which are important for cokemaking operation are given on Table 5.

process	emission	AEL/BAT	unit of measurement	remark
charging	dust	<5 or <50	g/t_{coke} or mg/Nm³	
	visible emission	< 30	sec	duration of visible emissions per charge
offgas from battery underfiring				
	SOx	<200 to 500 (as SO$_2$)	mg/Nm³	depending on the type of gas for underfiring
	NOx	<350 to 500 (as NO$_2$)	mg/Nm³	for new plants
	NOx	500 to 650 (as NO$_2$)	mg/Nm³	for existing plants which are equipped by primary measures for NOx reduction
	dust	< 1 to 20	mg/Nm³	
pushing	dust	< 10 to < 20	mg/Nm³	depending on filter type
quenching				
wet	dust	< 25	g/t_{coke}	existing plants
wet	dust	< 10	g/t_{coke}	new plants
dry	dust	20	mg/Nm³	
battery operation				
	visible emission	< 5 to 10	%	from leaks at doors
		adequate oven pressure regulation		
		work practice standards		
desulphurization of COG				
	H₂S	< 300 to 1000	mg/Nm³	applying absorption processes
	H₂S	< 10	mg/Nm³	applying wet oxidation processes

Table 4. BAT associated emission lewels (AEL) as described in the BREF document (EU, 2012)

emission	Limit value		remark
Benzene	5	$\mu g/m^3$	
Particulate Matter PM10	40	$\mu g/m^3$	
	50	$\mu g/m^3$	daily average for max. 35 days/a
Particulate Matter PM2.5	25	$\mu g/m^3$	from 2015
Benzo(a)pyrene *	1	ng/m^3	from 2012

Table 5. Ambient air quality standards (limit values) of the EU (EU, 2008) as an annual average with reference to coking plant operation; *: (EU, 2004)

4.3. USA

4.3.1. Clean Air Act

The Clean Air Act (CAA) of the United States of America was passed in the year 1990. This act of law describes standards for air quality, which exert a very strong influence on the requirements which have to be fulfilled for obtaining the permit to run an industrial plant. The so-called Residual Risk Standard (RRS) should provide an ample margin of safety to protect public health and to reduce the risk to cause cancer to a minimum.

In case of coking plants, amongst others, standards are set for the allowed number of visible emissions (leaking rates as %) from battery operation to reach this goal, as described by the US EPA (US-EPA, 1993a, 2005). For the construction of new coke plants at the green site, the CAA calls for zero visible emissions from battery operation. That means in practise, that in the USA, the non-cecovery technology is the only one, which is allowed by the US EPA for new green field plants because of the prevailing negative pressure and consequently of the prevention of leaks at the ovens.

For existing conventional coking plants the Residual Risk Standard, which is still open, has to be reached from 2020. It is to assume that the relevant legal demands will be very ambitious. During the recent 20 years the US coke oven plant operators had the chance to approach this target on different tracks, which specify different compliance timetables (Fig. 10) (Ailor, 2003; US-EPA, 1993a). While the MACT-track (Maximum Achievable Control Technology) allows less stringent standards for a long period to fulfill the highest lewel of emission standards already in 2005, operators who have chosen the LAER-track (Lowest Achievable Emissions Rate) got an extension to reach this standard only in the year 2010.

The relevant standards for the allowed visible emissions are shown on Table 6. Estimates of visible emissions should be based on the results of daily visible emission inspections using EPA Method 303 (US-EPA, 1993b).

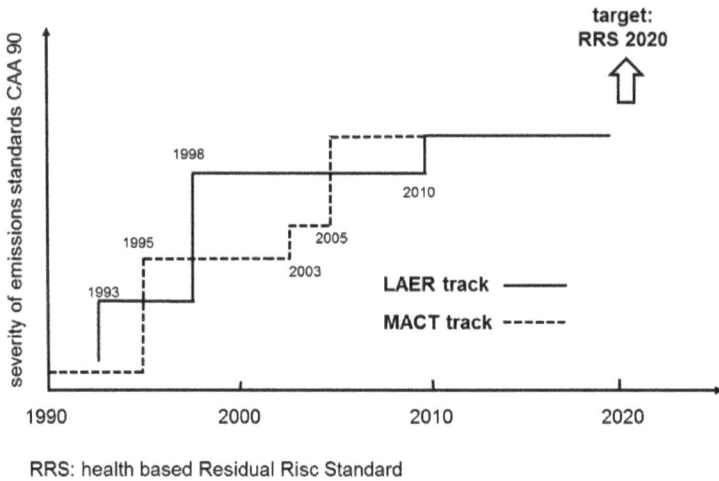

RRS: health based Residual Risc Standard

Figure 10. Timetable to comply with the legal demands of the US Clean Air Act

source	MACT	LAER	remark
	from 01.01.2003	from 01.01.2010	
doors	5.5 %	4 %	≥ 6 m
doors	5.0 %	4 %	foundry coke
doors	5.0 %	3.3 %	< 6 m
lids	0.6 %	0.4 %	all plants
offtakes	3.0 %	2.5 %	all plants
charging secs per charge	12	12	all plants

Table 6. Standards for visible emissions according MACT- and LAER-track respectively for conventional coking plants

It is easily to understand that operators of older plants would have preferentially followed the MACT track as their coking plants will be no longer in operation in the year 2010, probably. After all there were only 5 conventional batteries which have to comply with emission standards equivalent to the 2010-LAER-standard in 2005. On the other hand, operators of new plants, which were equipped with modern techniques for emission control on the date of their track choice, or for which a modernisation was planned, would have preferred the LAER-track supposably. Based on informations given in the year 2003 (Ailor, 2003) the LAER-track was chosen for 40 conventional batteries.

Emissions from pushing, quenching, and combustion stacks are adressed in (US-EPA, 2003a). The most relevant figures of this rule are given on Table 7. The local authority can make an order on more stringend limits than given on Table 7 on special reason, and can set emission standards for other emitted compounds than given on Table 7 with regard to the allowed annual mass flow, aditionally.

process	emission	limit value	unit of measurement	remark
pushing				
fugitive (not captured) emissions	opacity*	< 30/35	%	depending on oven hight *
outlet of dedusting device	dust	0.01 – 0.04 (5 – 20)	lb/t$_{short}$ coke (g/t $_{coke}$),	depending on type of control device
battery underfiring				
stack for offgas	opacity*	< 15/20 %	%	depending on coking time
quenching				
outlet of quench tower	dissolved solids	< 1.1	mg/l	quench water

Table 7. Emission standards for coking plants according (US-EPA, 2003a); *: determination of opacity is made by Method 9 given by US EPA (US-EPA, 1996)

German and European legal regulations set no standards for opacity. Therefore, only the 0.02 lb/t$_{short}$ (10 g/t) limit for pushing emissions from the stack when applying a moveable hood with a stationary control device can be compared with the relevant figure of 5 g/t coke set by German TA Luft for this technique.

In addition to the limit values as described before, the US environmental legislation sets work practice standards. These standards, for example, describe techniques which have to apply with regard to emission control and to emission monitoring, or how to operate the coking plant in a most environmental friendly way.

4.3.2. Quantification of visible emissions

The philosophy of EPA's rules for visible emissions caused from coke oven operation is based on a chain of causalities between:

- number of visible emissions, and
- mass flow of the emitted hazardous compound, and
- concentration of the emitted hazardous compound in ambient air, and
- ambient air quality and cancer risk

due to the usual practice when rating the health risk caused by air pollutants by dose/effect relations. This means, that, amongst others, there must be a quantitative correlation between the set standards for visible emissions and the emission mass flow (mass per time) of the hazardous compound.

The latter can be calculated on base of the frequency of the visible emissions (leaking rate) and of the source strength (emission mass flow) of the visible emission (US-EPA, 2008a, 2008b). Typical source strengths given as kg BSO/h/leak as derived from from page 4-30 of (US-EPA, 2008b) are listed on Table 8. BSO means the so-called Benzene soluble (BSO) portion of the emission. By using a conversion factor for BaP/BSO of 0,00836 (US-EPA, 2008b) the relevant BaP emissions can be calculated. They are given on Table 8 too.

type of leak	kg BSO/h/leak	mg BaP/h/leak
leaks observed according EPA 303 from the yard	0.019	159
leaks observed from the bench*	0.011	92
without visible emissions	0.002	17

Table 8. Emission mass flows of door leaks as given by US EPA (US-EPA, 2008b); *: for calculations according equ. 1 smaller leaks which cannot be observed from the yard but only from the bench are additionally taken into account; US EPA estimates the leaking rate of these emissions to 6 % as an average.

Applying a 4 % leaking rate (according EPA method 303) at the doors (post-NESHAP control standard according (US-EPA, 2008b)) the total BSO emissions of a model battery with 62 ovens (124 doors) can be calculatet as follows:

$$[(124 \times 0.04) \text{ method 303 leaks} \times 0.019 \text{ kg/h/leak} +$$

$$(124 \times 0.06) \text{ bench leaks} \times 0.011 \text{ kg BSO/h/leak} +$$

$$(124 \times 0.90) \text{ no visible leaks} \times 0.002 \text{ kg/h/leak}] \times 8760 \text{ h/a} = 3\ 498 \text{ kg BSO/a.} \qquad (1)$$

Considering a coke plant with a coal input of 492 000 t/a (344 000 t coke/a) a specific emission factor of 0.0071 kg BSO/t(coal) results for door emissions. By using a conversion factor for BaP/BSO of 0.00836 (US-EPA, 2008b) the specific BaP emissions from the doors amounts to 59.4 mg/t $_{\text{coal}}$ and 84.8 mg BaP/t$_{\text{coke}}$, resectively. By comparable evolutions emission factors for leaks at lids and offtakes as well as for charging can be received (Table 9; compare with Table 4-11 of (US-EPA, 2008b)). It is obvious that the doors are the dominant emission source out of all leaks at the battery.

US-EPA standard	BSO			
	charging	doors	lids	offtakes
	kg/t_{coal}	kg/t_{coal}	kg/t_{coal}	kg/t_{coal}
POST-NESHAP	0.00025	0.0071	0.000044	0.00015
	BaP			
	charging	doors	lids	offtakes
	mg/t_{coal}	mg/t_{coal}	mg/t_{coal}	mg/t_{coal}
POST-NESHAP	2.09	59.36	0.37	1.25
	BaP			
	charging	doors	lids	offtakes
	mg/t_{coke}	mg/t_{coke}	mg/t_{coke}	mg/t_{coke}
POST-NESHAP	2.99	84.79	0.53	1.79

Table 9. Specific emissions at doors according (US-EPA, 2008b)

Emission factors as given in (US-EPA, 2008b) are based on measurements carried out before the year 1980 on coking plants, which could not meet the emissions control standards of current plants. Thereby the coke-side dedusting facilities were used for capturing the emissions from the doors. The US EPA by itself designates the results of these measurements as highly uncertain.

5. Progress in emission control technologies – Best Available Techniques (BAT)

Environmental legislations for industrial plants, like the German TA Luft (TA-Luft, 2002) or IED of the EU (EU, 2010), demand very often for application of the so-called Best Available Techniques (BAT) for emission control according the state of the art in technology, (section

4.1/4.2). The following section will give a brief description of the most important techniques. Additional informations on the emission levels which can be achieved by the relevant technique are given on Table 4 (section 4.2).

5.1. Battery operation

5.1.1. Charging

BAT is an emission free charging by transfer of charging gases to the collecting main and into the neighbour oven, as an option (Fig. 11)

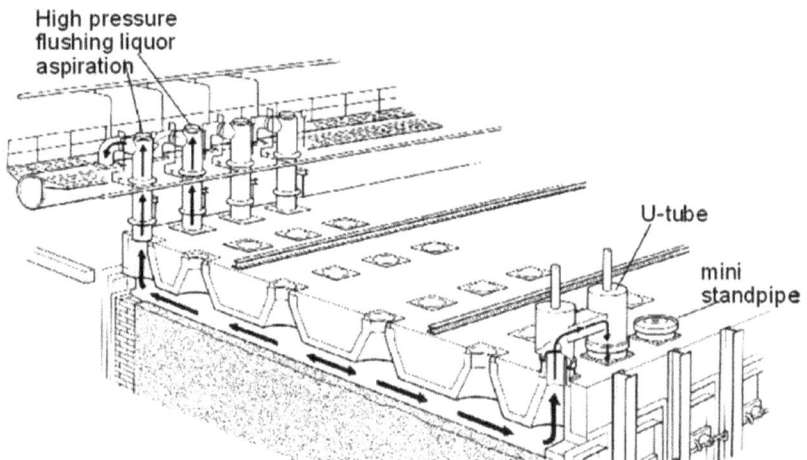

Figure 11. Principles of emission free charging of coke ovens

5.1.2. Larger oven chambers

A reduction of total fugitive emissions from battery operation can be achieved by lessening the sealing surfaces as well as the number of oven cycles. Naturally, such measures can be achieved only when building a new battery equipped with larger chambers as they were built by 7 to 8 meter ovens in the 1980th in Germany (section 6). Larger oven chambers provide less openings per t of produced coke due a reduction of the specific sealing surface. Fig. 12 shows (top side) the reduction of the number of closure facilities (openings) which was reached by a replacement of two smaller and older plants by the new coke plant Kaiserstuhl III, while the total capacity of both variants kept constant at 2 million tonnes coke per year. The drastic reduction of fugitive emissions of Benzo(a)pyrene and Benzene, caused by less openings but also by improved techniques, can be read from Fig. 12 (bottom side).

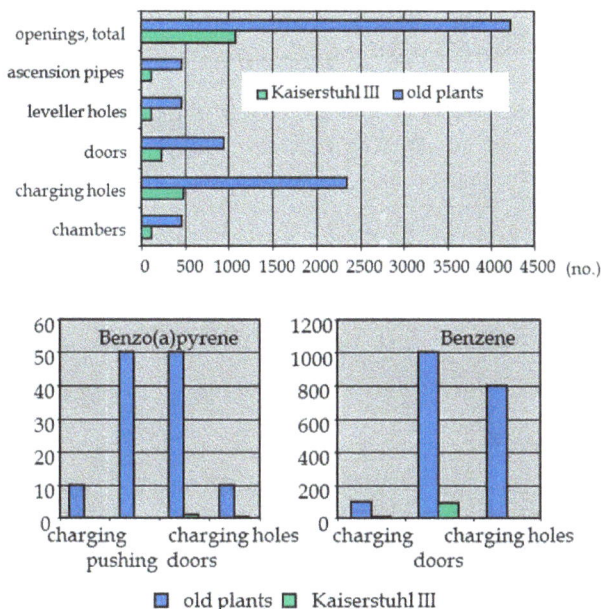

Figure 12. Emission reduction by lessening the sealing surface; top side: reduction of openings; bottom side: reduction of fugitive emissions by Benzo(a)pyrene and Benzene respectively (Hein, 2010)

Construction of larger oven chambers do not favour the intention of environmental control only, but also the economics of cokemaking. Desing data of the modern high capacity batteries as running in Germany today, can be received from Table 10 in section 6.

The development of chamber heights during the last 100 years is shown very arrestingly in Fig. 13.

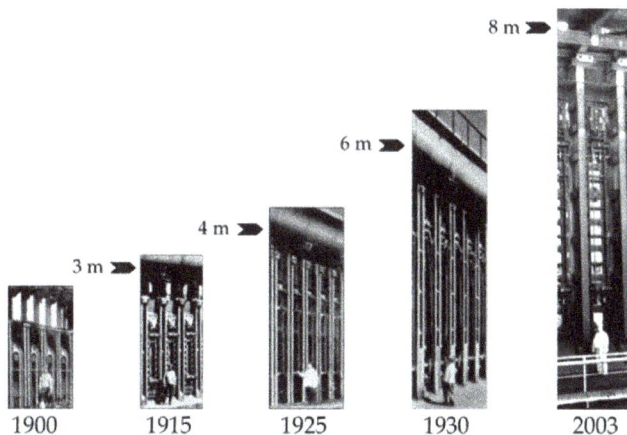

Figure 13. Development of typical heights of coke oven chambers (Hein, 2009)

5.1.3. Closure facilities

In order to improve the control of fugitive emissions from leaks at the battery, optimized closure facilities at doors, charging hole and offtakes have to be applied, and a good maintenance of them is demanded. BAT are flexible doors with springloaded sealings (Fig. 14, left side), for batteries higher than 6 m especially. An additional improvement is attainable if the pressure gradient at the sealing that constitutes the driving force for emissions could be lowered. This was done by the coke oven builders by means of gas channels in the door through which the escaping gas can flow into the direction of the gas space without greater flow resistance. All modern coke oven doors meanwhile have such gas channels as can be seen from Fig. 14, right side).

At the offtakes water sealed lids are BAT in order to reduce emissions.

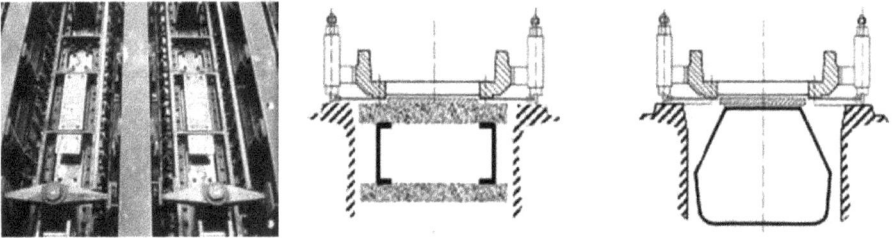

Figure 14. Modern door systems; left side: flexible doors (Krupp-Koppers, n.d.); right side: principle drawings of gas channels behind the door (Arendt et al., 2009)

5.1.4. Oven chamber pressure regulation

A reduction of fugitive emissions can be achieved by measures to regulate the chamber pressure within the coke ovens as function of progress in carbonization. BAT, e.g. is the PROven system (Pressure Regulated Oven), which was invented by DMT (Huhn, 1995). PROven regulates the pressure within each oven chamber at a constant and slight positive pressure during coking in order to eliminate fugitive emissions as much as possible. Fig. 15 shows on the left side principles of this system, and on the right side the reduction of PAH emissions by use of PROven in contrast to a non pressure regulated oven chamber (100 % PAH) (Spitz, 2005). In the year 2011 the PROven system was installed at 15 coking plants worldwide with more than 2100 ovens (Kaiser, 2011) including the new coking plant Schwelgern.

An alternative system has been developed by Paul Wurth and is called SOPRECO (Single Oven Pressure Control System). In 2011 the SOPRECO system was installed at the coking plant Dillingen, Germany, in 50 ovens, a second battery with 50 ovens is under construction (Faust, 2010).

5.1.5. Battery heating

Emissions from battery underfiring are limited by application of the following techniques: improved desulphurization of the used coke oven gas in order to a reach a remaining

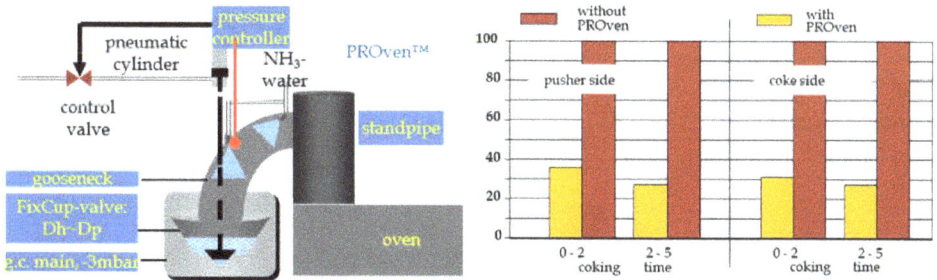

Figure 15. Left side: Principle of the ROven-system; right side: achievable emission reduction for PAH compounds

sulphur content of less than < 0.8 g /Nm³ and by special heating relevant technical measures in order to comply with a NOx standard of 500 mg/Nm³. While the desulphurization is achieved by absorption or by wet oxidation of H_2S (see section 5.2.1.), the NOx reduction is reached by waste gas recirculation and stage wise heating, in particular (Fig. 16). The latter was necessary anyway because of the taller becoming chamber heights.

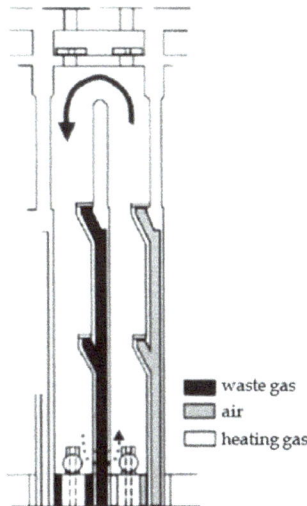

Figure 16. Principle scheme of stage wise heating

5.1.6. Coke pushing

In order to minimize emissions during coke pushing, an installation of a dedusting system is required, disposing of a hood, a suction device and of a filter system. The so-called "Bandschleifenwagen" (Fig. 17) with a subsequent stationary dedusting achieved acceptance.

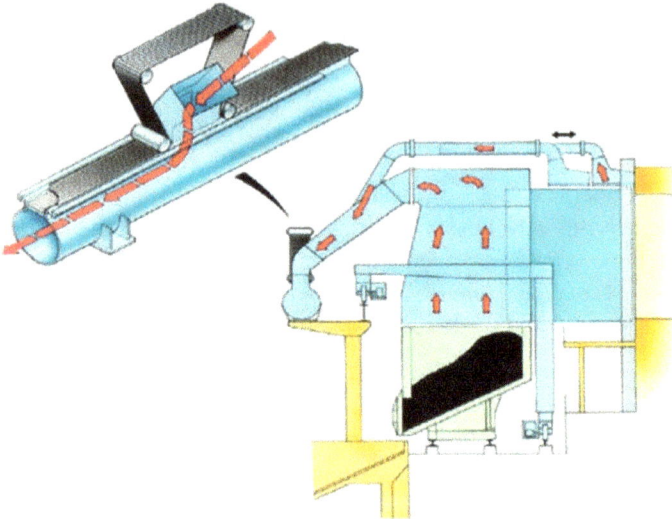

Figure 17. Drawing of the "Bandschleifenwagen" as part of the coke side dedusting device (Stoppa, 2003)

The efficiency of a modern coke side dedusting system is illustrated from Fig. 18.

without

with de-dedusting

Figure 18. Coke pushing without (left side) and with coke side dedusting (Coking Plant Prosper, Germany - right side)

5.1.7. Quenching

BATs are wet quenching as well as dry quenching.

Wet quenching

The hot coke is treated by water spraying under the quench tower to cool it down. The caused dust is hindered to leave the tower by special baffle constructions which are installed in the tower. The so-called Coke Stabilisation Quenching (CSQ) represents an advanced quenching technology comprising a combination of spray quenching and submerging in water. The CSQ tower contains a two set of baffles and comprises a hight of 70 m, in contrast to approx. 40 m which was the maximum hight of conventional quenching towers up to now (Fig. 19).

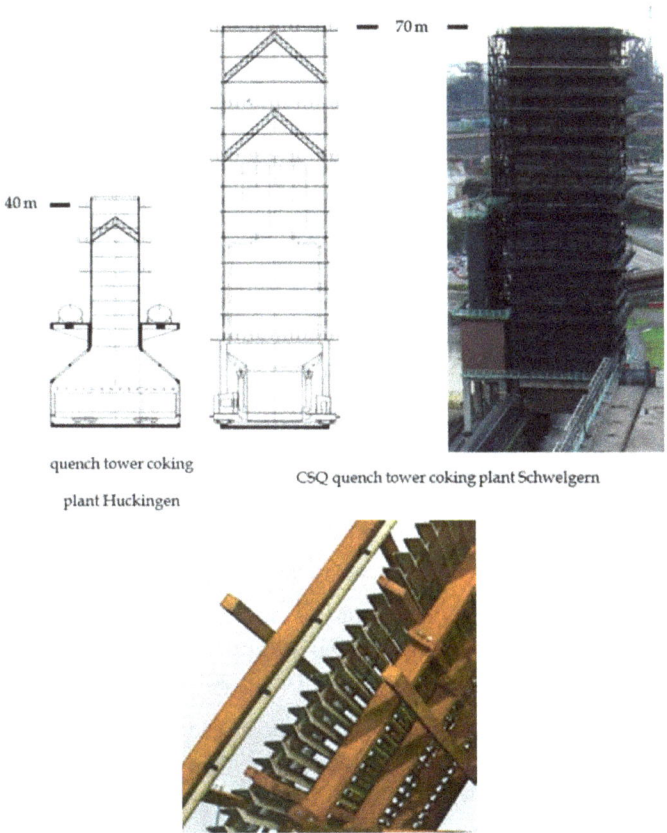

quench tower coking

plant Huckingen

CSQ quench tower coking plant Schwelgern

Figure 19. CSQ quench tower of the coking plant Schwelgern in contrast to the quench tower of the coking plant Huckingen (top side); bottom side: baffles (Nathaus, n.d.) for dust emission control before installation in a quench tower

Dry quenching

During dry quenching the hot coke is cooled down in a closed cooling chamber by use of an inert gas which is circulated and cooled down thereby within a heat exchanger. The produced steam can be used for electricity production. A scheme of a dry quenching plant is shown in Fig. 20.

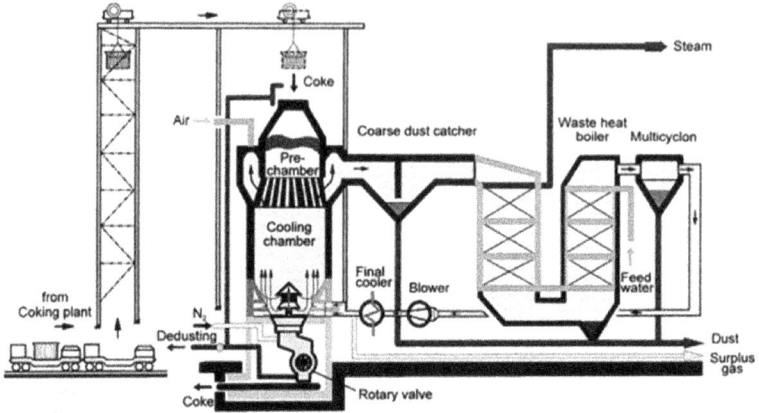

Figure 20. Schema of the dry quenching plant of the former coking plant Kaiserstuhl III (Stoppa et al., 1999)

Dry quenching is extended for application in countries, in which a water operated wet quenching is not possible because of meteorological reason, or which are characterized by water shortage. On the other hand, the use of dry quenching techniques is advantageously to operate in countries with high prizes for electricity.

5.2. By-product plant

5.2.1. Desulphurisation of coke oven gas

Because of its hydrogen sulphide (H_2S) content (up to 8 g/Nm³) unpurified coke oven gas (COG) is unsuited for use in many industrial applications. Typical desulphurisation processes according BAT to clean COG are (Sowa et al., 2011):

- absorption/stripping processes with subsequent conversion to sulphur containing compounds,
- wet oxidation processes with subsequent production of sulphur.

In Europe, the most commonly applied process is the absorptive process using a so-called ASK process (Ammonia-Sulphur cycle process, ASK; see Fig. 6 in section 2.2., too). It is a combination of H_2S and NH_3 removal. A first scrubber removes H_2S, using deacidified water providing from the distillation. A second scrubber is in combination with the first one for the removal of NH_3. The washer fluid which is loaded with H_2S and NH_3, respectively, is sent to a distillation unit (stripping/deacidification). This unit removes the adsorbed gases from the enriched solution; the water is mostly recirculated to the gas scrubbing. The H_2S/NH_3-vapours are led to the desulphurization unit, which is mostly a catalytic ammonia cracking combined with a sulphur recovery plant (Claus plant). A photo of a modern Claus plant can be seen in Fig. 21. Other options for desulphurization are the production of supheric acid or ammonia suphate. In all cases the produced chemicals are further by-products.

Figure 21. View on a modern Claus plant

The second absorptive process variant is the Vacuum Carbonate process commonly operated with potassium carbonate which has some tradition at West European and Asian coke plants.

The most commonly applied wet oxidative process (outside Europe) is the Stretford process. Wet oxidative processes possess a higher efficiency for H_2S removal than adsorption processes (see Table 4). However, they need the addition of specific chemicals, like vanadium compounds, quinone and hydroquinone compounds as catalysts, the wastes of which have to be discharged. Usually this waste water is treated separately owing to the presence of compounds that have a detrimental effect on the biological wastewater treatment plant.

5.2.2. Gas tight operation of the by-product plant

In modern by-product plants fugitive gaseous emissions are minimized by gas-tight operation of the gas treatment plant. The measures are, minimize the number of flanges, using of gas-tight flanges, or closed venting system for tanks and equipment containing aromatic hydrocarbons. By use of pumps and piping suitable to prevent leakages, a release of any effluent to the environment can be avoided.

5.2.3. Biological waste water treatment plant

BAT is a wastewater treatment by using efficient tar and PAH removal, using efficient ammonia stripping and biological waste water treatment with integrated nitrification and denitrification to fulfill the common local regulations for discharge water quality. Limiting values are existing for free ammonia, NH_3-N, BOD, COD, cyanides, hydrocarbons and phenol.

6. Situation of the German cokemaking industry

Today five modern coking plants comprising with high capacity batteries are in operation in Germany. These plants, the fotos of which are given on Figures 22 and 23, fulfill the highest standards for emission control techniques with regard to the state of the art. They are equipped with modern wet quenching systems in order to comply with the legal demands of the actual TA Luft (TA-Luft, 2002) while the former coking plants August Thyssen and Kaiserstuhl III have been provided with modern dry quenching facilities.

In 2011 battery no. 1 of the coking plant Dillingen is under construction; this is a replacement of an old battery. At Huckingen a second battery is under construction as an extension.

1983: Dillingen

1984: Huckingen

1985: Salzgitter

1985: Prosper

Figure 22. Coking plants currently in operation in Germany which were build in the 1980th, including date of commissioning

2003: Schwelgern   2010: Dillingen
                                              (battery no. 3)

Figure 23. Coking plants currently in operation in Germany which were commissioned under the influence of the TA Luft 2002, including date of commissioning

The most essential design data of the five coke plants operating today are summarized on Table 10.

		oven chamber dimensions				production rate	
		hight	length	width	volume		per oven
	commis-sioning	m	m	mm	m³	10^6 t/a	t/a
Central Coking Plant Saar	1983/2010	6.25	16.4	490	44	2.4	13300
Huckingen	1984	7.8	18	550	70	1.08	15430
Salzgitter	1985	6,2	16.6	465	43	1.42	13150
Prosper	1985	7	16.3	590	61	2	13700
Schwelgern	2003	8.4	20	590	93	2.64	18860

Table 10. Design data of the five German coke plants currently in operation (Hein, 2009)

The chamber height of 8.4 m of the new Schwelgern plant marked a new record for coke constructions. Now, the coke plant with the tallest chamber heights and the highest chamber volume worldwide is operating at the coking plant Schwelgern in Duisburg, Germany. The coking plant Saar in Dillingen is operated as stamp charging plant, and with 6.25 m hight the tallest for this technique.

The total production of the five plants was 8.15 mio. t coke in 2010. This is a sharp decrease when looking back to the year 1957 when approx. 50 mio. t coke were produced (Fig. 24).

The main cause for this change in Germany was the decline of coke sale for home firing and other applications than for pig iron making. On the other hand the coke need of the German iron and steel industry has fallen due to the reduction of the specific coke demand for the blast furnace as well as to the buying of coke from abroad, while the total hot metal production kept nearly constant since this time. The necessary adjustments in capacity were carried out in such way, that preferable older plants were shut down, which could not meet

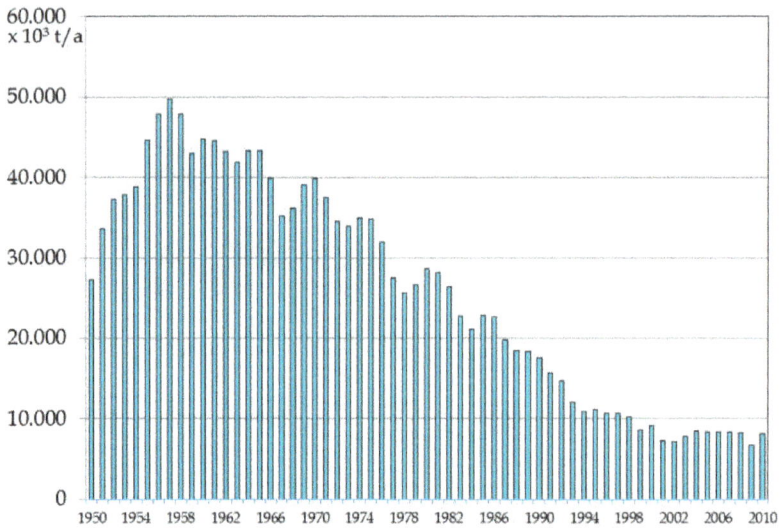

Figure 24. Annual coke production in Germany since the year 1950 (Kohlenstatistik, 2012)

the more stringent environmental standards, and which were not able to reach the economics which were typical for this time. This change has faced the mining industry, in particular, as this branche was the owner of nearly 75 % of the coking plants in Germany 50 years ago.

Due to the former dominance of the mining coking plants for the coke production the most sustainable impetus for new developments in cokemaking technology came from the German mining industry till the early 1990th years. Thereby, in particular, the basics were set for the construction of high capacity batteries as realized in the five coking plants running today, by research and development carried out in technical and semi-technical testing facilities for coking trials owned by the mining industry. The research in cokemaking technologies was centralized at the Bergbau-Forschung in Essen, the nucleus of the today´s DMT GmbH & Co. KG.

Progress made in further development of cokemaking technology and its implementation in practice, in particular, would not have been successful without the innovative legacy of the German coke oven constructor companies. Out of the four prosperous German companies Dr. C. Otto, Carl Still, Heinrich Koppers and Didier Kogag Hinselmann, today only one exists, the Uhde GmbH which took over their business activities during the last 30 years step by step. German cokemaking technique is accepted worldwide, and according to this it is not surprising that more than 100 000 coke ovens all over the world have been constructed by German companies.

Progress reached in emissions control on German coking plants can be described by a drastic reduction of production specific emissions caused by battery operation due to the more stringend becoming legal rules for environmental control (Fig. 25).

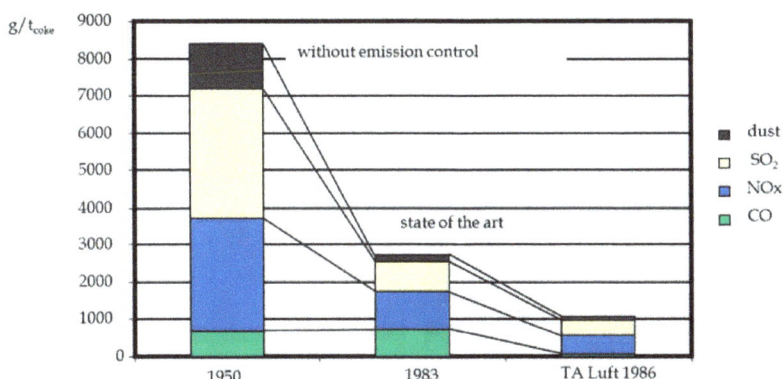

Figure 25. Reduction of specific emissions on German coking plants between 1950 and 1986

7. Determination of fugitive emissions of Benzo(a)pyrene from leaks at the battery

7.1. Measuring method

A quantitative method for measuring fugitive emissions from leakages at the battery was developed by Deutsche Montan Technologie GmbH (DMT) and its predecessor institute Bergbau-Forschung GmbH (BF) respectively. The relevant measurements included particle bound as well as gaseous compounds, and were carried out between 1980 and 2006 at various coking plants of different age in Europe which additionally were different in their design and in the state of maintenance of the closure facilities.

For the measurements a complete encapsulating of the relevant source is necessary as described in the following as an example of measurements at the coke oven doors. For this the outer door zone of the coke oven door is covered (see Fig. 26, left) by a thermo-stable transparent film (foil) in order to detect the strength of visible emissions, simultaneously. Preferentially the foil is fixed on the buck stays. The gas accumulated in the collecting space has to be withdrawn and analysed. For this, the foil at its bottom contains an opening while the top of the collecting space is combined with a vertical arranged tube. Because of thermal buoyancy clean air enters the opening at the bottom while the mixture of air and the emissions looked for leave through the pipe at the top of the collecting space. Typical volume flows are in the range between 50 and 200 Nm³/h depending on the design, the dimension of the door, the magnitude of the opening at the foil's bottom as well as on the meteorological marginal conditions. The relevant gas velocities range between 4 to 10 m/sec. From this main gas flow the sampling gas was sucked off isocinetically with a flow rate of about 2 Nm³/h.

For measurements of leakages at closed lids of the charging holes and of the offtakes, respectively, equipment for encapsulating was used, as shown on Fig. 26, (right). In order to get a constant gas-flow, pressured air as carrier gas was injected into the encapsulated space.

Figure 26. Equipment for measurements of fugitive emissions at doors (left side), lids (right side, top) and offtakes (right side, bottom)

In all cases the sampling gas is led via a dust filter and afterwards through an additional filter containing a synthetic resin for adsorption of still remaining gaseous PAH compounds. Sampling has to be done during the whole coking cycle, which was devided in several steps with separate sampling in some trials.

The taken samples are analysed in the laboratory for PAH-compounds by means of GC/MS and HPLC, respectively, in accordance with a national standard method (VDI, 1996).

7.2. Results from measurements at single leaks

Results from measurements at single leaks are given as emission mass flow mf (mg BaP/h/closure facility) as an average of the sampling time) in a first step. The relevant figures are derived from the initially measured mass concentration (mg BaP/Nm^3) in the sampling gas and the main gas volume flow (Nm^3/h). To make the results more comparable the emission mass flows are converted to product specific emissions (mg BaP/t$_{coke}$) by consideration of the production rate per oven and the coking time. This figure is typical for the closure facility under investigation.

Fig. 27 shows the distribution of BaP in the gaseous and on the particle phase of emissions from oven leaks, as function of total particle concentration and off-gas temperature, respectively. It could be shown, that with increasing temperature of the waste gas, the portion of BaP in the gaseous phase increases, too (Fig. 27, right). And one receives the result, also, that with increasing dust emission the portion of BaP in the gaseous phase descreases (Fig. 27, left).

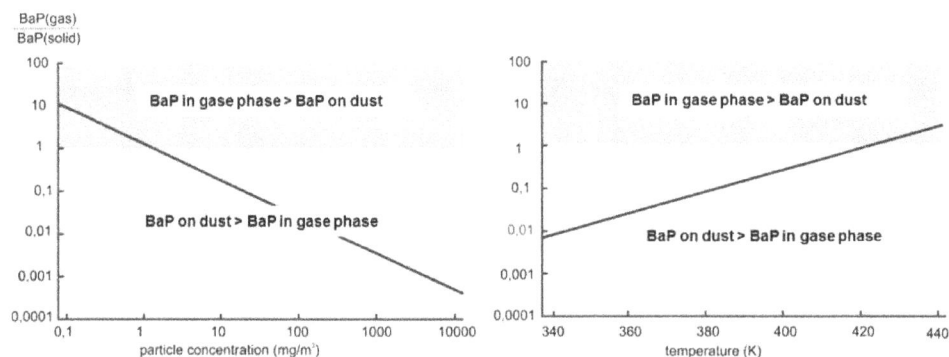

Figure 27. Proportion of BaP in gaseous and dust bound phase in emissions from coke oven leaks

Typical emission ranges for Benzo(a)pyrene as received by the measurements with concern to leaks at coke oven doors and chamber lids, respectively, are listed in Figures 28 and 29 (Eisenhut et al., 1990, 1992).

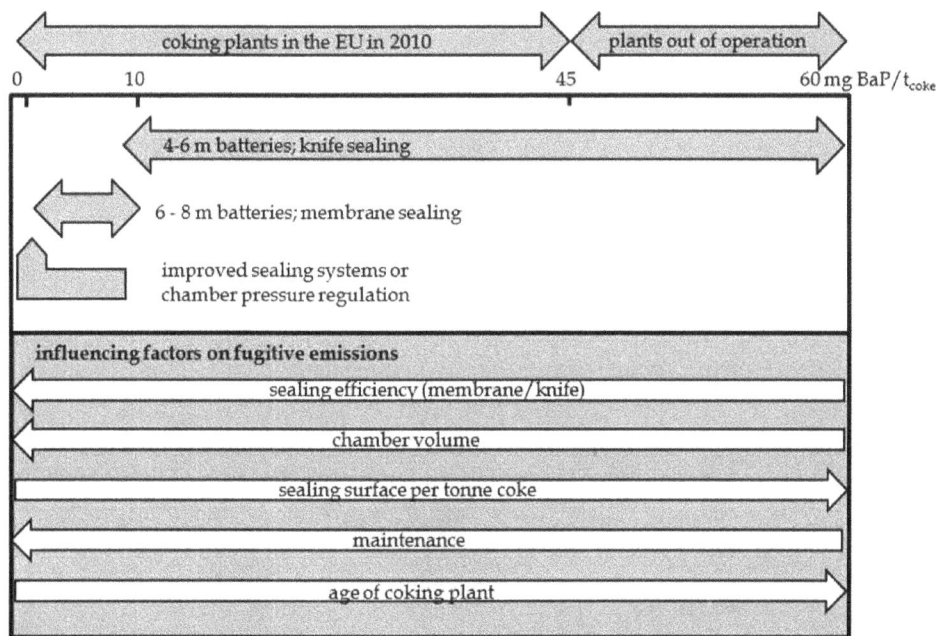

Figure 28. Typical ranges for Benzo(a)pyrene emissions (mg BaP/t$_{coke}$) from single leaks at coke oven doors as received from measurements

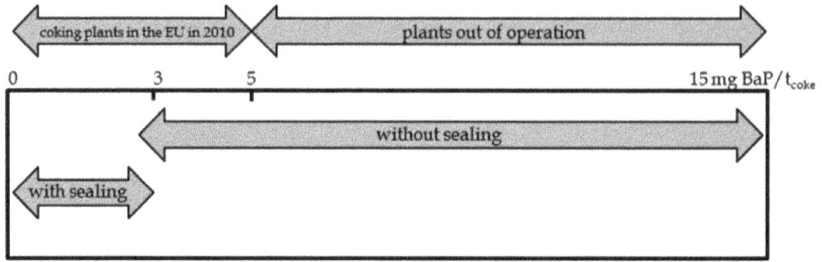

Figure 29. Typical ranges for Benzo(a)pyrene emissions (mg BaP/t$_{coke}$) from single leaks at charging lids as received from measurements

Fig. 28 shows also factors which have influenced the measurement results. These influence factors are valid for the results of measurements at the closed lids of the charging holes, too (Fig. 29). In both cases the age of the plants, the maintenance of them, the quality of the sealing facilities and the specific sealing surface per tonne of coke, which is in the opposite direction with the oven volume, have an impact on the amount of the emissions. As the measurement have started in early 1980th the shown ranges for emissions also include results from old plants with 4 m ovens in a bad condition and antiquated techniques for emission control. These plants are no longer in operation in Europa. And also in a more generalized view, one has to state that these plants are not typical for worldwide cokemaking operation of today. By consideration of this, Table 11 contains typical emission ranges for Benzo(a)pyren for coking plants caused by single leaks at the batteries which are still running today. Besides emissions from leakages at closed doors and lids, Table 11 contains also emissions from closed offtakes. Consequently it is to state that the lowest BaP emissions can be received at 6 to 8 m high flexible doors which are equipped by membrane sealings. The relevant emissions per door lie in the range between 1 to 10 mg BaP per t of coke. For new plants with an excellent maintenance, emissions at single doors go down to 1 mg/t$_{coke}$. Under optimal conditions, for example if a chamber pressure regulation system is installed (chapter 5.1.4.), BaP emissions are reduced below 1 mg/t$_{coke}$. BaP emissions at the chamber lids lie in a range between 0.3 and 5 mg/t$_{coke}$. The lowest emissions can be achieved at modern and well tended plants if the lids are sealed by special fluids or pastes after closing the relevant opening at the roof of the battery. In this case emission below 1 mg/t$_{coke}$ can be received. Typical BaP emissions from leaks at the offtakes are below 3 mg/t$_{coke}$. On modern plants with water sealed lids at the offtakes emissions go down below 1 mg/t$_{coke}$.

doors	control technique	lids	control technique	offtakes	control technique	unit of measurement
10 - 45	knife sealing	3 - 5	not sealed	< 3	metal/metal	mg BaP/t$_{coke}$
1 - 10	membrane sealing	0.3 - 3	sealed	< 1	water sealed	mg BaP/t$_{coke}$
< 1	improved techniques, like PROven					mg BaP/t$_{coke}$

Table 11. Product specific emissions for single leaks at the batteries of current coking plants

From Fig. 30 one can derive that over three-fourth of the fugitive BaP emissions from battery leaks in total is caused by emissions at the doors.

Figure 30. Spread of fugitive emissions from single leaks at the battery

This is in good correlation with the Figures given bei the US EPA (Table 9 of section 4.3.2.), and is the reason why in the following section emissions from coke oven doors are concerned, only, when discussing strength of leakages, as estimated by the US EPA and DMT, respectively.

7.3. Investigations at door leaks of definite strength

Normally, by use of only one emission figure, as received from Table 11, and multiplication with the annual coke production is not possible to estimate the annual BaP emissions of the total coke oven battery. The reason for this is the inequality of the strengths of the emissions at the various sources of one type (door, lid and offtake, respectively).

Analogously to the procedure from the US EPA (see section 4.3.2.) the total emissions of the plant should be calculated on base of the frequency of the visible emissions (leaking rate; section 7.4.) and of their strength (mass/h/leak), in the following. This will be done as an example for door emissions, as these emissions play the dominant role with regard to the total emissions caused by the battery (see Fig. 30 in section 7.2.)

To meet this goal varios door leaks, which strongly differ in their visible strength, were investigated as described in section 7.1., however by applying shorter sampling times (up to 5 h) with a nearly constant source strength over the sampling period. Typical strengths of visible emissions at doors are shown in Fig. 31. The emissions are categorized in:

-    strong (st),
-    medium (m),
-    slight (sl)
-    non visible emissions (n.v.e.)

For each category of visible strength typical BaP emission mass-flows (mf) could be determined, the ranges of which are shown on Table 12 (see also Fig. 32 in section 7.6.).

The specific mass-flows which are typical for visible emission strengths can be transfered to other plants where measurements have not been carried out. The assignment has to be done by an expert, on base of comparisons with results of measurements at comparable plants.

Figure 31. Four categories of visible strengths of door emissions

strength of visible emissions	strong (mf$_{st}$)	medium (mf$_m$)	slight (mf$_{sl}$)	n.v.e. (mf$_n$)	unit of measurement
all plants	150-600	50-150	10-40	< 10	mg BaP/h/leak
plants according state of the art (membrane sealings)	150-200	50-150	10-40	< 10	mg BaP/h/leak

Table 12. Typical emissions BaP mass flows mf for leaks of different visible strength at coke oven doors

7.4. Assessment of visible emissions and of leaking rates

The leaking rates at the different sources are determined by an inspection of the battery and counting the visible emissions according tho EPA method 303 (US-EPA, 1993b). A distinction from the EPA method is made with regard to the different strengths of the visible emission, as it is shown for door emissions in Fig. 31, as an example.

Thus, the result of the determination of visible emissions will be, in pinciple:

no. k of strong emission
no. l of medium emissions
no. of slight emissions, and
n-(k+l+m) no. of none visible emissions,

whereby k,l and m are the numbers of leaks with visible emissions of different strengths, and n ist the number of doors in total.

The DMT-method for inspection of the leaking rates differs from the US-EPA 303 method by its four categories for emission strength while the US EPA method only results in the decision on the existence of a visible emission or not.

7.5. Determination of the total emissions caused by the battery

By mathematical combination of the number of leaks with their relevant emission mass flow the total emission E (mg BaP/h) of the battery (plant) with regard to emissions from door leaks can be determined, according equation 2.

$$E = k \times mf_{st} + l \times mf_m + m \times mf_{sl} + (n-k-k-m) \times mf_n \qquad (2)$$

Where mf_{st}, mf_m, mf_{sl} and mf_n are the emission mass flows of different strengths of visible emission (Table 12), k,l and m are the numbers of visible emissions of different strengths at doors, and n ist the number of doors in total. Equation no. 2 is comparable to equation no. 1 (section 4.3.2.) by which relevant calculations are made by US EPA (US-EPA, 2008b). Product specific BaP emissions caused by door leaks, which are typical for the emissions of the total plant, can be derived by multiplication of the result of equation no. 2 with the annual operation time and dividing by the annual coke throughput. Results of these calculations, which often are called emission factors, are given in section 7.6., and are compared there with relevant emissions given by the US EPA.

7.6. Comparison of BaP emissions from own measurements with results given by US EPA

On base of equation 2, total BaP emissions caused by all doors of a modern high capacity battery (70 ovens, 7.8m hight, 1 mio. t coke per year) are calculated (line 8 and 9 of Table 13) by applying the extreme values of the given ranges for emission mass flows according Table 12 (line 3). Leaking rates (portion of no. of visible emissions (no. v. e.) of the total no. of openings in %) of 4 % (2 % slight and 2 % medium emissions) according the post-NESHAP standard and of 3.3 % (1 % slight and 2.2 % medium emissions) according the LAER standard are applied in order to make the results comparable with calculations of the US EPA (line 1 to 7 of Table 13).

Results given in lines 1 and 2 are derived on base of the model battery, as described in section 4.3.2. (62 4 m ovens per battery with a coke capacity of 344 000 t coke per year), and on leaking rates of 4 % and 3.3 % respectively, analogously to equation no. 1. These emissions will be reduced significantly when considering a high capacity battery with larger oven dimensions (line 3 and 4) due to the lower specific sealing lengths. Lines 5 to 7 contain ranges for BaP emissions caused at doors as given by a Risk Assessment Document of the US EPA (US-EPA, 2003b) for 5 US batteries which comply with the LAER standard (2010) already today (see section 4.3.1. also).

The origin of the applied emission mass-flows for the calculations according equation no. 1 and no. 2 one can read from column 8 of Table 13. To make data from US EPA comparable with own results, a conversion of BSO to BaP and t_{short} to t_{metric} was necessary.

		leaking rate		batt. height	capacity	BaP	ref. of emission mass flow
		no. v.e. (%)		(m)	(t/a x10³)	(mg/t_{coke})	
1	model batt.	4	post-NESHAP	4	344	84,8	(US-EPA, 2008b)
2	model batt	3.3	LAER	4	344	81,4	(US-EPA, 2008b)
3	high capacity oven	4	post-NESHAP	7.8	1000	30,6	(US-EPA, 2008b)
4	high capacity oven	3.3	LAER	7.8	1000	29,4	(US-EPA, 2008b)
5	5 US batt.	1.58 - 2.81	actual	3.4 - 5	65 - 589	22 - 57	(US-EPA, 2003b)
6	5 US batt.	5; (3.8)	MACT	3.4 - 5	65 - 589	25 - 88	(US-EPA, 2003b)
7	5 US batt.	3.3; (3.8)	LAER	3.4 - 5	65 - 589	25 - 78	(US-EPA, 2003b)
8	high capacity batt.	4	2sl+2m	7.8	1000	2.65 - 16.43	DMT/Table 11
9	high capacity batt.	3.3	1ss+2.2m	7.8	1000	2.66 - 16.41	DMT/Table 11
	*: non visible emissions are not considered						

Table 13. Comparison of product specific BaP emissions (emission factors) caused by door leaks from own measurements with figures given by the US EPA (US-EPA, 2008b, 2003b).

From Table 13 one can read that all data given by the US EPA for BaP emissions from door leaks are significantly higher than those calculated by DMT. The reason for this is easily to understand and can be caused back to the higher values for the emission strengths (emission mass-flows of the single leak) as given by the US EPA (see Fig. 32 and Table 8 in section 4.3.2., respectively), and to the extra addition of 6 % emissions which can be observed only from the bench according the procedure of the US EPA. And in addition, it is to remark that the total emissions of plants according the state of the art with visible emissions less than 4 % are predominantly influenced by the strength of the non visible emissions (< 10 against 17 mg BaP/h/leak). The quality of the DMT-values for BaP emission strength could be confirmed by several dispersion calculations, by which the additional load caused by coke plant emissions on the ambient air in the surrounding of the coke plant, where the actual BaP concentration was determined by measurements, could be forecasted sufficiently on base of the above mentioned emission factors. In this context, it is to remark, that emission, as published by the US EPA, will lead to an overestimation of the BaP concentration in the surrounding, if forecasting (section 8.2.) the addition load in ambient air caused by a planned coking plant, e. g. in the process for getting a license for operation.

An explanation for the differences in BaP emission strength as determined by the US EPA and DMT, respectively, can be found in the high uncertainty of the US EPA figures, and in their determination on old coking plants with low standards for emission control, according to the acertainment of the US EPA by itself (US-EPA, 2008b).

If emissions from charging lids and offtakes are taken into account, one can assert that there are only slight differences in the emissions determined by DMT and the US EPA, respectively.

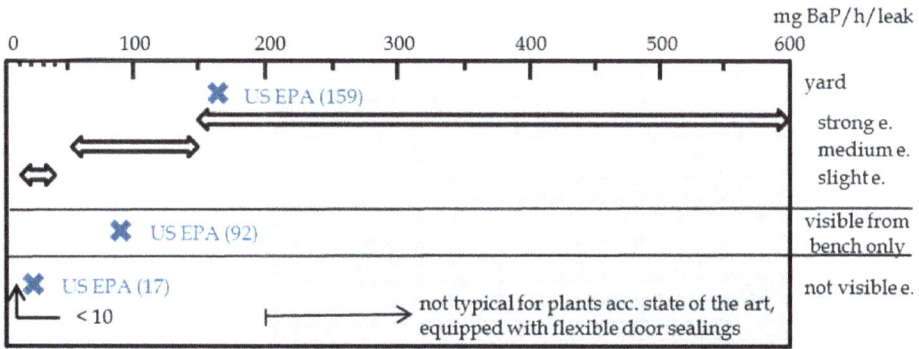

Figure 32. Ranges for emission strength (mg BaP/h/door leak) as determined by DMT and US EPA (cross marks), respectively (US-EPA, 2008b)

8. Benzo(a)pyrene in the vicinity of coking plants

A correlation between the Benzo(a)pyrene (BaP) emissions caused by a coking plant and the BaP concentration in ambient air in the surrounding of the plant could be shown by a lot of measurements. Measurements were made according (DIN-EN, 2008) by analysing the partice bound portion of the collected dust. Thereby factors could be determined, which influence the amount of concentrations, as given on Table 14, and which will be described in the following.

BaP in ambient air near coking plants is caused by:	-applied techniques for emission control on the plant
	-status of plant maintenance
	-age of the battery and of the closure facilities
	-local meteorological influences on spread of emissions
	- distance of the impacted area (measuring point) from the battery

Table 14. Factors influencing BaP concentrations in amient air caused by coking plant operation

Three coking plants, located in the Rhine-Ruhr area in Germany, were under investigation. In the following they are called coking plant A, B or C.

8.1. Results from measurements

For more than 20 years, ambient air has been examined for BaP in the surrounding of coking plant A (LANUV, 2012). The measuring station is located about 800 m away, in lee-side position to the coke plant. Measurements are taken two times or three times per week over a sampling period of 24 h. The coking plant is a modern plant yielding an annual coke production of approx. 2 million tons. The batteries are aged approx. 25 years, and fulfill the requirements imposed under the 2002 TA Luft for emission control.

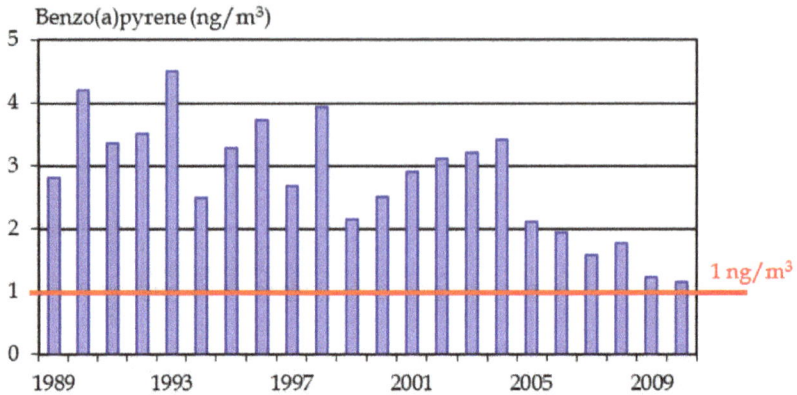

Figure 33. Benzo(a)pyrene in ambient air near (lee-side) coking plant A as annual mean (LANUV, 2012)

Fig. 33 shows the annual average concentration of BaP determined during the past years, that never fell under a BaP concentration of 1 ng/m³, which is set by the European Union as ambient air standard (EU, 2008).

The importance of coking plant´s emissions on the BaP burden in the vicinity can be proved by an evaluation of the measurements in front of the preferential wind direction at the measuring day within a two years' term (Fig. 34) (Hein et al, 2003; DWD, 2003; LANUV, 2003). The mean annual BaP concentrations for the period under evaluation (1999-2000) lie at 2.2 and 2.5 ng/m³, respectively. The highest BaP concentrations occur when the wind blows from the wind direction sector between 135 and 255°, with the maximum occurring during wind directions from approx. 200°. As the measuring station stands in a direct lee-side position to the coking plant in case of a wind direction from 195°, the inevitable conclusion is that the coke plant is mainly responsible for this burden during lee-side weather situations that reaches 3.7 ng/m³ on average. This conclusion can be confirmed by the absence of any other important BaP emitter in the weather-side of the coking plant. For measurements on days marked by wind directions falling outside the specified sector, it results a mean BaP concentration of 0.6 ng/m³. This BaP concentration is mainly congruent with the BaP background load which is typical for the industrial region where the coke plant is located, roughly amounting to 0.5 ng/m³.

When discussing the influence of meteorology on measured BaP concentrations, one should not ignore that other influential factors apart from the direction of wind are to be taken into account, for example the vertical exchange of air, which is typical for the season very often. A seasonal influence on the determined BaP concentrations can be clearly seen from measurements (LANUV, 2003) in the surrounding of coking plant B with an annual capacity of approx. 1.0 mio. t coke (Fig. 35). This plant is about 25 years old, and is equipped with techniques for emission control in compliance with the legal demands of the TA Luft 2002.

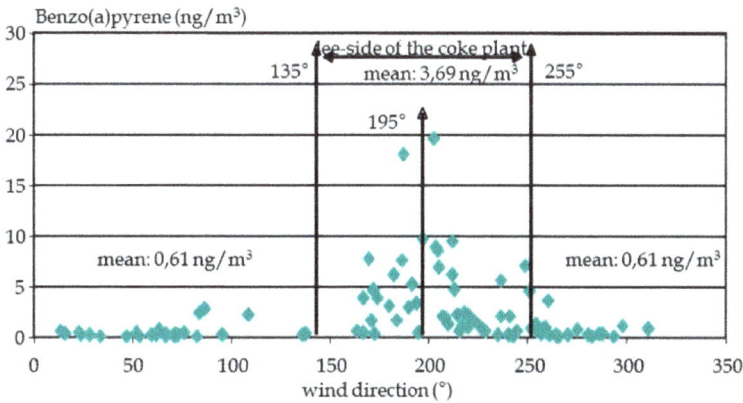

Figure 34. Influence of wind direction on Benzo(a)pyrene concentration in ambient air near coking plant A (DWD, 2003; Hein et al., 2003; LANUV, 2003)

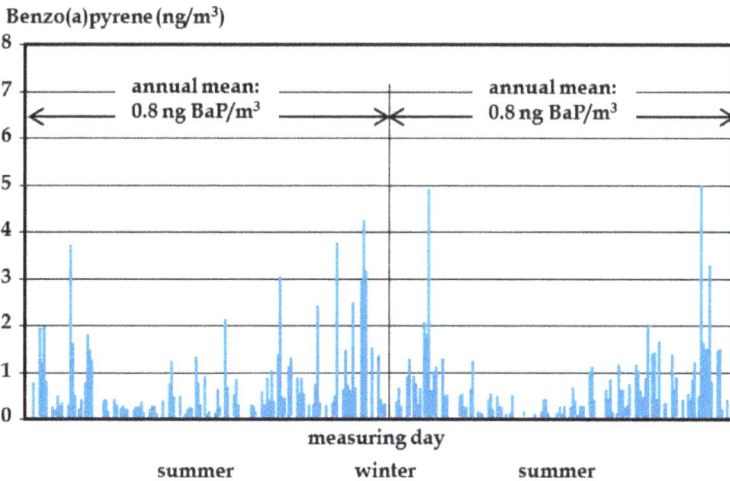

Figure 35. Benzo(a)pyrene in ambient air in lee-side of coking plant B during a two years period (LANUV, 2003)

The measurements were taken lee-side of the plant in a distance of 1000 m. Due to the larger distance of the measuring point from the plant, and to the less coking capacity, the BaP concentrations near coking plant B are lower than those near plant A. The annual BaP concentrations lie at 0.8 ng/m³ and complies with the relevant ambient air standard of 1 ng BaP/m³ as given by the EU (EU, 2004).

An influence caused by the seasonal effects, but also by improvements in the applied emission control techniques, can be also clearly seen from a three years measurement campaign in the surrounding of coking plant C, which composed of an annual capacity of approx. 1.5 mio. t coke (Fig. 36). The age of the various batteries of this coking plant, that has shutdown in 1999, was between 35 and 40 years on date of the measurements. However, the coke oven batteries including the oven machinery have been rehabilitated before the final measuring period such that they fulfilled the most essential demands imposed under the 1986 TA Luft. The measurements were carried out at a distance of approx. 250 m both on the lee-side and weather-side of the batteries. On the lee-side, the annual means for BaP ranges from 23 to 37 ng/m³, while more than 50 % of the measured values were above 10 ng/m³. The rehabilitation work carried-out during the measuring period led to a reduction in the BaP burden at the most strongly burdened measuring station on the lee-side by up to 20 % relative to the annual average. Fig. 36 shows the already known seasonal influence on measuring values which, like for coke plant B, is mainly attributable to the different meteorological conditions prevailing during the summer and winter term. However, a base load of up to 6 ng/m³ was determined for the winter months at both measuring positions. Presumably, coal fires in private households which were quite popular in this region at that time mainly caused this base load.

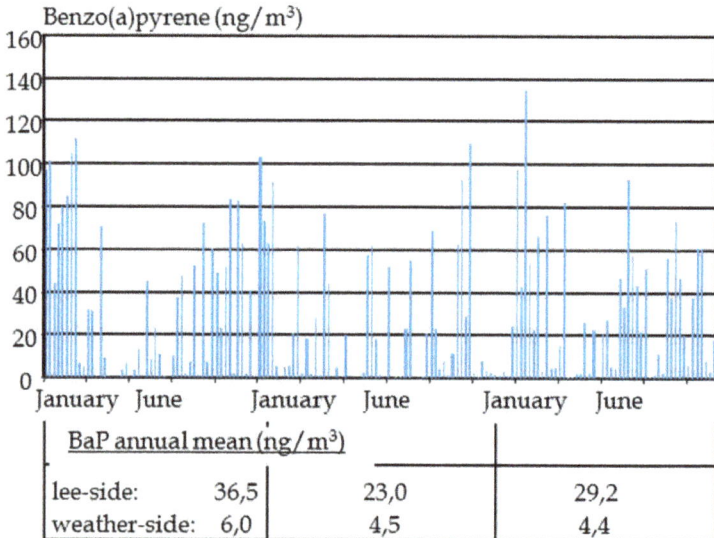

BaP annual mean (ng/m³)			
lee-side:	36,5	23,0	29,2
weather-side:	6,0	4,5	4,4

Figure 36. Benzo(a)pyrene in ambient air near coking plant C (shutdown in 1999) during a three years period

8.2. Calculated Benzo(a)pyrene concentrations

The additional burdens of BaP, caused by coke plant´s emissions, in the surrounding of coking plants A and C, respectively, were calculated by applying a spread model as per Gauß, without taking account of the influence exerted by buildings on the wind field (Hein et al., 2003). The wind field was just described by the spread class statistics for the site of the coking plant. The applied emission mass flow rates were based on those ranges given in section 7.

By evulation of existing measuring data on the overall BaP burden near both coking plants, it was possible to calibrate the mathematical assumptions, and the assumed emission mass-flows in particular, by a factor of 1.18. The corrected results from calculations for the site of coking plant A are reflected in Fig. 37 (top side) in a so-called iso-line representation, which gives the total load of BaP as an annual average (for the year under investigation) near this plant, assuming a base load of 0.5 ng/m³ which was typical for the Rhine-Ruhr area in the time under investigation.

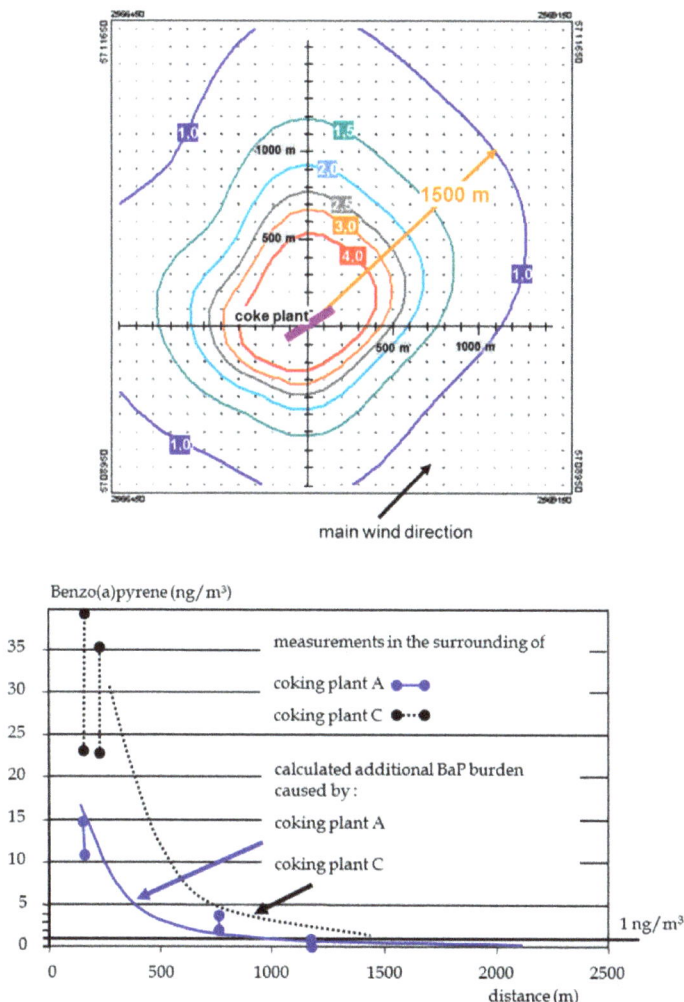

Figure 37. Calculated BaP concentrations (ng/m³ as an annual mean) in ambient air near German coking plants; top side: total BaP concentrations near coking plant A considering a base load of 0.5 ng BaP/m³; bottom side: calculated additional burden and measured concentrations in the surrounding of plant A and C, respectively

From Fig. 37 (top side) one may conclude that, in the period under investigation, the ambient air standard for BaP of 1 ng/m³ (as a sum of base load and additional burden) as demanded in (EU, 2008) will be complied with in north-east of the investigated coking plant A only from a distance of approx. 1,500 m onward away from the battery center, assuming a base load of 0.5 ng/m³. The graph in the bottom of Fig. 37 shows the nearly asymptotic decline of the additional burdens by BaP, caused by battery operation, in the main direction of wind in progressive distance from both coking plants under investigation. The spread characteristics shown here can be confirmed by BaP measurements (overall load) that were taken in the environment of these plants in the past.

Inasmuch as their meteorology as well as their coke throughput rates is comparable with the two investigated coking plants, a transfer of the outlined spread behaviour to other coking plants with comparable emission control standards should be possible.

9. Summary and conclusions

Coke will be an indispensable precursor for steel production worldwide, also in the future. A further extension of the current cokemaking capacities in the world will depend on the global economics and on the future behaviour of export willing countries to sell coke for reasonable prices, of China in particular. It is to assume, if there is need for building of additional cokemaking capacities, the relevant plants will be built in countries with an increasing steel demand. Besides for China, this will be the case for India, Southeast-Asia and South America. Another trend will be inevitable worldwide, that means the replacement of older and smaller plants by modern high capacities batteries for cokemaking. This will be necessary not only by economic but notably by ecological reasons. Worldwide the legal demands for improvements in emission control on coking plants have been tightened in the last years. Legislation for environmental control as given by the Clean Air Act in the US, by the German TA Luft or by the BREF document of the European Union are accepted as a standard for other countries. Improvements in emission control could be achieved by application of the Best Available Techniques for emission control on coking plants during the last years in Europea, and in Germany in particular. By consequent compliance with future standards, as described in the draft of the revised BREF document, further improvements in air quality in the surrounding of the plants will be achieved. At this, special importance is to be attached to emissions containing carcinogenic coumpounds, like Benzo(a)pyren (BaP), which are emitted during conventional cokemaking because of envitable leaks at the closure facilities of the oven chambers. Similar to the non-recovery technology for cokemaking, which operates under negative pressure, these fugitive emissions can be drastically reduced at conventional cokemaking, too, by application of techniques for control the pressure of the oven chamber.

In order to predict the impact of coke plant's emissions on the ambient air in the surrounding, it is necessary to quantify their amount. The paper describes methods for measuring fugitive emissions containing Benzo(a)pyrene at single closure facilities of the coke ovens, whereby emissions from doors play a dominant role. Based on these results, an estimation of the BaP emissions of the total plant is possible. It could be shown that so-called emission factors for BaP from doors, as an average of all doors of the battery, as given by the US EPA are higher than those from own measurements. By use of untypical high emission factors for a prognosis of the impact of coke plant's emissions on the ambient air and thus on the health risk for the people living nearby, it can happen that the importance of a coking plant is overestimated. By use of emission factors, which determination is described in this paper, for spread calculations, a sufficient forecast on the additional burden of BaP in ambient air in the surrounding of the coking plant is possible, when comparing with actual results of measurements. Additionally, parameters could be evolved which influence the impact of coke plant's emissions on ambient air. One of them is the location of the coking plant with regard to the relevant residential area where the ambient air measurements are carried out. In case that the coking plants are located mid of spacious industrial areas, the ambient air concentration for BaP of 1 ng/m^3 as set as a standard in Europe can be achieved in most cases, provided the relevant plant doesn't exceed a capacity of maximum 2 to 3 mio. t and is equipped with techniques for emission control according the state of the art.

Author details

Michael Hein and Manfred Kaiser
DMT GmbH and Co. KG, Cokemaking Technology Division, Essen, Germany

10. References

Ailor, David. (2003). Principal Environmental Issues Facing the U.S. Coke Industry in 2003, *Met World Coke Summit 2003*, Proceedings, Toronto 2003

Arendt, Paul; Hein, Michael; Huhn, Friedrich & Wanzl, Wolfgang. (2009). Kammerverkokung zur Erzeugung von Hochofenkoks, in *Die Veredlung und Umwandlung von Kohle Technologien und Projekte 1970 bis 2000 in Deutschland*, Abschnitt 3.2.1, Seite 13-50, Editor Jörg Schmalfeld, DGMK Deutsche Wissenschaftliche Gesellschaft für Erdöl, Erdgas und Kohle e.V., 2009

DIN-EN. (2008). Air quality - Standard method for the measurement of the concentration of benzo(a)pyrene in ambient air, *DIN EN 15549*, 2008

DWD. (2003). Informations given by Deutscher Wetterdienst, 2003

Eisenhut, Werner & Hein, Michael. (1994). Coke oven emission standards in the US - a discussion of the impact, significance and achievability, *Proceedings of the 53rd Ironmaking Conference*, Chicago March 20-23, 1994

Eisenhut, Werner, Hein, Michael & Friedrich, Frank. (1992). Aromatic Hydrocarbons in the Environment of Coke Oven Plants, *Proceedings of Clean Air Conference*, 9.Weltkongress der IUAPPA, Montreal 1992.

Eisenhut, Werner; Reinke, Martin & Friedrich, Frank. (1990). Coking Plant Environment in West-Germany, Coke Making International, Vol 1, 1, 1990, 74-77, ISSN 0937-9258

EU. (2004). *Directive 2004/107/EC* of the European Parliament and of the Council relating to arsenic, cadmium, mercury, nickel and polycyclic aromatic hydrocarbons in ambient air, Dezember 15, 2004

EU. (2008). *Directive 2008/50/EC* of the European Parliament and of the Council on ambient air quality and cleaner air for Europe, May 21, 2008

EU. (2010). *Directive 2010/75/EU* of the European Parliament and of the Council on industrial emissions (integrated pollution prevention and control), November 24, 2010

EU. (2012). Best Available Techniques Reference Document on the Production of Iron and Steel, *BREF*, European IPPC Bureau, Seville 2011

Faust, Winfried; Hansmann, Thomas; Lonardi, Emile & Stefano Pivot. (2010). Entwicklung einer Einzelkammerdruckregelung unter Berücksichtigung der Anforderungen einer Koksofenbatterie im Stampfbetrieb, *Proceedings of the Conference of Cokemaking Technologies*, April 29-30, 2010, Essen

Hein, Michael. (2002). Do coke oven plants meet the environmental requirements today? Comparison of different cokemaking systems, *Coke Making International*, Vol. 14, 1, 2002, 44-50, ISSN 0937-9258

Hein, Michael; Huhn, Friedrich & Sippel, Michael. (2003). Benzo(a)pyrene in ambient air near coke plants – a survey in view of the intended air quality standard of the EU, *Stahl und Eisen*, 123, 9, 2003, 61

Hein, Michael. (2009). Kokereitechnische Entwicklungen und Maßnahmen zu einer umweltverträglichen Koksproduktion nach 1945, in *Glückauf Ruhrgebiet, Der Steinkohlenbergbau nach 1945*, Katalog der gleichnamigen Ausstellung im Deutschen Bergbau Museum 2010, Editor Michael Farrenkopf et al, Bochum, 2009, ISBN 13 9783-937203-44-7

Hein, Michael. (2010). Kokereien im Blickfeld des Umweltschutzes - keine Rückschau ohne Ausblick, *Proceedings of the Conference of Cokemaking Technologies*, April 29-30, 2010, Essen

Huhn, Friedrich; Giertz, Hans-Josef & Hofherr, Klaus. (1995). New Process to avoid Emissions: Constant Pressure in Coke Ovens, *Proceedings of the 54th Ironmaking Conference*, 439-444, Nashville, TN, USA, 1995

Kaiser, Manfred. (2011). Emission reduction on coke oven and by-product plant, *Eurocoke Summit 2011*, Vienna, April 5-7, 2011

Krupp-Koppers. (n.d.). *Firmbrochure* of Krupp-Koppers GmbH, Essen, Germany

Kohlenstatistik. (2012). *Statistik der Kohlenwirtschaft e.V.*, Retrieved from http: <http://www.kohlenstatistik.de

LANUV. (2003). Informations given by Landesamt für Natur, Umwelt und Verbraucherschutz NRW, 2003

LANUV. (2012). Langjährige Entwicklungen ausgewählter Schadstoffkomponenten, Informations given by Landesamt für Natur, Umwelt und Verbraucherschutz NRW, 2012, Retrieved from
<http://www.lanuv.nrw.de/luft/immissionen/ber_trend/trends.htm

Nathaus. (n.d.). *Firmbrochure* of Kiro-Nathaus GmbH, Lüdinghausen, Germany

TA Luft. (2002). Technical Instruction on Air Quality Control (TA Luft), 2002

Re-Net. (2011). Research & Analysis of the Global Commodity Markets, Resource-Net, Brussels, 2011

Sowa, Frank, Kaiser, Manfred & Petzsch, Mario. (2011). Advanced Technologies for Desulphurisation of Coke Oven Gas, AIST Conference 2011, Proceedings, Indianapolis, May 2-5, 2011

Spitz, Joachim; Kochanski, Ulrich; Leuchtmann; Klaus-Peter, & Krebber, Frank. (2005). Operational Experiences gained with „PROven" in the new "Schwelgern" Coke Oven Plant, *Proceedings of the 5th International Cokemaking Congress*, Stockholm, 2005

Stoppa, Harald; Strunk, Joachim; Wuch, Gerd & Hein, Michael. (1999). Cost and Environmental Impact of Coke Dry and Wet Quenching, *Coke Making International*, Vol. 11, 1, 19990, 65-70, ISSN 0937-9258

Stoppa, Harald. (2003). Die Kokerei Kaiserstuhl, in *Koks Die Geschichte eines Wertstoffes*, Band 1, 62-79, Editor Michael Farrenkopf, Bochum, 2003, ISBN 3-921533-90-2

US-EPA. (1993a). National Emission Standards for Hazardous Air Pollutants for Source Categories, 40 CFR Part 63, subpart L, US Environmental Protection Agency, 1993

US-EPA. (1993b). Method 303-Determination of visible emissions from by-product coke oven batteries, US Environmental Protection Agency, 1993

US-EPA. (1996). Method 9-Visual determination of the opacity of emissions from stationary sources, US Environmetal Protection Agency, 1996

USE-PA. (2003a). National Emission Standards for Hazardous Air Pollutants for Coke Ovens: Pushing, Quenching, Battery Stack (66FR35326), US Environmental Protection Agency, 2003

US-EPA. (2003b). Risk Assessment Document for Coke Oven MACT Residual Risk, US Environmental Protection Agancy, 2003

US-EPA. (2005). National Emission Standards for Coke Oven Batteries, Final Rule, 40 CFR Part 63, US Environmental Protection Agency, 2005

US-EPA. (2008a). Chapter 12.2 (Coke production of AP 42); US Environmental Protection Agency, 2008

US-EPA. (2008b). Emission Factor Documentation for AP-42, Chapter 12.2, Coke Production, Final Report, US Environmental Protection Agency, 2008

VDI. (1996). Outdoor-air pollution measurement - Indoor-air pollution measurement - Measurement of polycyclic aromatic hydrocarbons (PAHs) - Gas-chromatographic determination, *VDI 3875, part 1*, Verein Deutscher Ingenieure, 1996.

Worldsteel. (2012). Informations given by WorldSteel Association, Retrieved from <www.worldsteel.org>

Contribution of the Compressed Air Energy Storage in the Reduction of GHG – Case Study: Application on the Remote Area Power Supply System

Hussein Ibrahim and Adrian Ilinca

Additional information is available at the end of the chapter

1. Introduction

There are many different interpretations and classifications in use today to describe rural and/or remote areas for the purposes of discussing methods of electrification. Some useful examples are as follows [1]:

1. By density and concentration or clustering – setting the context of the environment or geography:
   - Small communities, villages or even towns that are remote from other habitation,
   - Dispersed households, farms and enterprises of low density over wide areas or regions,
   - Community clusters or villages surrounded by lower density dispersed households,
   - Geographically on the same land mass, but separated by physical obstacles such as long distances, mountainous terrain, or possibly separated by water such as island communities,
2. By energy use:
   - By power and energy (or load factor=f(energy/power)) and load profile,
   - By application: household, commercial enterprise, institution, agricultural processing, etc.
3. By choice and method of energy provision:
   - Reticulated electricity, connected to some form of larger grid, or a local micro grid,
   - Reticulated/piped fuel such as natural gas, LPG, fuel oil, diesel,

- Transported fuel such as natural gas, LPG, fuel oil, diesel, by land or sea transport,
- Reliance on renewable energy products such as hydro, solar photovoltaic (PV), wind, waves, tides,

The most suitable method of electricity provision (technology, business model, etc.) will usually depend on the combination of the geographic context, the consumer need, and the possibilities that are available and affordable to provide the energy requirements. Therefore, the most appropriate solutions in one place might be quite unsuitable in another [1].

Clusters and communities that are very remote from other habitation will generally be supplied by some form of centralised local generation, or via a connection to a larger but somewhat remote grid.

2. Challenges related to the electrification of remote areas

Today, diesel generators are mainly used, around the world, as emergency supply sets in telecommunication, public buildings, hospitals, or other technical installations (meteorological systems, tourist facilities, farms, etc.), and as standalone military and marine power plants, as well as the reliable isolated power source for islands or remote villages placed far from the power network [2]. In fact, there are two general methods of supplying electricity to remote areas: grid extension and the use of diesel generators. Grid extension can be very expensive in many locations. Diesel generators are therefore the only viable option for remote area electrification [3].

Classic gensets based on internal combustion engines are equipped with synchronous generators, therefore fixed speed operation is required. It gives low efficiency during low load operation (figure 1). It is not critical in emergency case operated sets, but very important in continuously operated systems, where fuel consumption is significant economic and logistic aspect. In fact, remote areas with relatively small communities generally show significant variation between the time of peak loads and the time of minimum loads. A typical example of a load profile of a remote community in Western Australia is shown below in figure 2. Diesel-powered electric generators are typically sized to meet the peak demand during the evening but must run at very low loads during "off-peak" hours during the day and night. This low-load operation results in poor fuel efficiency and increased operation and maintenance costs [3].

Moreover, low load operation of diesel genset at synchronous speed reduces the engine lifetime, by incomplete combustion of the fuel, therefore an additional dump load is required to improve the combustion process. The efficiency and fuel combustion at low load conditions can be improved by use of load adaptive adjustable speed operation of the genset [4]. In some remote locations, a dual diesel generator system is employed. When the load is light, the smaller generator is used; as the load increased, the manual switch is transferred to the larger generator. This approach results in some fuel savings, however managing this dual system is time consuming and impractical [3].

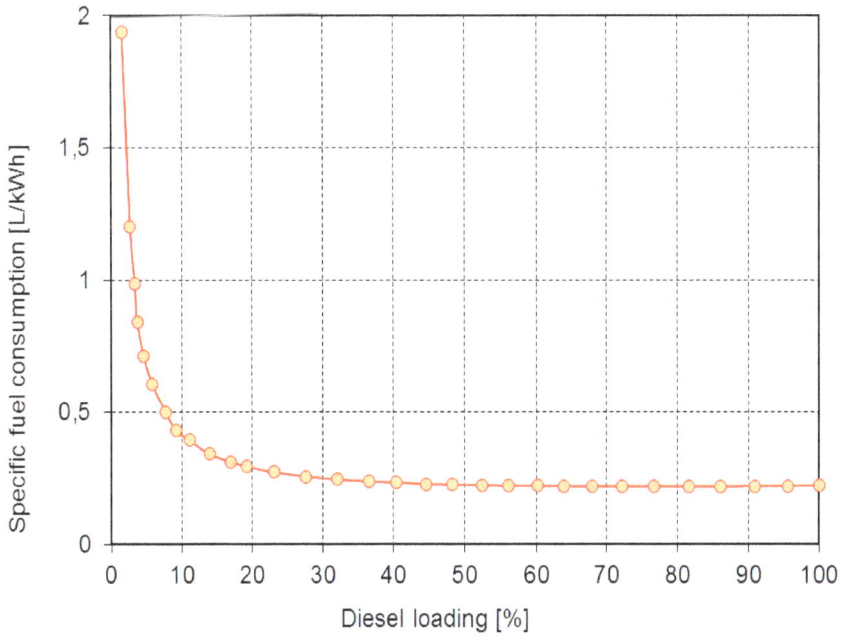

Figure 1. Example of a variation of diesel fuel consumption with loading

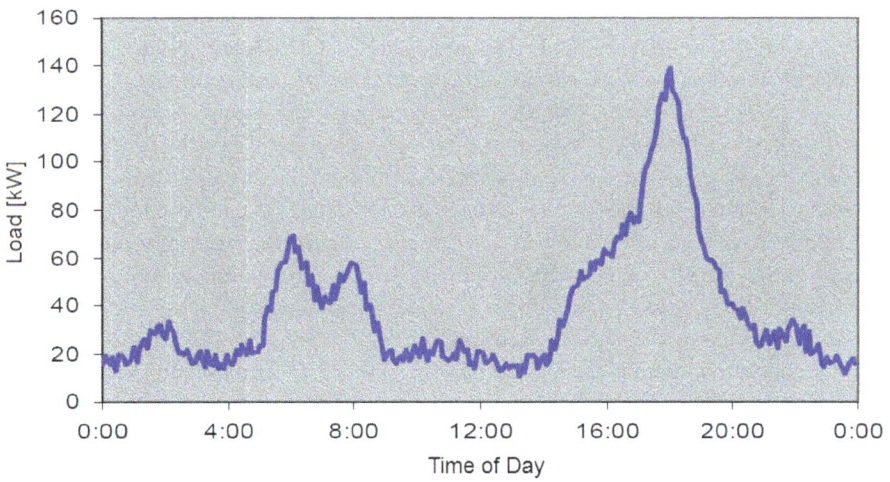

Figure 2. Typical load profile of a remote community [3]

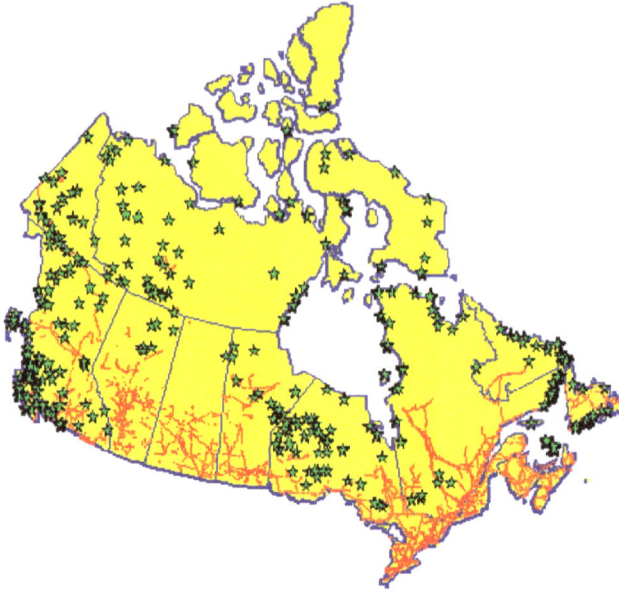

Figure 3. Canadian remote communities [5]

In Canada, approximately 200,000 people live in more than 300 remote communities (Yukon, TNO, Nunavut, islands) (figure 3) and are using diesel-generated electricity, responsible for the emission of 1.2 million tons of greenhouse gases (GHG) annually [6]. In Quebec province, there are over 14,000 subscribers distributed in about forty communities not connected to the main grid. Each community constitutes an autonomous network that uses diesel generators.

In Quebec, the total production of diesel power generating units is approximately 300 GWh per year. In the meantime, the exploitation of the diesel generators is extremely expensive due to the oil price increase and transportation costs. Indeed, the communities are dependent on imported fossil fuels for most of their energy requirements. Also, there are exposed to diesel fuel price volatility, frequent fuel spills and high operation and maintenance costs including fuel transportation and bulk storage. Having said this however, in the past decade, diesel prices have more than doubled. High fuel costs have translated into tremendous increases in the cost of energy generation [3]. In Quebec for example, as the fuel should be delivered to remote locations, some of them reachable only during summer periods by barge, the cost of electricity produced by diesel generators reached in 2007 more than 50 cent/kWh in some communities, while the price for selling the electricity is established, as in the rest of Quebec, at approximately 6 cent/kWh [7]. The deficit is spread among all Quebec population as the total consumption of the autonomous grids is far from being negligible. In 2004, the autonomous networks represented 144MW of installed power, and the consumption was established at 300 GWh. Hydro-Quebec, the provincial utility, estimated at approximately 133 million CAD$ the annual loss, resulting from the difference between the diesel electricity production cost and the uniform selling price of electricity [7].

Moreover, the electricity production by the diesel is ineffective, presents significant environmental risks (spilling), contaminates the local air and largely contributes to GHG emission. In all, we estimate at 140,000 tons annual GHG emission resulting from the use of diesel generators for the subscribers of the autonomous networks in Quebec. This is equivalent to GHG emitted by 35,000 cars during one year.

The use of diesel engines to supply power to rural communities has provided light and energy services to places where previously there has only been darkness. However, the rising cost of diesel fuel (brought on by higher oil prices and the environmental regulation for its transportation, use, and storage) combined with carbon emissions concerns is driving remote communities to look at alternative methods to supplement this power source. During the past few years, wind energy is increasingly used to reduce diesel fuel consumption, providing economic, environmental, social, and security benefits [8].

Wind-diesel systems have been the most successfully and widely hybrid power systems applied up to date. These systems are designed to use as much as possible wind power in order to lower diesel consumption. The challenge is to keep the power quality and stability of the system besides the variability of the wind power generation and diesel operational constraints [9]. Indeed, one of the disadvantages is the intermittent nature of wind power generation. Diesel engine driven synchronous generators operating in parallel with wind turbine must maintain a good voltage and frequency regulation against active and reactive load variations and wind speed changes [10]. Integration of a storage element with diesel and wind turbine is necessary in order to get a smooth power output from a wind turbine and to optimize energy use to further reduce consumption of diesel fuel [11]. The next sections present an overview of technical challenges of wind-diesel hybrid system (WDHS), the justification of the choice of compressed air as device of energy storage to be used with WDHS and the impact of using of this storage energy system on the fuel consumption of diesel generators and on the GHG emissions.

3. Overview of wind-diesel hybrid system

3.1. Description of wind-diesel hybrid system

A wind-diesel hybrid system is any autonomous electricity generating system using wind turbine(s) with diesel generator(s) to obtain a maximum contribution by the intermittent wind resource to the total power produced, while providing continuous high quality electric power [12]. Overview of typical wind-diesel installations can be found in [13]. In the most cases, the power of installed diesel gensets is much higher than the power of wind turbines. In peak, the wind turbines can cover even more than 90% of demanded power, but in long term the fuel saving is 10-15%. The same level of fuel saving can be obtained by gensets based on power electronics, load adaptive, adjustable speed diesel without use of wind turbines. It is used in light mobile power gensets [14].

Figure 4 presents a schematic diagram of a generalized wind diesel system. As shown, this system consists of the following major components:

- One or more wind turbines
- One or more diesel generator sets
- A consumer load
- An additional controllable or dump load
- A storage system
- A control unit (including possible load management)

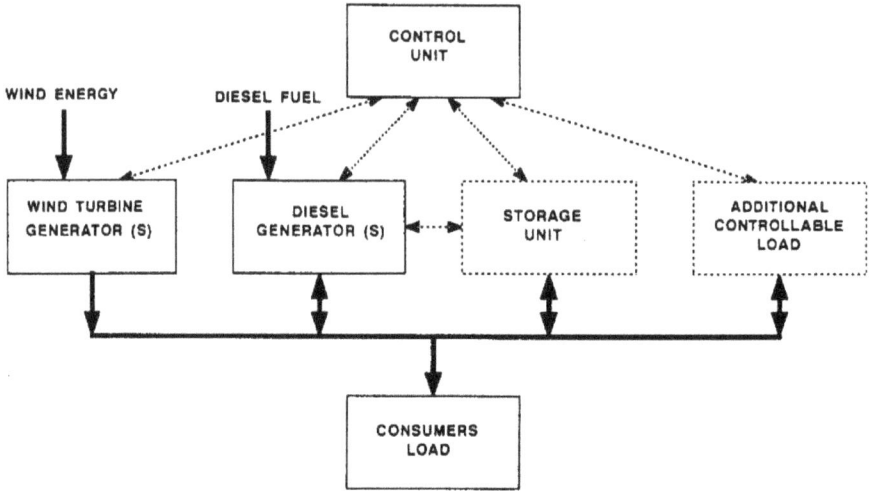

Figure 4. Schematic of generalized wind-diesel system [15]

3.2. Classification of wind-diesel systems versus wind penetration rate

Wind-Diesel hybrid power systems are particularly suited for locations where wind resource availability is high and the cost of diesel fuel and generator operation control the cost of electrical energy supplied. As a result of turbine developments the economics of wind power have now become competitive with conventional power source. The economy of operation of wind turbines is critically dependent on the wind speeds at the site. If the wind turbine is used along (high penetration of wind energy) with a diesel engine, the cost of power generation could be reduced, in addition to reducing greenhouse gas emission problems.

Penetration here is defined as the ratio of rated capacity of the wind energy source to the total system-rated capacity. It should also be noted that load patterns may also significantly affect system operation [15].

A classification system is used when discussing the amount of wind that is being integrated into the grid system (Table 1). A system is considered to be a high penetration system (figure 5) when the amount of wind produced at any time versus the total amount of energy produced is over 100%. Low penetration systems (figure 6) are those with less than 50% peak instantaneous penetration and medium penetration systems have between 50%-100%

of their energy being produced from wind at any one time. Low and medium penetration systems are a mature technology. High penetration systems, however, still have many problems, especially when installed with that capacity to operate in a diesel-off mode.

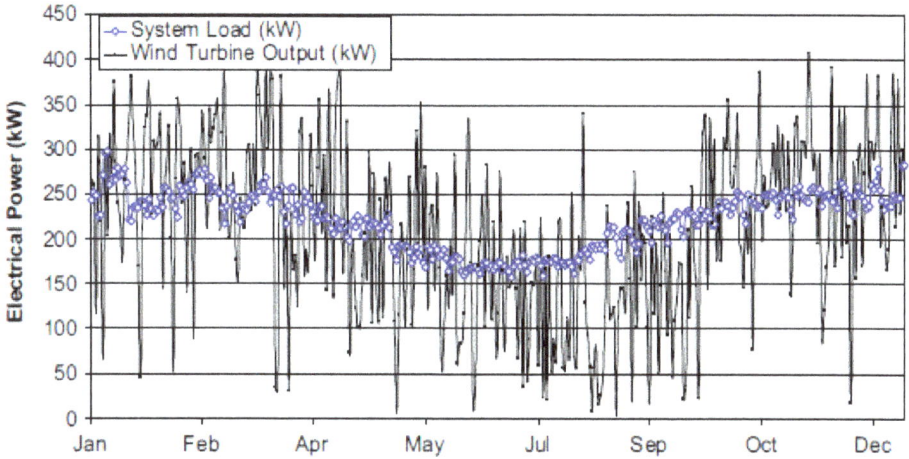

Figure 5. Example of a high-penetration wind–diesel system outputs [16]

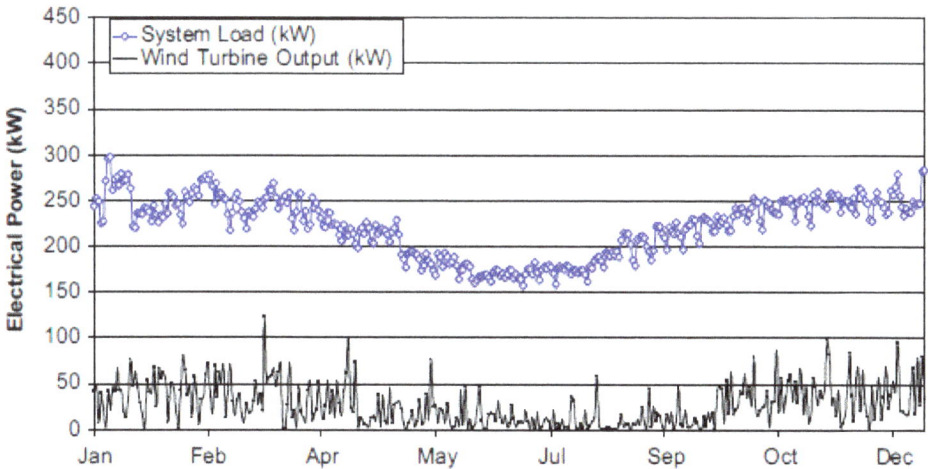

Figure 6. Example of a low-penetration wind–diesel system outputs [16]

Penetration class	Operating characteristics	Penetration	
		Peak instantaneous	Annual average
Low	Diesel runs full-time. Wind power reduces net load on diesel. All wind energy goes to primary load. No supervisory control system	< 50%	< 20%
Medium	Diesel runs full-time. At high wind power levels, secondary loads dispatched to ensure sufficient diesel loading or wind generation is curtailed. Requires relatively simple control system.	50% - 100%	20% - 50%
High	Diesel may be shut down during high wind availability. Auxiliary components required to regulate voltage and frequency. Requires sophisticated control system.	100% - 400%	50% - 150%

Table 1. Wind-diesel classification [15]

3.3. Technical challenges of wind-diesel hybrid system

Hybrid wind-diesel systems with high penetration of wind power have three plant modes: diesel only (DO), wind-diesel (WD) and wind only (WO). In DO mode, the maximum power from the wind turbine generator (WTG) is always significantly less than the system load. It is the mode of classical diesel power plant. In this case, the diesel generators (DG) never stop operation and supply the active and reactive power demanded by the consumer load. Frequency regulation is performed by load sharing and speed governors controlling each diesel engine and voltage regulation is performed by the synchronous voltage regulators in each generator. The main goals of maximize fuel savings or minimize generation costs to supply the actual load [17] is achieved by careful planning/scheduling of the DG having into account factors such as their specific fuel consumption, their rated power, etc.

Wind-Diesel mode can be considered as a diesel plant with the wind turbine as a negative load. It is the mode of many low/medium wind penetration power wind-diesel systems already implemented in Nordic communities in Yukon [18], Nunavut [19] and in Alaska [20]. In this case, the WTG power is frequently approximately the same as the consumer load and in addition to DG(s), WTG(s) also supply active power. Some new problems appear in this mode like to determine the diesel spinning reserve (the wind power can disappear in any moment due to the unpredictable wind resource and the current load can overload the diesel(s) currently supplying), or to assure a minimum diesel load needed by some engines (this situation can happen at high wind power levels and low loads). Under these conditions, two operating modes are possible: (1) the diesel can be allowed to run continuously, or (2) the diesel can be stopped and started, depending on the instantaneous power from the wind and the requirements of the load. Running the diesel continuously decreases the load factor, with an increase in the aforementioned diesel operating costs.

Using the energy storage unit showed in figure 4 can solve both problems. The second problem can be solved by the use of the dump load showed in figure 4 or reducing the power coming from the wind turbine. Some variable speed wind turbines have this possibility [21]. Also in this mode additional reactive power must be generated, because wind turbines are normally reactive power consumers, although adding capacitor banks or overexciting the synchronous generators can solve this. However, the first obstacle with this perspective results from the operation constraints of diesels. Beyond a certain penetration, the obligation to maintain idle the diesel at any time, generally around 25-30 % of its nominal output power, forces the system to function at a very inefficient regime. Also, this limits the wind energy to a level of too weak penetration and the wind turbines act only as a negative charge for the network. Indeed, for low and medium penetration systems, the diesel consumes, even without load, approximately 50% of the fuel at nominal power output. These systems are easier to implement but their economic and environmental benefits are marginal [22].

The use of high penetration systems allows the stop of the thermal groups, ideally as soon as the wind power equals the instantaneous charge, to maximize the fuel savings. This is the wind only mode. The WO mode is only possible if the power coming from the wind turbine(s) is greater than the consumed power by the load (with a safety margin). Because no diesel generators run in this mode, auxiliary components are required to regulate voltage and frequency. The frequency is controlled through the active power balance. To accomplish this active power balance, the energy storage system can be added to store the surplus active power from the wind turbine or retrieve power in the periods when the wind power is less than current load; also the surplus wind power can be consumed by dump loads. The voltage is controlled by the reactive power balance and it is normally achieved through synchronous condensers which deliver the reactive power needed by the loads and the wind turbine. To supply power uninterruptedly, the size of the energy storage has to be big enough to assure power to the load during transitions from the wind power source to the diesel power source when there is a failure or absence of wind energy. In the meantime, the high-penetration wind diesel systems without storage (WDHPWS) is subject to complex technical problems [23], [24] which did that a single project of this type, without any storage, is presently operational in Alaska [20].

During time intervals when the excess of wind energy over the charge is considerable the diesel engine must still be maintained on standby so that it can quickly respond to a wind speed reduction (reduce the time of starting up and consequent heating of the engine). This is an important source of over consumption because the engine could turn during hours without supplying any useful energy. Assuming optimum exploitation conditions [25], the use of energy storage with wind-diesel systems can lead to better economic and environmental results , allows reduction of the overall cost of energy supply and increase the wind energy penetration rate (i.e., the proportion of wind energy as the total energy consumption on an annual basis) [16].

4. Choice of the energy storage device for a high penetration wind-diesel hybrid system

Presently, the excess wind energy is stored either as thermal potential (hot water), an inefficient way to store electricity as it cannot be transformed back in electricity when needed or in batteries which are expensive, difficult to recycle, a source of pollution (lead-acid) and limited in power and lifecycle. The fuel cells propose a viable alternative but due to their technical complexity, their prohibitive price and their weak efficiency, their appreciation in the market is still in an early phase. The required storage system should be easily adaptable to the hybrid system, available in real time and offer smooth power fluctuations.

Due to technical, economical and energetic advantages demonstrated by the compressed air energy storage (CAES) in hybrid systems at large scale (figure 7) use in the USA and Germany, we investigated the possibility to associate the wind-diesel with compressed air energy storage system for medium and small scale applications (isolated sites).

The choice of this system was not only based on the successes of large scale CAES system. The energy storage in the form of compressed air is suitable for both wind and diesel applications. Moreover, the CAES presents an interesting solution for the problem of strong stochastic fluctuations in wind power by offering a high efficiency conversion rate (60-70% for a complete charge-discharge cycle). It, also, uses conventional materials that are easy to recycle and can support an almost unlimited number of cycles [26].

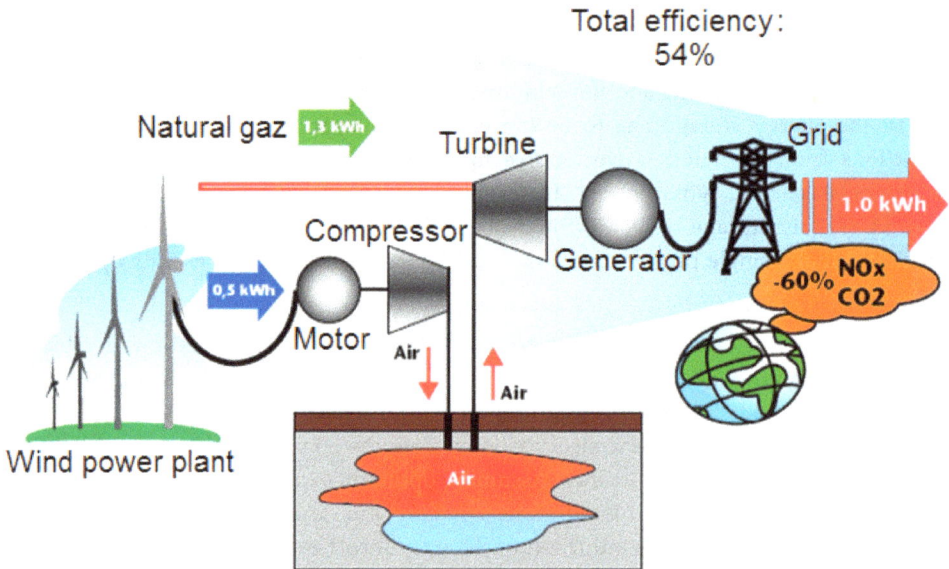

Figure 7. Illustration of the large-scale wind-compressed air hybrid system

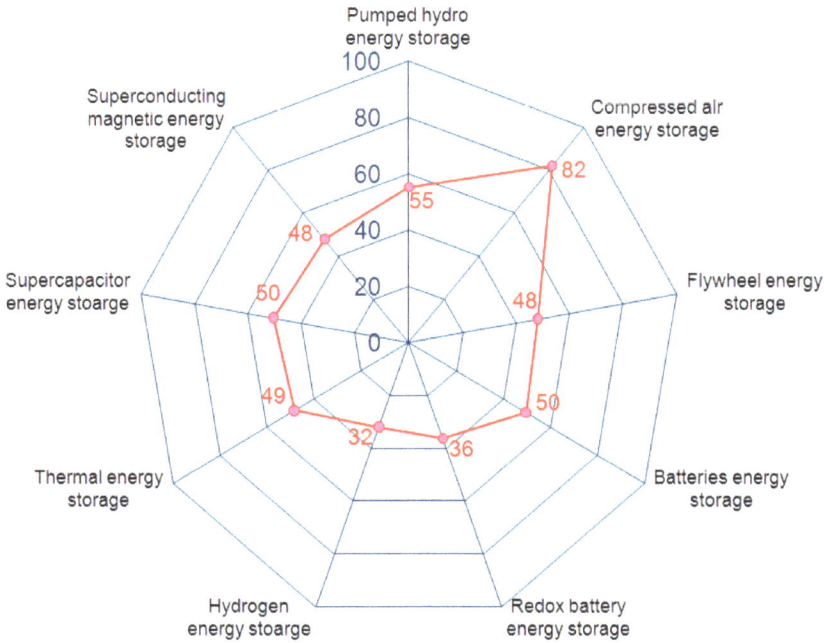

Figure 8. Performance index of different energy storage systems

A detailed study based on a critical analysis of all techno-economical characteristics of the possible energy storage technologies (for example, cost, efficiency, simplicity, life time, maturity, self-discharging, reliability, environmental impact, operation constraints, energy and power capacity, adaptability with wind-diesel system, contribution to reduce of fuel consumption, etc.); it was proposed a solution that meets all the technical and financial requirements while ensuring a reliable electricity supply of these sites. It is the wind-diesel hybrid system with compressed air energy storage (WDCAHS). This study demonstrates the value of compressed air storage for a high penetration wind-diesel hybrid system and its advantages with regard to the other energy storage technologies. It was based on the aggregation in a «performance index» of technical, economic and environmental characteristics of various storage methods [27]. The results of this analysis and the values of the performance index are illustrated in the figure 8 for different possible strategies of storage.

The performance index is the measure of the applicability of a technique of storage to a specified application [27]. For another application than the power supply of a remote area, the values of the performance index can be different. The determination of the indication of performance is done using a decision matrix that helps to balance the importance of each characteristic (15 criteria, for example, cost, self-discharging, reliability, time response, efficiency, simplicity, life time, maturity, environmental impact, operation constraints, energy and power capacity, adaptability with wind-diesel system, operational constraints, contribution to reduce of fuel consumption, etc.) of the storage system with regard to the specific requirements of the envisaged application.

It is easy to establish from the figure 8 that the compressed air energy storage system (CAES) answers the choice criteria with a performance index of approximately 82 %.

5. Medium-scale wind-diesel-compressed air hybrid system

5.1. Operation principle

The medium scale wind-diesel-compressed air hybrid system (MSWDCAHS) (figure 9) can be used, for example, in the case of remote villages or islands with important level of local electrical load (few hundred kilowatts to few tens of MW). MSWDCAHS combined with diesel engine supercharging, will increase the wind energy penetration rate. Supercharging is a process consisting of a preliminary compression that aims to increase the density of the engine's air intake, in order to increase the specific power (power by swept volume). During periods of strong wind (when wind power penetration rate – WPPR, defined as the quotient between the wind generated power and the charge is greater than 1; WPPR>1), the wind power surplus is used to compress the air via a compressor and store it in a tank. The compressed air is then used to supercharge the diesel engine with the two-fold advantage of increasing its power and decreasing its fuel consumption. The diesel generator works during periods of low wind speed, i.e., when the wind power is not sufficient to sustain the load.

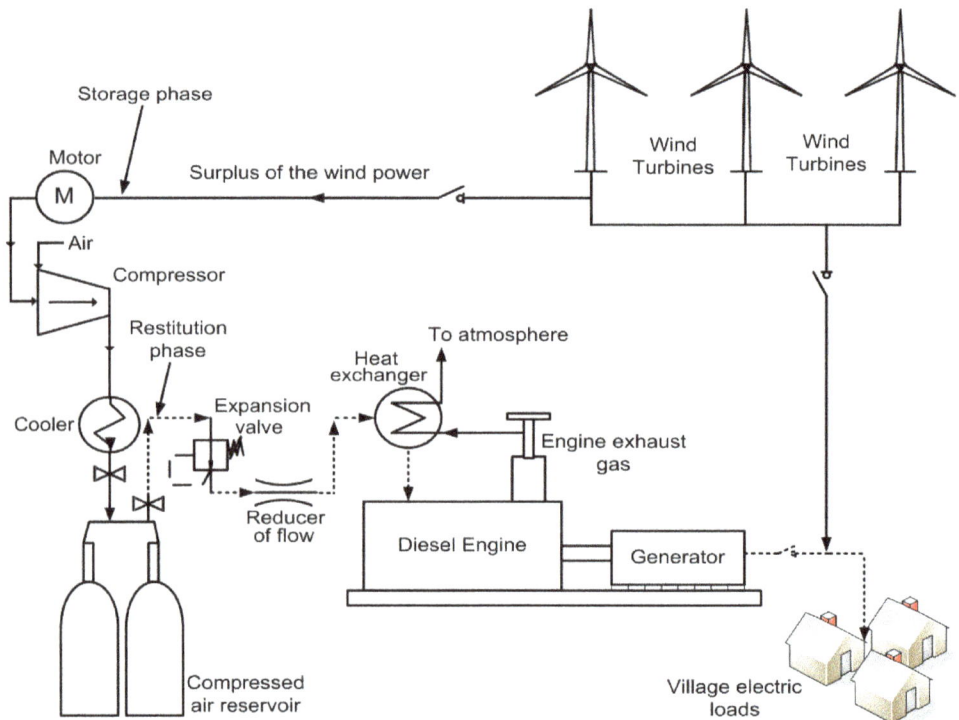

Figure 9. Illustration of the medium-scale wind-diesel compressed air hybrid system

5.2. Technical advantages of an additional supercharging of diesel generator with stored compressed air

Most diesel generators used in remote areas (medium-scale case) are already equipped with a turbocharging system via a turbocharger. However, this type of system loses its advantages during operation at low regime because its efficiency is directly related to the quantity of exhaust gases. To understand the advantage of an additional turbocharging of diesel engine and the operation limits of a turbocharger, we present in figure 10 an example which compares a diesel engine in two functioning modes: atmospheric (without turbocharger) and turbocharged.

Figure 10 shows that as compared to an atmospheric diesel engine with an engine capacity of 10 L, supercharging can increase the values of the indicated efficiency of the engine (maximal efficiency = 45%) and extend the operating range in the area of high efficiency thanks to the large permissible quantity of air into the engine. For a load of 600 N.m, the efficiency of the supercharged engine is about 38% compared with that of atmospheric engine (14%), i.e., an increase about 170%. On the other hand, increasing the applied load on the engine triggers a degradation of the diesel performance due to the operation limits of the turbocharger and to the increase of the heat loss through the cylinder walls. However, this does not exclude the fact that the efficiency for high loads are better through supercharging as compared to the efficiency obtained with atmospheric engine (an increase about 64% for a load of 1200 N.m).

The figure 10 also shows that the compression ratio reaches its maximal value (figure 10) only for the highest loads (this corresponds to high flow and pressure of exhaust gas). This delay to reach the maximal pressure of the compressed air at the engine intake will delay the achievement of the maximal power of the turbocharged engine. The objective of the additional supercharging via the stored compressed air is, then, to maximize the overall efficiency of the diesel engine (figure 10), by several improvements:

- Improving the combustion efficiency by operating the engine at all times with an optimal air/fuel ratio, which does not allow the turbocharger to operate alone.
- Reducing the pumping losses for the low pressure loop of the thermodynamic cycle of diesel engine to increase the work supplied for the same quantity of burned fuel.
- Increasing the specific power (power per swept volume unit) of the diesel engine and its performance.
- Increase the intake pressure at a level which allows a decrease of the fuel quantity injected while maintaining the same maximal pressure in the cylinder of the engine. This allows decreasing the mechanical and thermal constraints due to the supercharging.

6. Small-scale wind-diesel-compressed air hybrid system

The small scale wind-diesel-compressed air hybrid system (SSWDCAHS) (figure 11) can be used, for example, in the case of remote telecom infrastructures that the level of electrical load is not very high (few tens of kW). These infrastructures require continuous, stable, and safe energy supply to maximize the deployment, signal strength, and coverage of the cellular phone. The difference between MSWDCAHS and the SSWDCAHS is the utilization method of the stored compressed air. Indeed, when the output energy of a wind turbine is more than energy demand at the load side (WPPR>1), the excess energy will be converted into the mechanical form using high pressure compressors. The energy is stored in a high pressure reservoir as potential energy. In a case that the wind turbine cannot deliver the required energy at the load side (WPPR<1), the stored mechanical energy will be converted to the electrical energy. In this case, the stored compressed air will be expanded into a pneumatic generator that supply the load. In this step, the diesel generator is stopped. The genset works during periods of low wind speed and only if the compressed air energy storage capacity is not sufficient to supply the pneumatic generator.

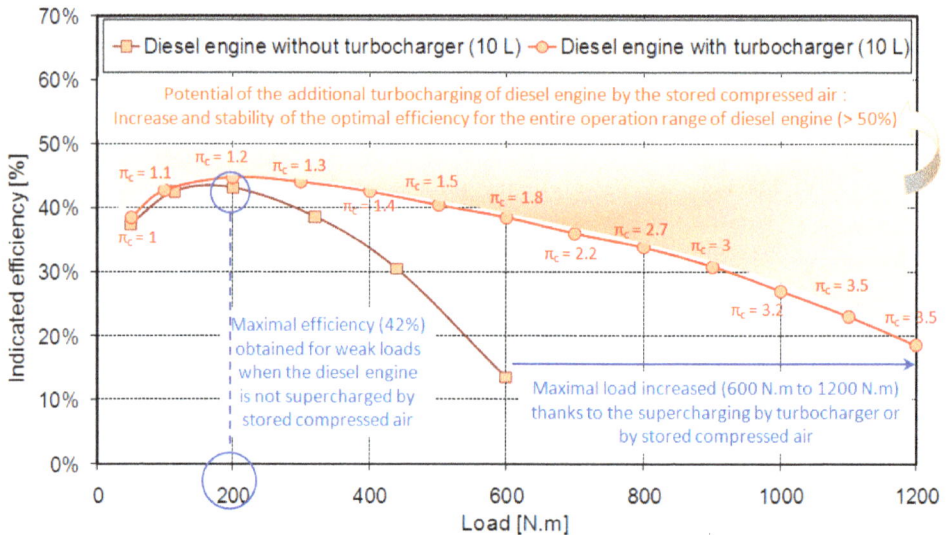

Figure 10. Potential of the additional supercharging of diesel engine by the stored compressed air

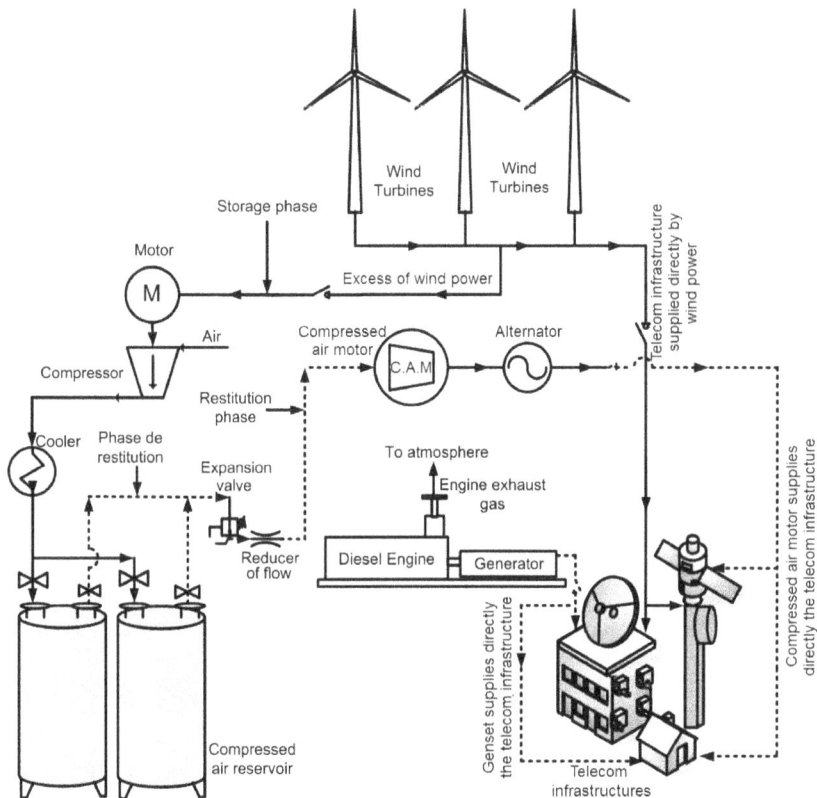

Figure 11. Illustration of the small-scale wind-diesel compressed air hybrid system

7. Advantages of wind-diesel-compressed air hybrid system

WDCAHS represents an innovative concept and it has a very important commercial potential for remote areas as it is based on the use of diesel generators already in place. To our knowledge, the type of WDCAHS that is, proposed in this paper was never the object of a commercial application or an experimental project, and we did not find studies relative to the design or performance of such a system in the scientific literature.

The lack of information on the economics, as well as on performances and reliability data of such systems is currently the main barrier to the acceptance of wind energy deployment in the remote areas. WDCAHS is designed to overcome most of the technical, economic and social barriers that face the deployment of wind energy in isolated sites [28]. Indeed, implementation costs are minimized and reliability is increased by using the existing diesel generators. Our WDCAS solution is threefold: modification and adaptation of the existing engines at the intake level (for medium-scale), addition of a generator that functions with compressed air (for small-scale), addition of a wind power plant and addition of an air compression and storage system.

Using information available [29-31], and performance analysis [32-33], we estimate that on a site with appreciable wind potential, the return on investment (ROI) for such installation is between 2 and 5 years, subject to the costs of fuel transport. For sites accessible only by helicopters the ROI can be less than a year [25].

This analysis does not take into account the raising prices of fuel, nor GHG credit which only tend to reduce the ROI [34].

8. Case study applied to the medium scale wind-diesel-compressed air system

To estimate the potential gain of the MSWDCAS on a target site, we recovered the hourly wind speed data and the hourly electrical load of the diesel engine on the site of the village of Tuktoyaktuk in the Northwest Territories of Canada on the Arctic coast. The maximum and average electric loads of this village are respectively 851 kW and 506 kW. Initially, the village's electricity is supplied by 2 diesel generators, each having 544 kW as maximal power. To these generators a wind plant composed of 4 wind turbines of type Enercon, each having a nominal power equal to 335 kilowatts, a total power equal to 1340 kW was added. We estimated fuel consumption, greenhouse gases (GHG) emissions and maintenance cost of diesel engines for different scenarios: diesel only, wind–diesel hybrid system (WDHS) without CAES and wind–diesel hybrid system with CAES, over a period of 1 year (2007 year's). Figures 12–18 illustrate the results.

Figures 12 and 13 represent the profile of the average wind speed corrected to hub height of wind turbines, the profile of the monthly electric load of the village, the variations of power supplied by wind turbines, the variations of power supplied by diesel generators before and after hybridization with the wind turbines, the operation frequency of diesel engines after hybridization and the profiles of the power directed toward the storage system and that absorbed by the compressor. These figures show that the maximal average consumption of the village occurs during the fall and winter seasons due to the increase of the electric load for the heating. Unfortunately, the highest wind speed is registered during the spring and summer seasons where the average electricity consumption decreases approximately 200 kW in comparison with that of the winter.

Figure 14 shows the operation frequency of diesel engines after the hybridization with the wind turbines. The number of functioning hours of diesel engines depends strongly of the availability of the wind power and the level of the electric load of the village. During 2007, the hybridization would have allowed the operation of a single engine during 5628 h (64%), of two engines during 1766 h (20%) and stop both diesel generators approximately 1366 h (16%).

Figure 15 represents the operation frequency of diesel engines according to their supercharging mode (with or without CAES). This figure shows that the hybridization allows the functioning of diesel engines supercharged by stored compressed air during 3608 h (41%). During 3786 h (43%), the diesel engines are operating without CAES and they are stopped for 1366 h.

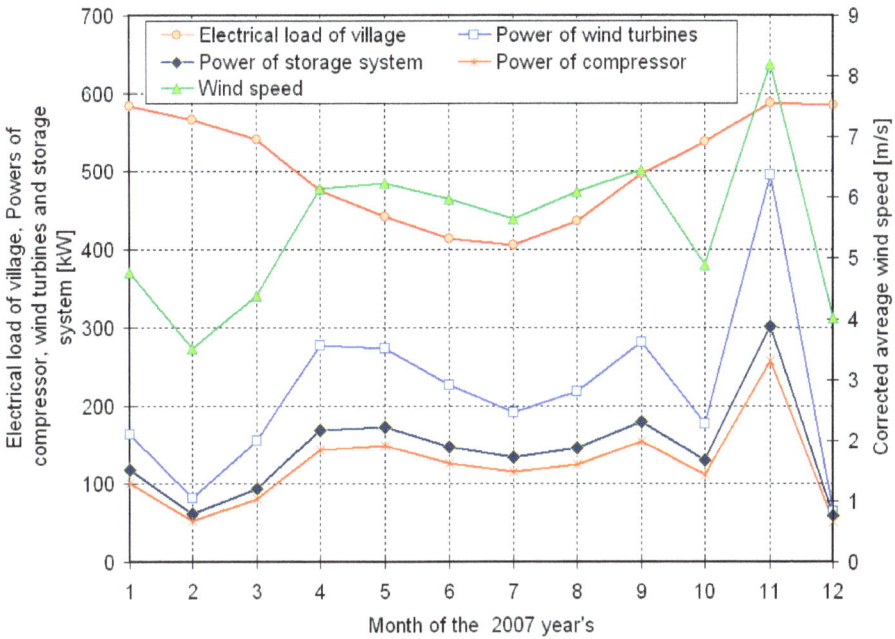

Figure 12. Average wind speed and power profiles of the: electrical load, wind turbines, compressor and energy storage system

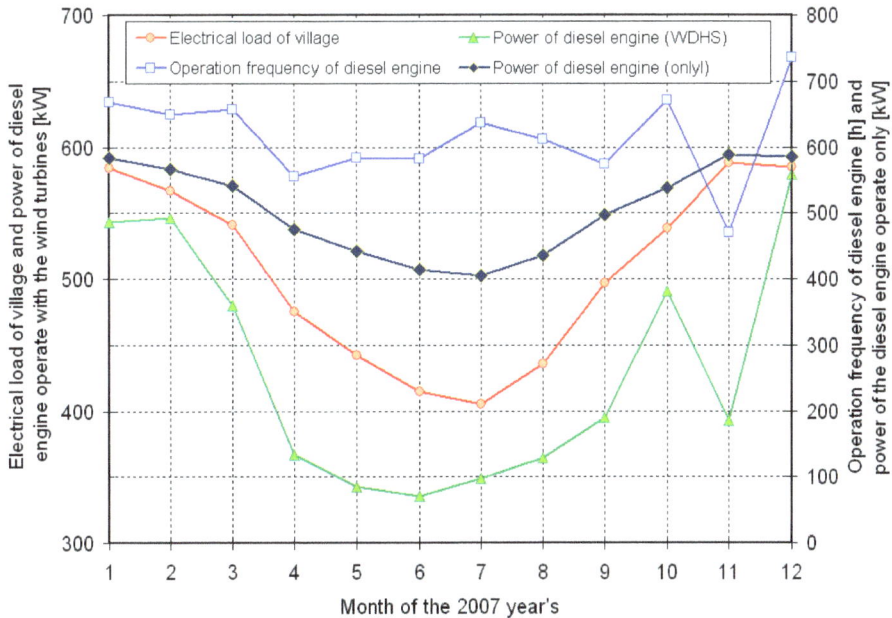

Figure 13. Operation frequency and power curve of diesel engines and profile of the electrical load

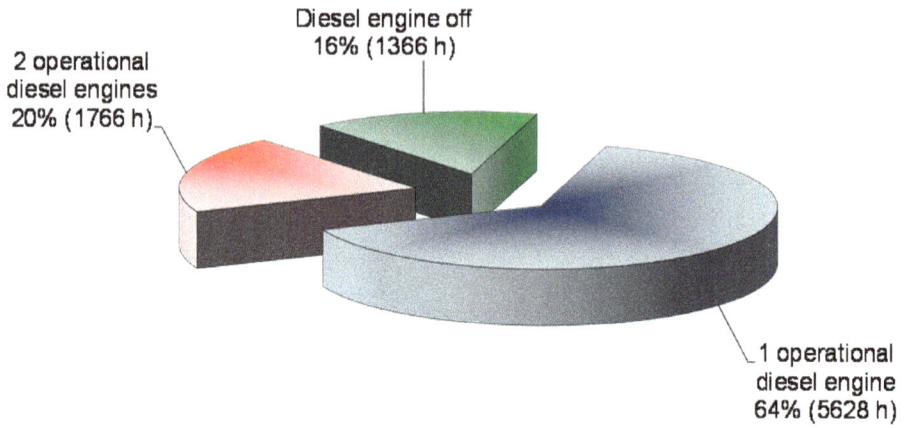

Figure 14. Operation frequency of diesel engines after the hybridization with the wind turbines

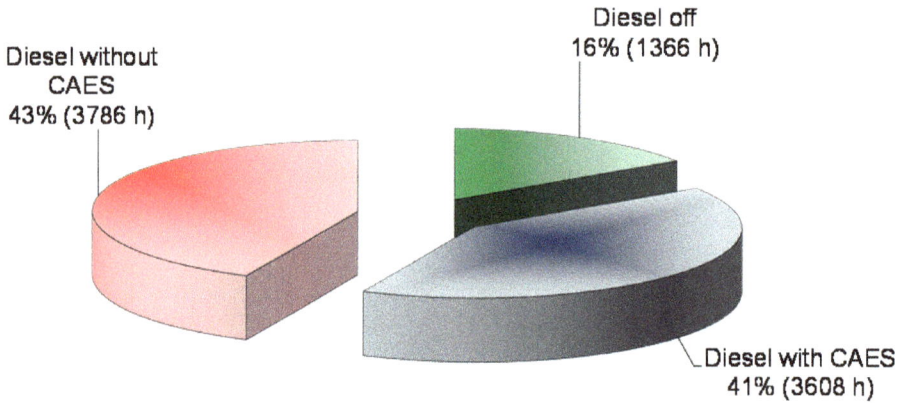

Figure 15. Operation frequency of diesel engines according to their supercharging mode

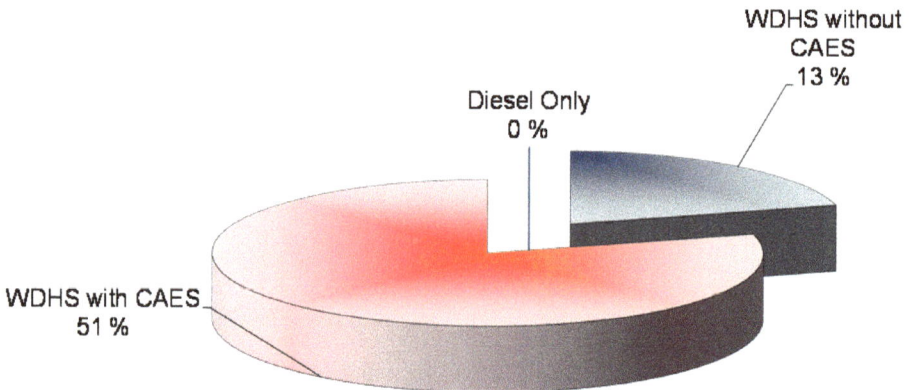

Figure 16. Annual reduction of maintenance and operation costs

The estimation of the annual reduction of maintenance and operation costs, based on the reduction of operation time of the two diesel engines, is represented in figure 16. The base line for comparison is the scenario without hybridization, where no savings in the cost of maintenance can be realized. The WDHS without CAES allows 13% reduction while with CAES, this rate increases to 51%. It is important to mention that the supercharged diesel engine by compressed air stored allows operating with a single diesel engine, whatever the load of the village. On the other hand, a permutation between the two supercharged engines will be necessary to avoid the blocking of some mechanical moving pieces of the engine.

Figure 17 shows the monthly consumption of fuel along a year (2007). Compared with the base line scenario, the use of WDHS without CAES allows fuel reduction varying from 3,000 litres (minimal value) in February to 36,000 litres (maximal value) in November. On the other hand, a WDHS with CAES will significantly increase this economy with a minimum fuel saving of 10,000 litres (February) and a maximum of 53,000 litres (November).

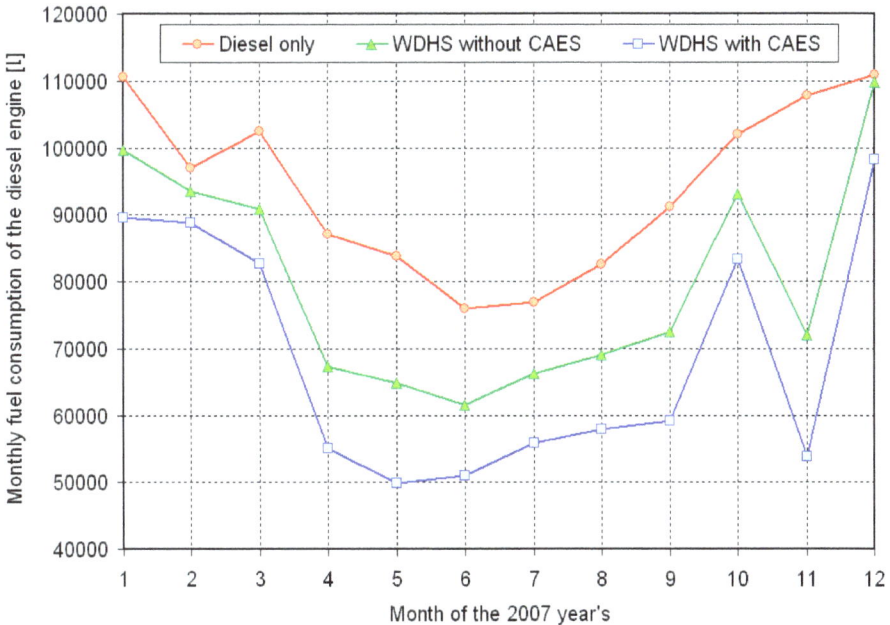

Figure 17. Monthly consumption of fuel along a 2007 year's

Figure 18 illustrates the annual fuel savings. The hybridization between wind energy and diesel engines without CAES reduces by 168,324 L the annual fuel consumption (15%) while with CAES, this reduction increases to 27% (303,143 L). This quantity (27%) is equivalent to 848.8 tons of CO_2 or the annual emission of 167 automobiles and light trucks traveling 15,000 km per year. In Table 2, we review the quantity of greenhouse gases (GHG) avoided thanks to the use of MSWDCAHS.

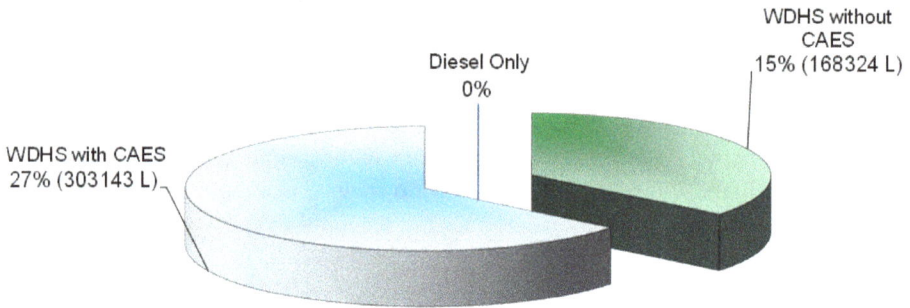

Figure 18. Annual fuel savings

Name of the substance	Emission Factor (kg/m³)	Total value of emissions (tones)
Carbon dioxide (CO_2)	2800	848.8
Carbon monoxide (CO)	13.954	4.23
Sulfur dioxide (SO_2)	0.083	0.025
Oxides of nitrogen (NOx)	52.532	15.925
Volatile organic Compounds (VOC)	1.344	0.408
Total suspended Particles (TSP)	1.018	0.309
Particles with diameters ≤ 10 μm (P_{10})	0.814	0.247
Particles with diameters ≤ 2.5 μm ($P_{2.5}$)	0.786	0.238

Table 2. Quantities of GHG avoided by MSWDCAHS

9. Case study applied to the small scale wind-diesel-compressed air system

To estimate the potential gain of the SSWDCAS on a target site, we recovered the hourly wind speed data (for one month, April 2005) on the site of the telecom station of Bell-Canada situated in Kuujjuarapik (North of Quebec) at 1130 kilometers from Montreal (figure 19). The wind speed data of this site for the month April 2005 are shown in figures 20.

The electrical load of the station is considered constant, about 5kW, including the secondary load of heating. The diesel generator guarantees the supply's continuity of the station by providing exactly the power level consumed by the load. The case study was conducted using two types of wind turbines: the first is a Bergey [35] (10kW, already installed on site) and the second is a PGE (currently named Endurance, 35kW) [36] that we propose to be able to increase the penetration of wind energy and use the excess of this energy to produce the compressed air. Figures 21-24 illustrate the obtained results.

Figure 19. Telecom station of Bell-Canada at Kuujjuarapik [37]

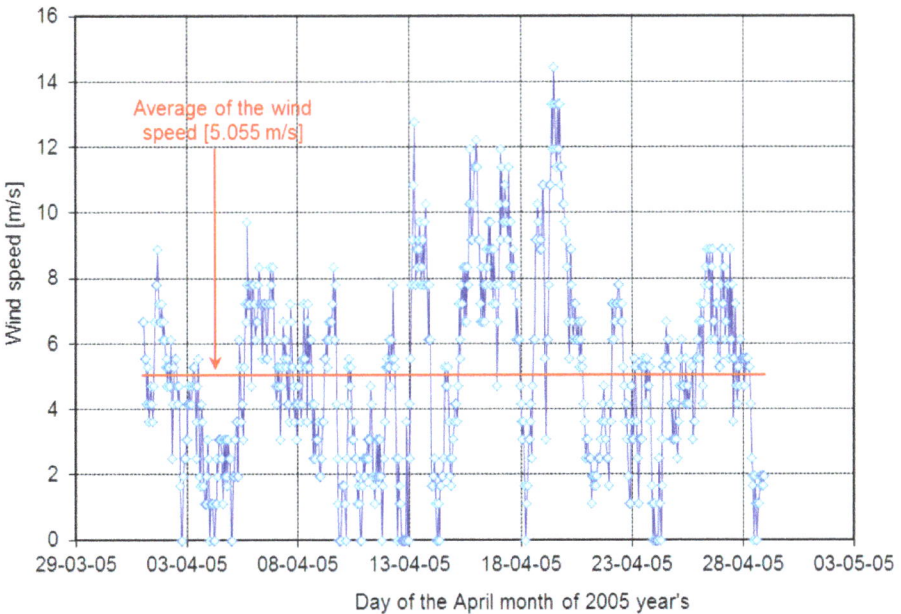

Figure 20. Wind speed data of the Kuujjuarapik site along an April month of 2005 year's

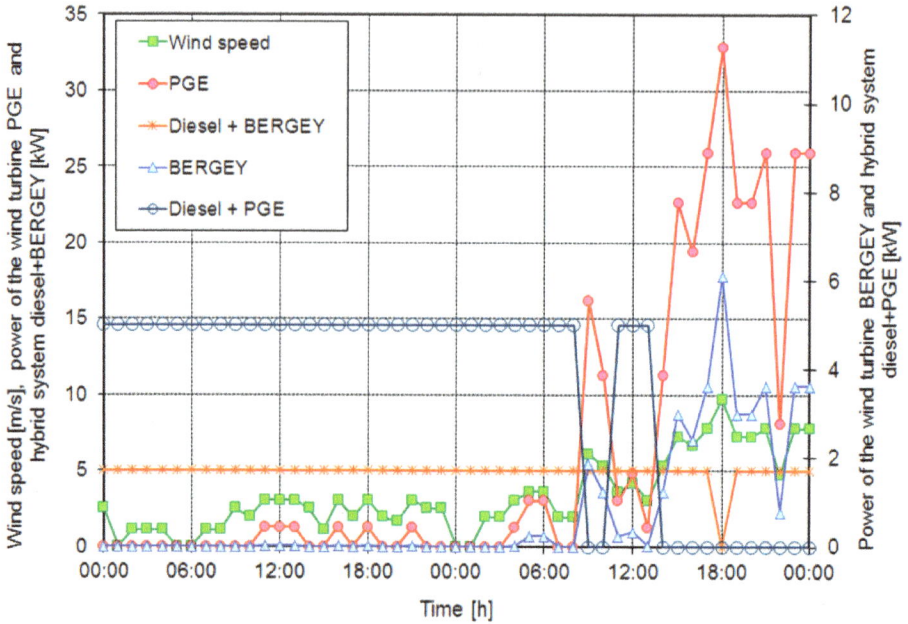

Figure 21. Operating modes of the studied systems along period from 4 to 5 April 2005

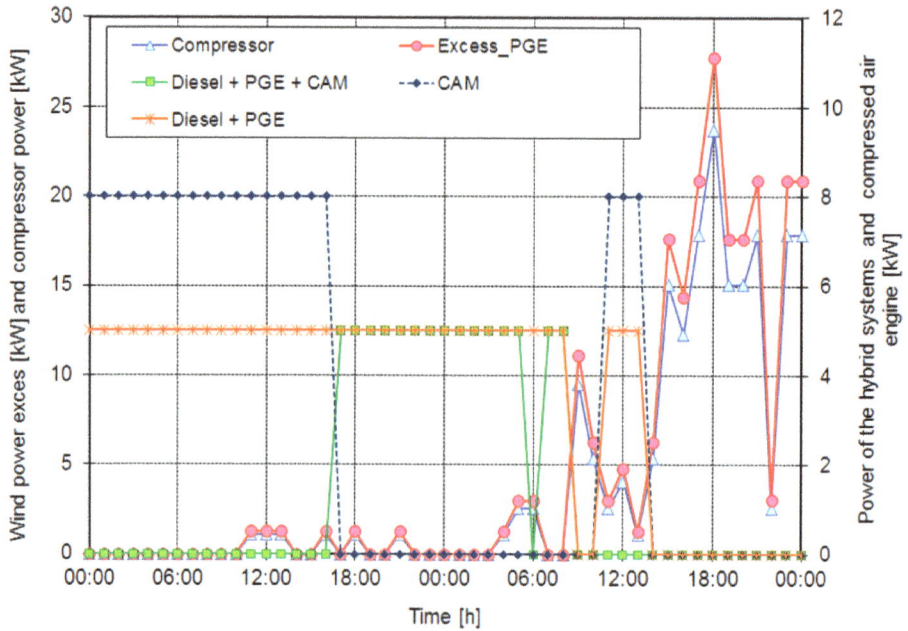

Figure 22. Operating modes of the studied systems along period from 4 to 5 April 2005

In figures 21 and 22, the legends (Diesel + Bergey), (Diesel + PGE) and (diesel + PGE + CAM) represent the power supplied by diesel generator according to the different types of hybridizations with two models of wind turbines and a compressed air energy storage system, respectively.

It is interesting to observe, in figures 21 et 22, the advantages of hybridization (Diesel + PGE + CAM) that appears in the short duration of the diesel operation time (DOT). Indeed, it can stop completely the diesel generator for 33 hours during two days of operation (saving of 69% of the DOT) compared to 13 hours of shutdown of the diesel (27% of DOT is avoided) obtained through the system (Diesel + PGE) and 1 hour (saving of 2% of the DOT) during which the diesel will be stopped thanks to the hybridization between it and Bergey wind turbine..

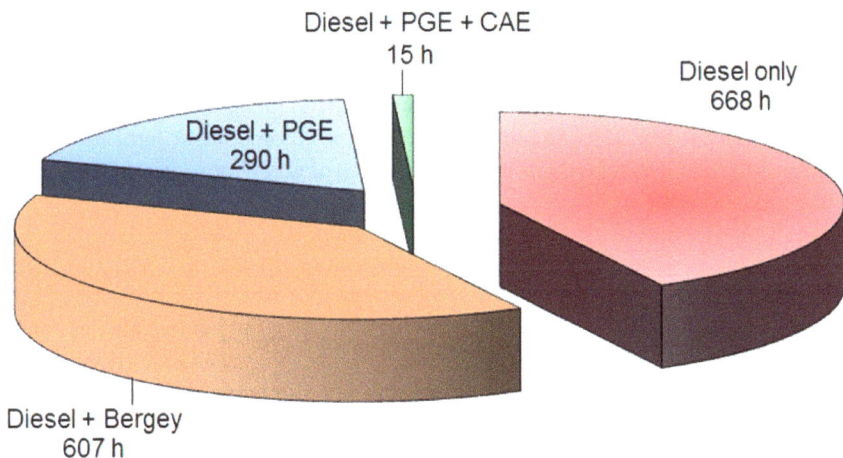

Figure 23. Operating time of diesel generator according to the exploitation scenarios

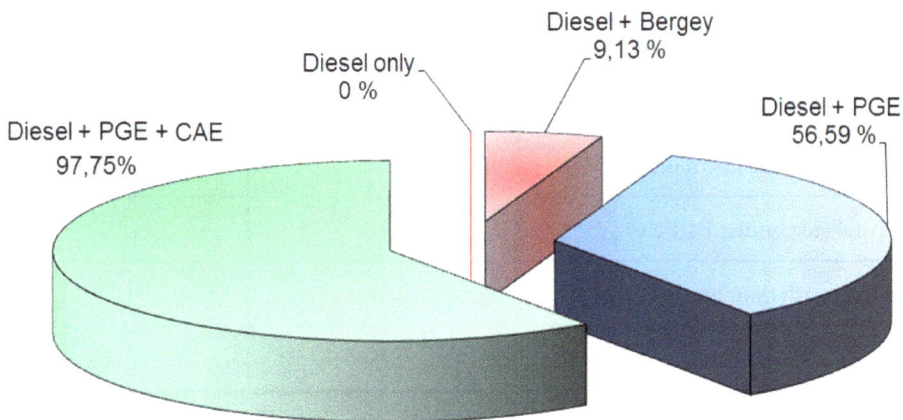

Figure 24. Fuel saving according to the exploitation scenarios

Figure 23 represents the operating time of diesel generator according to the functioning scenario of the system (diesel only, Bergey + diesel, PGE + diesel or PGE + diesel + CAM). Figure 23 shows that hybridization between the Bergey wind turbine and diesel does not allow a remarkable decrease in the operation frequency of the diesel generator (DG) that runs about 91% of operating time in April 2005 (607 h). But by combining the DG to a PGE wind turbine, the DG will work almost 43% (290 h) of time during the month of April and 15 h (2%) if it works in hybridization with the CAM and a PGE wind turbine.

Figure 24 represents the fuel saving according to the functioning scenario of the system (diesel only, Bergey + diesel, PGE + diesel or PGE + diesel + CAM), Figure 24 shows that WDHS avoids approximately 139 liters of fuel (9% saving in fuel consumption) if the diesel is associated with a Bergey wind turbine. However, this rate increases to 57% (863 liters), if the hybridization of the diesel generator is done with the PGE wind turbine. On the other hand, the hybridization between diesel generator, compressed air generator and PGE wind turbine increases this fuel saving very significantly where the amount of fuel avoided is approximately 1491 liters (98%). The fuel saved thanks to SSWDCAHS (during the month of April 2005), allows to reduce the greenhouse gases (GHG) emission approximately 4 tons, which is equivalent to the GHG amount emitted by one automobile or light truck traveling 15,000 km per year. In Table 3, we review the quantity of greenhouse gases (GHG) avoided thanks to the use of WDCAHS.

Name of the substance	Emission Factor (kg/m³)	Total value of emissions (tones)
Carbon dioxide (CO_2)	2800	4.174
Carbon monoxide (CO)	13.954	0.023
Sulfur dioxide (SO_2)	0.083	0.007
Oxides of nitrogen (NO_x)	52.532	0.108
Volatile organic Compounds (VOC)	1.344	0.009
Total suspended Particles (TSP)	1.018	0.008
Particles with diameters $\leq 10\ \mu m$ (P_{10})	0.814	0.008
Particles with diameters $\leq 2.5\ \mu m$ ($P_{2.5}$)	0.786	0.008

Table 3. Quantities of GHG avoided by SSWDCAHS during one month (April 2005)

10. Conclusion

WDCAHS (medium and small scales) represents an innovative concept designed to overcome most of the technical, economic and social barriers that face the deployment of wind energy in isolated sites. Indeed, implementation costs are minimized and reliability is increased by using the existing diesel generators. Thus, the results, theoretical and experimental, obtained have demonstrated the great potential of wind-diesel-compressed air energy storage system for two types of applications: small and medium scale. The application of these systems on real case studies has demonstrated that the fuel economy and the saved GHG obtained with MSWDCAHS (for medium-scale) and SSWDCAHS (small-scale) is about 30% and 98% respectively.

Despite the low average wind speed that characterize the sites chosen for the case studies, remarkable savings may be obtained through the use of compressed air, by avoiding the consumption of large fuel quantities and by allowing the use of a single diesel engine (instead of two: medium-scale case) or by stopping the diesel thanks to the compressed air motor (small-scale case). This allows not only reducing the exploitation deficit of diesel engines supplying the autonomous networks in remote areas, but also to prolong engine life-cycle and reduce maintenance costs. These percentages, especially for medium-scale application, can be increased if the wind-diesel-compressed air hybrid system is used in the sites characterized by a good wind energy potential (average wind speed about 8-9 m/s).

However, in future works, the calculation of the energy cost (kW/h), based on the investment cost and the purchase of new equipment (wind turbines, CAES equipment, etc.) will allow determining the system's economic viability for remote applications.

Finally, further investigation and analysis, as well as building and testing a prototype, are required to validate the present conclusions. This can be validated on the future bed-test of TechnoCentre éolien at Rivière-au-Renard (Quebec, Canada).

Author details

Hussein Ibrahim
TechnoCentre Éolien, Gaspé, QC, Canada

Adrian Ilinca
Université du Québec à Rimouski, Rimouski, QC, Canada

11. References

[1] G. Iwanski, W. Koczara, "Isolated Wind-Diesel Hybrid Variable Speed Power Generation System", presented at EVER 2009, Monaco 2009, May 26-29.

[2]   Ken Ash, Trevor Gaunt, Sissy-qianqian Zhang, Erkki Lakervi, Innovative solutions and best practices for electrification of remote areas, CIGRE SC-C6 (COLL 2007), Working Group C6-13 "Rural Electrification", Topic B.

[3]   Chemmangot Nayar, High Renewable Energy Penetration – Diesel Generator Systems, - Electrical India Vol. 50, N°6, June 2010.

[4]   L. Grzesiak, W. Koczara, M. da Ponte, "Power Quality of the Hygen Autonomous Load – Adaptive Adjustable Speed Generating System", Proc. of Applied Power Electronics Conf. APEC'99. Dallas, USA, March 1999, pp. 398 – 400.

[5]   Kim Ah-You, Greg Leng, Énergies renouvelables dans les communautés éloignées du Canada, Programme des énergies renouvelables pour les communautés éloignées, Ressources Naturelles Canada.

[6]   Weis TM, Ilinca A. The utility of energy storage to improve the economics of wind-diesel power plants in Canada. Renewable Energy 2008;33(7):1544e57.

[7]   La stratégie énergétique du Québec 2006e2015. L'énergie pour construire le Québec de demain. http://www.mrnf.gouv.qc.ca/energie/eolien.

[8]   I. Baring-Gould, M. Dabo, "Technology, Performance, and Market Report of Wind-Diesel Applications for Remote and Island Communities", Proc. of WINDPOWER 2009, Chicago, Illinois May 4–7, 2009.

[9]   Chedid, R. B., S. H. Karaki et C. EI-Chamali. ,"Adaptive fuzzy control for wind-diesel weak power systems", Energy Conversion, IEEE Trans on, vol. 15, nO I, p. 71-78.

[10] Saha, T. K., et D. Kastha. ,"Design Optimization and Dynamic Performance Analysis of a Stand-Alone Hybrid Wind-Diesel Electrical Power Generation System ", Energy Conversion, IEEE Trans on, vol. 25, no 4, p. 1209-1217.

[11] Abbey, C.," A Stochastic Optimization Approach to Rating of Energy Storage Systems in Wind-Diesel Isolated Grids ", Power Systems, IEEE Trans. on, vol. 24, no I, p. 418-426.

[12] Wind/Diesel Systems Architecture Guidebook, AWEA, 1991.

[13] http://en.wikipedia.org/wiki/Wind-Diesel_Hybrid_Power_Systems

[14] W. Koczara, Z. Chlodnicki, E. Ernest, N. Brown, "Hybrid Adjustable Speed Generation System", proceedings on 3rd International Conference on Ecological Vehicles & Renewable Energies, Monaco 2008, March 27-30.

[15] J.G. McGowan, J.F. Manwella and S.R. Connors, "Wind/diesel energy systems: Review of design options and recent developments", Solar Energy, Volume 41, Issue 6, Pages 561-575, 1988.

[16] Timothy M. Weis, Adrian Ilinca, The utility of energy storage to improve the economics of wind–diesel power plants in Canada, Renewable Energy, Volume 33, Issue 7, July 2008, Pages 1544–1557.

[17] J. Kaldellis et al, "Autonomous energy systems for remote islands based on renewable energy sources", in Proceedings of EWEC 99, Nice 1999.

[18] J.F. Maisson « Wind Power Development in Sub-Arctic Conditions with Severe Rime Icing », Presented at the Circumpolar Climate Change Summit and Exposition, Whitehorse, Yukon, 2001

[19] www.nunavutpower.com

[20] B. Reeves « Kotzebue Electric Association Wind Projects », Proceedings of NREL/AWEA 2002 Wind-Diesel Workshop, Anchorage, Alaska, USA, 2002

[21] P. Ebert P and J. Zimmermann, "Successful high wind penetration into a medium sized diesel grid without energy storage using variable speed wind turbine technology", in Proceedings of EWEC 99, Nice 1999.

[22] Singh V. Blending Wind and Solar into the Diesel Generator Market. Renewable Energy Policy Projet (REPP) Research Report, Winter 2001, No. 12, Washington, DC.

[23] Y. Jean, P. Viarouge, D. Champagne, R. Reid, B. Saulnier, «Perfectionnement des outils pour l'implantation des éoliennes à Hydro-Québec», rapport IREQ-92-065, 1992

[24] R. Gagnon, A. Nouaili, Y. Jean, P. Viarouge; «Mise à jour des outils de modélisation et de simulation du Jumelage Éolien-Diesel à Haute Pénétration Sans Stockage et rédaction du devis de fabrication de la charge de lissage», Rapport IREQ-97-124-C, 1997.

[25] Ilinca A, Chaumel JL. Implantation d'une centrale éolienne comme source d'énergie d'appoint pour des stations de télécommunications. Colloque international sur l'énergie éolienne et les sites isolés, Îles de la Madeleine, 2005.

[26] H. Ibrahim, R. Younès, A. Ilinca, J. Perron, Investigation des générateurs hybrides d'électricité de type éolien-air comprimé. Numéro spécial CER'2007 de la Revue des énergies renouvelables, Parrainée par l'UNESCO, Éditée par le CDER, Algérie, Août 2008.

[27] H. Ibrahim, A. Ilinca, J. Perron, Investigations des différentes alternatives renouvelables et hybrides pour l'électrification des sites isolés, rapport interne, UQAR, LREE–03, 2008.

[28] T.M. Weis, A. Ilinca, J.Paul. Pinard, Stakeholders' perspectives on barriers to remote wind–diesel power plants in Canada Energy Policy, Volume 36, Issue 5, May 2008, Pages 1611-1621

[29] Hunter R, Elliot G. Windediesel systems e a guide to the technology and its implementation. Cambridge (UK): Cambridge University Press; 1994.

[30] HOMER v2.0 e the optimisation model for distributed power. NREL. www.nrel.org.

[31] Robb D. Making a CAES for wind energy storage. North American Wind Power, June 2005.

[32] Ibrahim H, Younès R, Ilinca A. Optimal conception of a hybrid generator of electricity. CANCAM02007 ETS-39, Toronto, Canada. p. 358 - 359.

[33] Ibrahim H, Ilinca A, Perron J. Moteur diesel suralimenté, bases et calculs, cycles réel, théorique et thermodynamique. Rapport interne, UQAR-UQAC, LREE-02; Novembre 2006.

[34] Ibrahim H, Ilinca A, Younès R, Perron J, Basbous T. Study of a hybrid wind-diesel system with compressed air energy storage. IEEE Canada, electrical power conference 2007, "Renewable and alternative energy resources", EPC2007, Montreal, Canada, October 25e26, 2007.

[35] http://www.bergey.com/

[36] http://www.endurancewindpower.com/

[37] Bell-Canada, www.bell.ca

Advances in Spatio-Temporal Modeling and Prediction for Environmental Risk Assessment

S. De Iaco, S. Maggio, M. Palma and D. Posa

Additional information is available at the end of the chapter

1. Introduction

Meteorological readings, hydrological parameters and many measures of air, soil and water pollution are often collected for a certain span, regularly in time, and at different survey stations of a monitoring network. Then, these observations can be viewed as realizations of a random function with a spatio-temporal variability. In this context, the arrangement of valid models for spatio-temporal prediction and environmental risk assessment is strongly required. Spatio-temporal models might be used for different goals: optimization of sampling design network, prediction at unsampled spatial locations or unsampled time points and computation of maps of predicted values, assessing the uncertainty of predicted values starting from the experimental measurements, trend detection in space and time, particularly important to cope with risks coming from concentrations of hazardous pollutants. Hence, more and more attention is given to spatio-temporal analysis in order to sort out these issues.

Spatio-temporal geostatistical techniques provide useful tools to analyze, interpret and control the complex evolution of various variables observed by environmental monitoring networks. However, in the literature there are no specialized monographs which contain a thorough presentation of multivariate methodologies available in Geostatistics, especially in a spatio-temporal context. Several authors have developed different multivariate models for analyzing the spatial and spatio-temporal behavior of environmental variables, as it is clarified in the following brief review.

In multivariate spatial analysis, direct and cross correlations for the variables under study are quantified by estimating and modeling the matrix variogram. The difficulty in modeling this matrix function, especially the off diagonal entries of the same matrix, has been first faced by using the linear coregionalization model (*LCM*), proposed by [45].

For matrix covariance functions, [28] constructed a parametric family of symmetric covariance models for stationary and isotropic multivariate Gaussian spatial random fields, where both the diagonal and off diagonal entries are of the Matérn type. In the bivariate case, they provided necessary and sufficient conditions for the positive definiteness of the second-order structure, whereas for the other multivariate cases they suggested a parsimonious model which imposes restrictions on the scale and cross-covariance smoothness parameters. In the bivariate case, where the smoothness parameter is the same for both covariance functions, the Gneiting model is a simplified LCM. The Gneiting cross-covariance model also assumes that the scale parameter is the same for all the covariance functions and the cross-covariance functions. Both the LCM and Gneiting constructions for cross-covariances result in symmetric models; however, no distributional assumptions are required for using a LCM, which can easily incorporate components with compact support and multiple ranges and an unbounded variogram component.

Although models for multivariate spatial data have been extensively explored [25, 54, 55], models for multivariate spatio-temporal data have received relatively less attention. In the literature, it is common to use classical techniques for multivariate spatial and temporal analysis [8, 54]. Recently, canonical correlation analysis was combined with space-time geostatistical tools for detecting possible interactions between two groups of variables, associated with pollutants and atmospheric conditions [6]. In the dynamic modeling framework, there are some results in studying the spatio-temporal variability of several correlated variables: [26], for example, extended univariate spatio-temporal dynamic models to multivariate dynamic spatial models. Moreover, [38] proposed a methodology to evaluate the appropriateness of several common assumptions, such as symmetry, separability and linear model of coregionalization, on multivariate covariance functions in the spatio-temporal context, while [4] proposed a spatio-temporal LCM where the multivariate spatio-temporal process was expressed as a linear combination of independent Gaussian processes in space-time with mean zero and a separable spatio-temporal covariance. [1] considered some solutions to the symmetry problem; moreover, they proposed a class of cross-covariance functions for multivariate random fields based on the work of [27]. The maximum likelihood estimation of heterotopic spatio-temporal models with spatial LCM components and temporal dynamics was developed by [22]. A $GSLib$ [19] routine for cokriging was properly modified in [12] to incorporate the spatio-temporal LCM, previously developed using the generalized product-sum variogram model [10]. Recently, in [15] an automatic procedure for fitting the spatio-temporal LCM using the product-sum variogram model has been presented and some computational aspects, analytically described by a main flow-chart, have been discussed. In [16] simultaneous diagonalization of the sample matrix variograms has been used to isolate the basic components of a spatio-temporal LCM and it has been illustrated how nearly simultaneous diagonalization of the cross-variogram matrices simplifies modeling of the matrix variogram.

In the following, after an introduction of the theoretical framework of the multivariate spatio-temporal random function and its features (Section 2), a review of recent techniques for building admissible models is proposed (Section 3). Successively, the spatio-temporal LCM, its assumptions and appropriate statistical tests are presented (Section 4) and techniques for prediction and risk assessment maps are introduced (Section 5). Some critical aspects regarding sampling, modeling and computational problems are discussed (Section 6). Finally, a case study concerning particle pollution (PM_{10}) and two atmospheric variables (Temperature and Wind Speed) in the South of Apulian region (Italy), has been presented (Section 7). Before using the spatio-temporal LCM to describe the spatio-tempor multivariate correlation structure among the variables under study, its adequacy with respect to the data has been analyzed; in particular, the assumption of symmetry of the cross-covariance function, has been properly tested [38]. By using a recent fitting procedure [16], based on the simultaneous diagonalization of several symmetric real-valued matrix variograms, the basic structures of the spatio-temporal LCM which describes the spatio-temporal correlation among the variables, have been easily detected. Predictions of the primary variable (PM_{10}) are obtained by using a modified $GSLib$ program, called "COK2ST" [12]. Then, risk maps showing the probability that the particle pollution exceeds the national law limit have been associated to predition maps and the estimation of the probability distributions for two sites of interest have been produced.

2. Multivariate spatio-temporal random function

Let $\mathbf{Z}(\mathbf{u}) = [Z_1(\mathbf{u}), \ldots, Z_p(\mathbf{u})]^T$, be a vector of p spatio-temporal random functions ($STRF$) defined on the domain $D \times T \subseteq \mathbb{R}^{d+1}$, with $(d \leq 3)$, then

$$\{\mathbf{Z}(\mathbf{u}), \ \mathbf{u} = (\mathbf{s}, t) \in D \times T \subseteq \mathbb{R}^{d+1}\},$$

represents a multivariate spatio-temporal random function ($MSTRF$), where $\mathbf{s} = (s_1, \ldots, s_d)$ are the coordinates of the spatial domain $D \subseteq \mathbb{R}^d$ and t the coordinate of the temporal domain $T \subseteq \mathbb{R}$.

Afterwards, the $MSTRF$ will be denoted with $\mathbf{Z}$ and its components with Z_i. The p $STRF$ $Z_i, i = 1, \ldots, p$, are the *components* of $\mathbf{Z}$ and they are associated to the spatio-temporal variables under study; these components are called *coregionalized variables* [29]. The observations $z_i(\mathbf{u}_\alpha), i = 1, \ldots, p, \ \alpha = 1, \ldots, N_i$, of the p variables Z_i, at the points $\mathbf{u}_\alpha \in D \times T$, are considered as a finite realization of a $MSTRF$ $\mathbf{Z}$.

2.1. Moments of a $MSTRF$

Given a $MSTRF$ $\mathbf{Z}$, with p components, we define, if they exist and they are finite:

- the *expected value*, or *first-order moment* of each component Z_i,

$$E[Z_i(\mathbf{u})] = m_i(\mathbf{u}), \quad \mathbf{u} \in D \times T, \ i = 1, \ldots, p; \tag{1}$$

- the *second-order moments*,
  1. the *variance* of each component Z_i,

$$Var[Z_i(\mathbf{u})] = E[Z_i(\mathbf{u}) - m_i(\mathbf{u})]^2, \quad \mathbf{u} \in D \times T, \ i = 1, \ldots, p; \tag{2}$$

  2. the *cross-covariance* for each pair of $STRF$ $(Z_i, Z_j), i \neq j$,

$$Cov[Z_i(\mathbf{u}), Z_j(\mathbf{u}')] =$$

$$= C_{ij}(\mathbf{u}, \mathbf{u}') = E\left[(Z_i(\mathbf{u}) - m_i(\mathbf{u}))(Z_j(\mathbf{u}') - m_j(\mathbf{u}'))\right], \tag{3}$$

$\mathbf{u}, \mathbf{u}' \in D \times T, \ i, j = 1, \ldots, p, \ i \neq j;$

3. the *cross-variogram* for each pair of STRF (Z_i, Z_j), $i \neq j$,

$$2\gamma_{ij}(\mathbf{u}, \mathbf{u}') = Cov\left[(Z_i(\mathbf{u}) - Z_i(\mathbf{u}')), (Z_j(\mathbf{u}) - Z_j(\mathbf{u}'))\right], \tag{4}$$

$\mathbf{u}, \mathbf{u}' \in D \times T$, $i, j = 1, \ldots, p$, $i \neq j$.

Note that for $i = j$, we obtain:

- the covariance of the STRF Z_i, called *direct covariance*, or simply covariance,

$$C_{ii}(\mathbf{u}, \mathbf{u}') = E\left[(Z_i(\mathbf{u}) - m_i(\mathbf{u}))(Z_i(\mathbf{u}') - m_i(\mathbf{u}'))\right],$$

with $\mathbf{u}, \mathbf{u}' \in D \times T$;
- the *direct variogram* of the STRF Z_i,

$$2\gamma_{ii}(\mathbf{u}, \mathbf{u}') = Var\left[(Z_i(\mathbf{u}) - Z_i(\mathbf{u}')\right], \quad \mathbf{u}, \mathbf{u}' \in D \times T.$$

These moments describe the basic features of a *MSTRF*, such as the spatio-temporal correlation for each variable and the cross-correlation among the variables.

2.2. Admissibility conditions

In multivariate Geostatistics, admissibility conditions concern both the cross-covariances and the cross-variograms, as described in the following.

Let $\mathbf{Z}$ be a *MSTRF*, with components Z_i, $i = 1, \ldots, p$, and let $\{\mathbf{u}_1, \ldots, \mathbf{u}_N\}$ a set of N points of a spatio-temporal domain $D \times T$; the direct and cross-covariances of the *MSTRF* must satisfy the following inequality:

$$\sum_{i=1}^{p} \sum_{j=1}^{p} \sum_{\alpha=1}^{N} \sum_{\beta=1}^{N} \lambda_{\alpha i} \lambda_{\beta j} C_{ij}(\mathbf{u}_\alpha - \mathbf{u}_\beta) \geq 0,$$

for any choice of the N points $\mathbf{u}_\alpha$ and for any choice of the weights $\lambda_{\alpha i}$. Using the matrix notation, the $(p \times p)$ matrices $\mathbf{C}(\mathbf{u}_\alpha - \mathbf{u}_\beta) = \left[C_{ij}(\mathbf{u}_\alpha - \mathbf{u}_\beta)\right]$ of the direct and cross-covariances of the STRF $Z_i(\mathbf{u}_\alpha)$ and $Z_j(\mathbf{u}_\beta)$ will be admissible if they satisfy the following condition:

$$\sum_{\alpha=1}^{N} \sum_{\beta=1}^{N} \vec{\lambda}_\alpha^T \mathbf{C}(\mathbf{u}_\alpha - \mathbf{u}_\beta) \vec{\lambda}_\beta \geq 0, \tag{5}$$

where $\vec{\lambda}_\alpha = \left[\lambda_{\alpha 1}, \ldots, \lambda_{\alpha p}\right]^T$ is a $(p \times 1)$ vector of weights $\lambda_{\alpha i}$.

As in the univariate case, the $(p \times p)$ matrices $\mathbf{\Gamma}(\mathbf{u}_\alpha - \mathbf{u}_\beta) = \left[\gamma_{ij}(\mathbf{u}_\alpha - \mathbf{u}_\beta)\right]$ of the direct and cross-variograms of the STRF $Z_i(\mathbf{u}_\alpha)$ and $Z_j(\mathbf{u}_\beta)$ will be admissible if, for any choice of the N points $\mathbf{u}_\alpha$, they satisfy the following condition

$$-\sum_{\alpha=1}^{N} \sum_{\beta=1}^{N} \vec{\lambda}_\alpha^T \mathbf{\Gamma}(\mathbf{u}_\alpha - \mathbf{u}_\beta) \vec{\lambda}_\beta \geq 0,$$

under the constraint: $\quad \sum_{\alpha=1}^{N} \vec{\lambda}_{\alpha} = 0.$

2.3. Stationarity hypotheses

Stationarity hypotheses allow to make inference on the $MSTRF$. In particular, second-order stationarity and intrinsic hypotheses concern the first and second-order moments of the $MSTRF$.

2.3.1. Second-order stationarity

A $MSTRF$ **Z**, with p components, is *second-order stationary* if:

- for any $STRF$ Z_i, $i, \dots, p$,

$$E[Z_i(\mathbf{u})] = m_i, \qquad \mathbf{u} \in D \times T, i = 1, \dots, p; \tag{6}$$

- for any pair of $STRF$ Z_i and Z_j, $i, j = 1, \dots, p$, the cross-covariance C_{ij} depends only on the spatio-temporal separation vector $\mathbf{h} = (\mathbf{h_s}, h_t)$ between the points $\mathbf{u}$ and $\mathbf{u} + \mathbf{h}$:

$$C_{ij}(\mathbf{h}) = E[(Z_i(\mathbf{u} + \mathbf{h}) - m_i)(Z_j(\mathbf{u}) - m_j)] =$$
$$= E[Z_i(\mathbf{u} + \mathbf{h})\, Z_j(\mathbf{u})] - m_i\, m_j, \tag{7}$$

where $\mathbf{u}, \mathbf{u} + \mathbf{h} \in D \times T, i, j = 1, \dots, p$. For $i = j$, the direct covariance function of the $STRF$ Z_i is obtained.

There exist several physical phenomena for which neither variance, nor the covariance exist, however it is possible to assume the existence of the variogram.

2.3.2. Intrinsic hypotheses

A $MSTRF$ **Z**, with p components, satisfies the *intrinsic hypotheses* if:

- for any $STRF$ Z_i, $i = 1, \dots, p$,

$$E\left[Z_i(\mathbf{u} + \mathbf{h}) - Z_i(\mathbf{u})\right] = 0, \qquad \mathbf{u}, \mathbf{u} + \mathbf{h} \in D \times T, i = 1, \dots, p; \tag{8}$$

- for any pair of $STRF$ Z_i and Z_j, $i, j = 1, \dots, p$, the cross-variogram exists and it depends only on the spatio-temporal separation vector $\mathbf{h}$:

$$2\, \gamma_{ij}(\mathbf{h}) = Cov[(Z_i(\mathbf{u} + \mathbf{h}) - Z_i(\mathbf{u})), (Z_j(\mathbf{u} + \mathbf{h}) - Z_j(\mathbf{u}))], \tag{9}$$

where $\mathbf{u}, \mathbf{u} + \mathbf{h} \in D \times T, i, j = 1, \dots, p$.

Second-order stationarity implies the existence of the intrinsic hypotheses, however the converse is not true. Intrinsic hypotheses imply that the cross-variogram can be expressed as the expected value of the product of the increments:

$$\gamma_{ij}(\mathbf{h}) = \frac{1}{2} E\{[Z_i(\mathbf{u} + \mathbf{h}) - Z_i(\mathbf{u})][Z_j(\mathbf{u} + \mathbf{h}) - Z_j(\mathbf{u})]\}, \tag{10}$$

$\mathbf{u}, \mathbf{u} + \mathbf{h} \in D \times T$, $i, j = 1, \ldots, p$. For $i = j$, the direct variogram of the STRF Z_i is obtained.

2.3.3. Properties of the cross-covariance for second-order stationary MSTRF

Given a second-order stationary MSTRF, the cross-covariance satisfies the properties listed below.

The cross-covariance is not invariant with respect to the exchange of the variables:

$$C_{ij}(\mathbf{h}) \neq C_{ji}(\mathbf{h}), \qquad i \neq j, \tag{11}$$

as well as it is not invariant with respect to the sign of the vector $\mathbf{h}$:

$$C_{ij}(-\mathbf{h}) \neq C_{ij}(\mathbf{h}), \qquad i \neq j. \tag{12}$$

However, the cross-covariance is invariant with respect to the joint exchange of the variables and the sign of the vector $\mathbf{h}$:

$$C_{ij}(\mathbf{h}) = C_{ji}(-\mathbf{h}). \tag{13}$$

2.3.4. Properties of the cross-variogram for intrinsic MSTRF

Afterwards, the main properties of the cross-variogram for intrinsic MSTRF are given.

1. The cross-variogram vanishes at the origin, that is:

$$\gamma_{ij}(\mathbf{0}) = 0. \tag{14}$$

2. The cross-variogram is invariant with respect to the exchange of the variables:

$$\gamma_{ij}(\mathbf{h}) = \gamma_{ji}(\mathbf{h}). \tag{15}$$

3. The cross-variogram is invariant with respect to the sign of the vector $\mathbf{h}$:

$$\gamma_{ij}(-\mathbf{h}) = \gamma_{ij}(\mathbf{h}). \tag{16}$$

From (15) and (16) follows that the cross-variogram is completely symmetric, as it will be pointed out in the next sections.

2.3.5. Separability for a MSTRF

The cross-covariance C_{ij} for a second-order stationary MSTRF $\mathbf{Z}$ is *separable* if:

$$C_{ij}(\mathbf{h}) = \rho(\mathbf{h}) a_{ij}, \qquad \mathbf{h} = (\mathbf{h}_s, h_t) \in D \times T, \; i, j = 1, \ldots, p,$$

where a_{ij} are the elements of a $(p \times p)$ positive definite matrix and $\rho(\cdot)$ is a correlation function. In this case, it results:

$$\frac{C_{ij}(\mathbf{h})}{C_{ij}(\mathbf{h}')} = \frac{\rho(\mathbf{h})}{\rho(\mathbf{h}')}, \quad \mathbf{h}, \mathbf{h}' \in D \times T, \; i, j = 1, \ldots, p,$$

hence the changes of the cross-covariance functions, with respect to the changes of the vector $\mathbf{h}$, do not depend on the pair of the STRF Z_i, Z_j.

The cross-covariance C_{ij} for a second-order stationary MSTRF $\mathbf{Z}$ is *fully separable* if:

$$C_{ij}(\mathbf{h}_s, h_t) = \rho_S(\mathbf{h}_s)\,\rho_T(h_t)a_{ij}, \quad (\mathbf{h}_s, h_t) \in D \times T, \ i, j = 1, \ldots, p,$$

where a_{ij} are the elements of a $(p \times p)$ positive definite matrix, $\rho_S(\cdot)$ is a spatial correlation function and $\rho_T(\cdot)$ is a temporal correlation function. In the literature, many statistical tests for separability have been proposed and are based on parametric models [2, 32, 51], likelihood ratio tests and subsampling [46] or spectral methods [23, 50].

2.3.6. Symmetry for a MSTRF

The cross-covariance C_{ij} of a second-order stationary MSTRF $\mathbf{Z}$, with p components, is *symmetric* if:

$$C_{ij}(\mathbf{h}) = C_{ij}(-\mathbf{h}), \quad \mathbf{h} \in D \times T, \ i, j = 1, \ldots, p,$$

or, equivalently, if:

$$C_{ij}(\mathbf{h}) = C_{ji}(\mathbf{h}), \quad \mathbf{h} \in D \times T, \ i, j = 1, \ldots, p.$$

The cross-covariance C_{ij} of a second-order stationary MSTRF $\mathbf{Z}$, with p components, is *fully symmetric* if:

$$C_{ij}(\mathbf{h}_s, h_t) = C_{ij}(\mathbf{h}_s, -h_t), \quad (\mathbf{h}_s, h_t) \in D \times T, \ i, j = 1, \ldots, p,$$

or, equivalently,

$$C_{ij}(\mathbf{h}_s, h_t) = C_{ij}(-\mathbf{h}_s, h_t), \quad (\mathbf{h}_s, h_t) \in D \times T, \ i, j = 1, \ldots, p.$$

Atmospheric, environmental and geophysical processes are often under the influence of prevailing air or water flows, resulting in a lack of full symmetry [18, 27, 52].

Fig. 1 summarizes the relationships between separability, symmetry, stationarity and the LCM in the general class of the cross-covariance functions of a MSTRF $\mathbf{Z}$. If a cross-covariance is separable, then it is symmetric, however, in general, the converse is not true. Moreover, the hypothesis of full separability is a special case of full symmetry.

Several tests to check symmetry and separability of cross-covariance functions can be found in the literature [38–40, 50].

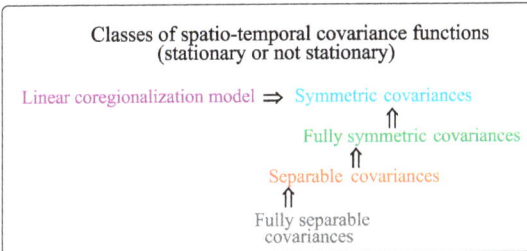

Figure 1. Relationships among different classes of spatio-temporal covariance functions

3. Techniques for building admissible models

In the following, a brief review of the most utilized techniques to construct admissible cross-covariance models is presented.

1. For the *intrinsic correlation model*, the matrices $\mathbf{C}$ are described by separable cross-covariances $C_{ij}, i, j = 1, \ldots, p$ [44], that is:

$$C_{ij}(\mathbf{u}_\alpha, \mathbf{u}_\beta) = \rho(\mathbf{u}_\alpha, \mathbf{u}_\beta)a_{ij},$$

where the coefficients a_{ij} are the elements of a $(p \times p)$ positive definite matrix, and $\rho(\cdot, \cdot)$ is a correlation function. However, this model is not flexible enough to handle complex relationships between processes, because the cross-covariance function between components measured at each location always has the same shape regardless of the relative displacement of the locations. As it will be discussed in the next section, the *LCM* is a straightforward extension of the intrinsic correlation model.

2. In the kernel convolution method [55] the cross-covariance functions is represented as follows:

$$C_{ij}(\mathbf{u}_\alpha, \mathbf{u}_\beta) = \int_{\mathbb{R}^{d+1}} \int_{\mathbb{R}^{d+1}} k_i(\mathbf{u}_\alpha - \mathbf{u})k_j(\mathbf{u}_\beta - \mathbf{u}')\rho(\mathbf{u} - \mathbf{u}')d\mathbf{u}d\mathbf{u}',$$

where the k_i are square integrable kernel functions and ρ is a valid stationary correlation function. This approach assumes that all the variables $Z_i, i = 1, \ldots, p$, are generated by the same underlying process, which is very restrictive. Moreover, this model and its parameters lack interpretability and, except for some special cases, it requires Monte Carlo integration.

3. In the covariance convolution for stationary processes [24, 43] the cross-covariance functions is represented as follows:

$$C_{ij}(\mathbf{h}) = \int_{\mathbb{R}^{d+1}} C_i(\mathbf{h} - \mathbf{h}')C_j(\mathbf{h}')d\mathbf{h}',$$

where C_i are second-order stationary covariances. The motivation and interpretation of the resulting cross-dependency structure is rather unclear. Although some closed-form expressions exist, this method usually requires Monte Carlo integration.

4. Recently, an approach based on latent dimensions and existing covariance models for univariate random fields, has been proposed; the idea is to develop flexible, interpretable and computationally feasible classes of cross-covariance functions in closed form [1].

4. Linear coregionalization model

LCM is based on the hypothesis that each direct or cross-variogram (covariogram) can be represented as a linear combination of some basic models and each direct or cross-variogram (covariogram) must be built using the same basic models [33].

LCM is utilized in several applications because of its flexibility, moreover it encouraged the development of algorithms able to estimate quickly the parameters of the selected model, assuring the admissibility conditions, even in presence of several variables [29–31].

Let $\mathbf{Z}$ be a second-order stationary $MSTRF$ with p components, Z_i, $i = 1, \ldots, p$, the direct and cross-covariances for the spatio-temporal LCM are defined as follows:

$$C_{ij}(\mathbf{h}) = Cov\left[Z_i(\mathbf{u}), Z_j(\mathbf{u} + \mathbf{h})\right] = \sum_{l=1}^{L} b_{ij}^l \, c_l(\mathbf{h}), \qquad i, j = 1, \ldots, p, \tag{17}$$

where c_l are covariances, called *basic structures*, and the non-negative coefficients b_{ij}^l satisfy the following property:

$$b_{ij}^l = b_{ji}^l, \quad i, j = 1, \ldots, p;$$

hence, in the LCM, it is assumed that:

$$C_{ij}(\mathbf{h}) = C_{ij}(-\mathbf{h}),$$

$$C_{ij}(\mathbf{h}) = C_{ji}(\mathbf{h}),$$

with $i, j = 1, \ldots, p, \ i \neq j$.

The matrix $\mathbf{C}$ for the second-order stationary $\mathbf{Z}$ is built as follows:

$$\mathbf{C}(\mathbf{h}) = \sum_{l=1}^{L} \mathbf{B}_l \, c_l(\mathbf{h}). \tag{18}$$

Analogously, it is also possible to introduce the LCM for a $MSTRF$ which satisfies the intrinsic hypotheses. In such a case, the direct and cross-variograms are built as follows:

$$\gamma_{ij}(\mathbf{h}) = \sum_{l=1}^{L} b_{ij}^l \, g_l(\mathbf{h}), \qquad i, j = 1, \ldots, p, \tag{19}$$

where each basic structure g_l is a variogram and the L matrices $\mathbf{B}_l$ of the coefficients b_{ij}^l, corresponding to the sill values of the basic models g_l, are positive definite.

Then, for a $MSTRF$ which satisfies the intrinsic hypotheses the matrix $\boldsymbol{\Gamma}$ of the direct and cross-variograms is:

$$\boldsymbol{\Gamma}(\mathbf{h}) = \sum_{l=1}^{L} \mathbf{B}_l \, g_l(\mathbf{h}). \tag{20}$$

The necessary and sufficient conditions because the model defined in (17) and (20) are admissible are:

1. $c_l(g_l)$ must be covariances (variograms),

2. the matrices $\mathbf{B}_l = \left[b_{ij}^l\right]$, called *coregionalization matrices*, must be positive definite.

The necessary, but not sufficient conditions, for the coefficients b_{ij}^l are the following:

a) $b_{ii}^l \geq 0 \qquad i = 1, \ldots, p, \ l = 1, \ldots, L;$

b) $|b_{ij}^l| \leq \sqrt{b_{ii}^l \, b_{jj}^l} \qquad i, j = 1, \ldots, p, \ l = 1, \ldots, L.$

The basic structures $g_l(\mathbf{h}) = g_l(\mathbf{h}_s, h_t)$ of the spatio-temporal *LCM* (20) can be modelled by using several spatio-temporal variogram models known in the literature, such as the *metric model* [20], *Cressie-Huang models* [5], *Gneiting models* [27] and many others [9, 35, 41, 42, 49, 52].

As discussed in [10], each basic spatio-temporal structure, $g_l(\mathbf{h}_s, h_t)$, $l = 1, \ldots, L$, can be modelled as a generalized product-sum variogram [7], that is

$$g_l(\mathbf{h}_s, h_t) = \gamma_l(\mathbf{h}_s, 0) + \gamma_l(\mathbf{0}, h_t) - k_l \, \gamma_l(\mathbf{h}_s, 0) \, \gamma_l(\mathbf{0}, h_t), \qquad l = 1, \ldots, L, \qquad (21)$$

where

- $\gamma_l(\mathbf{h}_s, 0)$ and $\gamma_l(\mathbf{0}, h_t)$ are, respectively, the marginal spatial and temporal variograms at lth scale of variability;
- k_l, defined as follows

$$k_l = \frac{sill[\gamma_l(\mathbf{h}_s, 0)] + sill[\gamma_l(\mathbf{0}, h_t)] - sill[g_l(\mathbf{h}_s, h_t)]}{sill[\gamma_l(\mathbf{h}_s, 0)] \, sill[\gamma_l(\mathbf{0}, h_t)]}, \qquad (22)$$

is the parameter of generalized product-sum variogram model and it is such that

$$0 < k_l \leq \frac{1}{max\{sill[\gamma_l(\mathbf{h}_s, 0)]; \, sill[\gamma_l(\mathbf{0}, h_t)]\}}. \qquad (23)$$

The inequality (23) represents a necessary and sufficient condition in order that each basic structure $g_l(\mathbf{h}_s, h_t)$, with $l = 1, \ldots, L$, is admissible. Recently, it was shown that strict conditional negative definiteness of both marginals is a necessary as well as a sufficient condition for the generalized product-sum (21) to be strictly conditionally negative definite [13, 14].

Substituting (21) in (20), the spatio-temporal *LCM* can be defined through two marginals: one in space and one in time, i.e.:

$$\Gamma(\mathbf{h}_s, 0) = \sum_{l=1}^{L} \mathbf{B}_l \, \gamma_l(\mathbf{h}_s, 0), \qquad \Gamma(\mathbf{0}, h_t) = \sum_{l=1}^{L} \mathbf{B}_l \, \gamma_l(\mathbf{0}, h_t). \qquad (24)$$

Using the generalized product-sum variogram model it is possible:

1. to identify the different scales of variability and build the matrices $\mathbf{B}_l$, $l = 1, \ldots, L$, by means of the direct and cross marginal variograms;

2. to describe the correlation structure of processes characterized by a different spatial and temporal variability.

4.1. Assumptions in the spatio-temporal *LCM*

Fitting a spatio-temporal *LCM* to the data requires the identification of the spatio-temporal basic variograms and the corresponding positive definite coregionalization matrices, however this is often a hard step to tackle. A recent approach [16], based on the simultaneous diagonalization of a set of matrix variograms computed for several spatio-temporal lags, allows to determine the spatio-temporal *LCM* parameters in a very simple way.

In several environmental applications [54], the cross-covariance function is not symmetric, as for example, in time series in presence of a delay effect, as well as in hydrology, for the cross-correlation between a variable and its derivative, such as water head and transmissivity [53]. Hence, this assumption should be tested before fitting a spatio-temporal *LCM*.

A useful hint to verify the symmetry of the cross-covariance can be given by estimating all the pseudo cross-variograms [47] of the standardized variables $\breve{Z}_i, i = 1, \ldots, p$, i.e., $\tilde{\gamma}_{ij}(\mathbf{h}) = 0.5 \, \mathrm{Var}[\breve{Z}_i(\mathbf{u}) - \breve{Z}_j(\mathbf{u} + \mathbf{h})], i, j = 1, \ldots, p, i \neq j$. If the differences between the estimated pseudo cross-variograms $\tilde{\gamma}_{ij}(\mathbf{h})$ and $\tilde{\gamma}_{ji}(\mathbf{h}), i, j = 1, \ldots, p$, are zero or close to zero, then it could be assumed that the cross-covariances are symmetric.

The appropriateness of the assumption of symmetry of a spatio-temporal *LCM* can be tested by using the methodology proposed by [38], based on the asymptotic joint normality of the sample spatio-temporal cross-covariances estimators. Given a set Λ of user-chosen spatio-temporal lags and the cardinality c of Λ, let $\mathbf{G}_n = \{C_{ji}(\mathbf{h}_s, h_t) : (\mathbf{h}_s, h_t) \in \Lambda, i, j = 1, \ldots, p\}$ be a vector of cp^2 cross-covariances at spatio-temporal lags $k = (\mathbf{h}_s, h_t)$ in Λ. Moreover, let $\hat{C}_{ji}(\mathbf{h}_s, h_t)$ be the estimator of $C_{ji}(\mathbf{h}_s, h_t)$ based on the sample data in the spatio-temporal domain $D \times T_n$, where D represents the spatial domain and $T_n = \{1, \ldots, n\}$ the temporal one, and define $\{\hat{C}_{ji}(\mathbf{h}_s, h_t) : (\mathbf{h}_s, h_t) \in \Lambda, i, j = 1, \ldots, p\}$. Under the assumptions given in the above paper, $|T_n|^{1/2}(\hat{\mathbf{G}}_n - \mathbf{G}) \xrightarrow{d} N_{cp^2}(0, \Sigma)$, where $|T_n|\Sigma$ converges to $Cov(\hat{\mathbf{G}}_n, \hat{\mathbf{G}}_n)$. Then the tests for symmetry properties can be based on the following statistics

$$TS = |T_n|(\mathbf{A}\hat{\mathbf{G}}_n)^T (\mathbf{A}\Sigma\mathbf{A}^T)^{-1}(\mathbf{A}\hat{\mathbf{G}}_n) \xrightarrow{d} \chi_a^2, \tag{25}$$

where a is the row rank of the matrix $\mathbf{A}$, which is such that $\mathbf{A}\mathbf{G} = 0$ under the null hypothesis. Moreover, the choice of modeling the *MSTRF* $\mathbf{Z}$ by a spatio-temporal *LCM* is based on the prior assumption that the multivariate correlation structure of the variables under study is characterized by $L(L \geq 2)$ scales of spatio-temporal variability. On the other hand, if the multivariate correlation of a set of variables does not present different scales of variability ($L = 1$), then the cross-covariance functions are separable, i.e.,

$$C_{ij}(\mathbf{h}) = \rho(\mathbf{h}) \, b_{ij}, \qquad i, j = 1, \ldots, p, \tag{26}$$

where b_{ij} are the entries of a $(p \times p)$ positive definite matrix $\mathbf{B}$ and $\rho(\cdot)$ is a spatio-temporal correlation function. Hence, as in the spatial context [54], a spatio-temporal intrinsic coregionalization model can be considered.

Obviously, this last model is just a particular case ($L = 1$) of the spatio-temporal *LCM* defined in (17) and it is much more restrictive than the linear model of coregionalization since it requires that all the variables have the same correlation function, with possible changes in the sill values. Note that, if a cross-covariance is separable, then it is symmetric.

Remarks

- In the spatio-temporal *LCM*, each component is represented as a linear combination of latent, independent univariate spatio-temporal processes. However, the smoothness of any component defaults to that of the roughest latent process, and thus the standard approach does not admit individually distinct smoothness properties, unless structural zeros are imposed on the latent process coefficients [28].

- In most applications, the wide use of the spatio-temporal *LCM* is justified by practical aspects concerning the admissibility condition for the matrix variograms (covariances). Indeed, it is enough to verify the positive definiteness of the coregionalization matrices, $\mathbf{B}_l$, at all scales of variability.

- The spatio-temporal *LCM* allows unbounded variogram components to be used [54].

5. Prediction and risk assessment in space-time

For prediction purposes, various cokriging algorithms can be found in the literature [3, 33]. As a natural extension of spatial ordinary cokriging to the spatio-temporal context, the linear spatio-temporal predictor can be written as

$$\widehat{\mathbf{Z}}(\mathbf{u}) = \sum_{\alpha=1}^{N} \Lambda_{\alpha}(\mathbf{u}) \mathbf{Z}(\mathbf{u}_{\alpha}), \tag{27}$$

where $\mathbf{u} = (\mathbf{s}, t) \in D \times T$ is a point in the spatio-temporal domain, $\mathbf{u}_{\alpha} = (\mathbf{s}, t)_{\alpha} \in D \times T$, $\alpha = 1, \ldots, N$, are the data points in the same domain and $\Lambda_{\alpha}(\mathbf{u})$, $\alpha = 1, \ldots, N$, are $(p \times p)$ matrices of weights whose elements $\lambda_{\alpha}^{ij}(\mathbf{u})$ are the weights assigned to the value of the jth variable, $j = 1, \ldots, p$, at the αth data point, to predict the ith variable, $i = 1, \ldots, p$, at the point $\mathbf{u} \in D \times T$.

The predicted spatio-temporal random vector $\widehat{\mathbf{Z}}(\mathbf{u})$ at $\mathbf{u} \in D \times T$, is such that each component $\widehat{Z}_i(\mathbf{u})$, $i = 1, \ldots, p$, is obtained by using all information at the data points $\mathbf{u}_{\alpha} = (\mathbf{s}, t)_{\alpha} \in D \times T$, $\alpha = 1, \ldots, N$.

The matrices of weights $\Lambda_{\alpha}(\mathbf{u})$, $\alpha = 1, \ldots, N$, are determined by ensuring the unbiased condition for the predictor $\widehat{\mathbf{Z}}(\mathbf{u})$ and the efficiency condition, obtained by minimizing the error variance [29].

The new *GSLib* routine "COK2ST" [12] produces multivariate predictions in space-time, for one or all the variables under study, using the spatio-temporal *LCM* (20) where the basic spatio-temporal variograms are modelled as generalized product-sum variograms. An application is also given in [11].

Similarly, for environmental risk assessment, the formalism of multivariate spatio-temporal indicator random function (*MSTIRF*) and corresponding predictor, have to be introduced. Let

$$\mathbf{I}(\mathbf{u}, \mathbf{z}) = [I_1(\mathbf{u}, z_1), \ldots, I_p(\mathbf{u}, z_p)]^T,$$

be a vector of p spatio-temporal indicator random functions (*STIRF*) defined on the domain $D \times T \subseteq \mathbb{R}^{d+1}$, with $(d \leq 3)$, as follows

$$I_i(\mathbf{u}, z_i) = \begin{cases} 1 & \text{if } Z_i \text{ is not greater (or not less) than the threshold } z_i, \\ 0 & \text{otherwise} \end{cases}$$

where $\mathbf{z} = [z_1, \ldots, z_p]^T$. Then

$$\{\mathbf{l}(\mathbf{u}, \mathbf{z}), \; \mathbf{u} = (\mathbf{s}, t) \in D \times T \subseteq \mathbb{R}^{d+1}\},$$

represents a $MSTIRF$. In other words, for each coregionalized variable Z_i, with $i = 1, \ldots, p$, a $STIRF$ I_i can be appropriately defined. Then the linear spatio-temporal predictor (27) can be easily written in terms of the indicator random variables I_i, $i = 1, \ldots, p$. If the spatio-temporal correlation structure of a $MSTIRF$ is modelled by using the spatio-temporal LCM, based on the product-sum, the new $GSLib$ routine "COK2ST" [12] can be used to produce risk assessment maps, for one or all the variables under study. If $p = 1$, the dependence of the indicator variable is characterized by the corresponding indicator variogram of I: $2\gamma_{ST}(\mathbf{h}) = \text{Var}[I(\mathbf{s} + \mathbf{h}_s, t + h_t) - I(\mathbf{s}, t)]$, which depends solely on the lag vector $\mathbf{h} = (\mathbf{h}_s, h_t)$, $(\mathbf{s}, \mathbf{s} + \mathbf{h}_s) \in D^2$ and $(t, t + h_t) \in T^2$. After fitting a model for γ_{ST}, which must be conditionally negative definite, ordinary kriging can be applied to generate the environmental risk assessment maps. In this case, the $GSLib$ routine "K2ST" [17] can be used for prediction purposes in space and time.

6. Some critical aspects

Multivariate geostatistical analysis for spatio-temporal data is rather complex because of several problems concerning:

a) sampling,

b) the choice of admissible direct and cross-correlation models,

c) the definition of automatic procedures for estimation and modeling.

Sampling problems

There exist several sampling techniques for multivariate spatio-temporal data, as specified herein. Let

$$U_i = \{\mathbf{u}_\alpha \in D \times T, \alpha = 1, \ldots, N_i\}, \qquad i = 1, \ldots, p,$$

be the sets of sampled points in the spatio-temporal domain for the p variables under study. It is possible to distinguish the following situations:

1. *total heterotopy*, where the sets of the sampled points are pairwise disjoint, that is

$$\forall \, i, j = 1, \ldots, p, \quad U_i \cap U_j = \varnothing, \, i \neq j; \tag{28}$$

2. *partial heterotopy*, where the sets of the sampled points are not pairwise disjoint, that is

$$\exists \, i \neq j \mid U_i \cap U_j \neq \varnothing, \quad i, j = 1, \ldots, p; \tag{29}$$

3. *isotopy*, where the sets of the sampled points coincide, that is

$$\forall \, i, j = 1, \ldots, p, \quad U_i \equiv U_j. \tag{30}$$

A special case of partial heterotopy is the so-called *undersampling*: in such a case, the points where a variable, called *primary* or *principal*, has been sampled, constitute a subset of the points where the remaining variables, called *auxiliary* or *secondary*, have been observed. The secondary variables provide additional information useful to improve the prediction of the primary variable.

Modeling problems

In Geostatistics, the main modeling problems concern the choice of an admissible parametric model to be fitted to the empirical correlation function (covariance or variogram). In particular, in multivariate analysis for spatio-temporal data it is important to identify an admissible model able to describe the correlation among several variables which describe the spatio-temporal process. In this context, it is suitable to underline that

- it is not enough to select an admissible direct variogram to model a cross-variogram;
- the direct variograms are positive functions, on the other hand cross-variograms could be negative functions;
- only some necessary conditions of admissibility are known to model a cross-variogram, as the Cauchy-Schwartz inequality, however sufficient conditions cannot be easily applied [54].

The use of multivariate correlation models well-known in the literature, such as the *LCM* [54], the class of non-separable and asymmetric cross-covariances, proposed by [1], or the parametric family of cross-covariances, where each component is a Matérn process [28], requires the identification of several parameters, especially in presence of many variables.

Moreover, estimation and modeling the direct and cross-correlation functions could be compromised by the sampling plan.

Computational problems

For spatial multivariate data different algorithms for fitting the *LCM* have been implemented in software packages. [30] and [31] described an iterative procedure to fit a *LCM* using a weighted least-squares like technique: this requires first fitting the diagonal entries, i.e. the basic variogram structures must be determined first. In contrast, [56] and [57] developed an alternative method for modeling the matrix-valued variogram by near simultaneously diagonalizing the sample variogram matrices, without assuming any model for the variogram matrix; [36] proposed estimating the range parameters of a *LCM* using a non-linear regression method to fit the range parameters; [37] used simulated annealing to minimize a weighted sum of squares of differences between the empirical and the modelled variograms; [48] modified the Goulard and Voltz algorithm to make it more general and usable for generalized least-squares or any other least-squares estimation procedure, such as ordinary least-squares; [58] developed an algorithm for the maximum-likelihood estimation for the purely spatial *LCM* and proved that the EM algorithm gives an iterative procedure based on quasi-closed-form formulas, at least in the isotopic case. Significant contributions concerning estimation and computational aspects of a *LCM* can be found in [21]. Unfortunately, for multivariate spatio-temporal data there does not exist software packages which perform in an unified way a) the structural analysis, b) a convenient graphical representation of the covariance (variogram) models fitted to the empirical ones and c) predictions. One of the solutions could be to extend the above mentioned techniques to space-time. Indeed, some *routines* already implemented in the *GSLib software* [19] or in the module *gstat* of *R*, can only be applied in a multivariate spatial context.

In the last years, a new *GSLib* routine, called "COK2ST", can be used to make predictions in the domain under study, utilizing the spatio-temporal *LCM*, based on the generalized product-sum model [12]. This routine could be merged with the automatic procedure for fitting the spatio-temporal *LCM* using the product-sum variogram model, presented in [15], in order to provide a complete and helpful package for the analyst who needs to obtain predictions in a spatio-temporal multivariate context. This is certainly the first step for other developments and improvements in this field.

7. Case study

In the present case study, the environmental data set, with a multivariate spatio-temporal structure, involves PM_{10} daily concentrations, Temperature and Wind Speed daily averages measured at some monitored stations located in the South of Apulia region (Italy), from the 1st to the 23rd of November 2009. In particular, there are 28 PM_{10} survey stations, 60 and 54 atmospherical stations for monitoring Temperature and Wind Speed, respectively. As it is highlighted in Fig. 2(a), over the domain of interest, almost all the PM_{10} monitoring stations are either traffic or industrial stations, depending on the area where they are located (close to heavy traffic area or to industrialized area). The remaining monitoring stations are called peripheral. In Fig. 2(b), box plots of PM_{10} daily concentrations classified by typology of survey stations are illustrated. Fig. 3 shows the temporal profiles of the observed values. It is

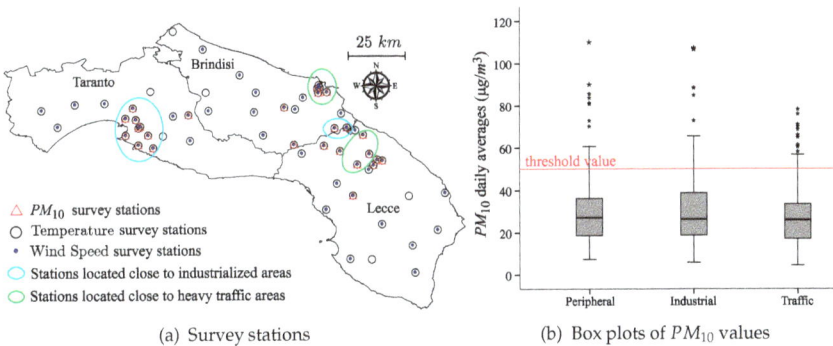

(a) Survey stations   (b) Box plots of PM_{10} values

Figure 2. Posting map and box plots of PM_{10} daily concentrations classified by typology of survey stations

evident that low (high) values of Temperature and Wind Speed are associated with high (low) values of PM_{10}.

Note that, during the period of interest, the PM_{10} threshold value fixed by National Laws for the human health protection (i.e. 50 $\mu g/m^3$ which should not be exceeded more than 35 times per year) has been overcome the 13rd, the 14th, the 15th and the 23rd of November 2009. Indeed, as regards this last aspect, it is worth noting that the highest PM_{10} daily averages have

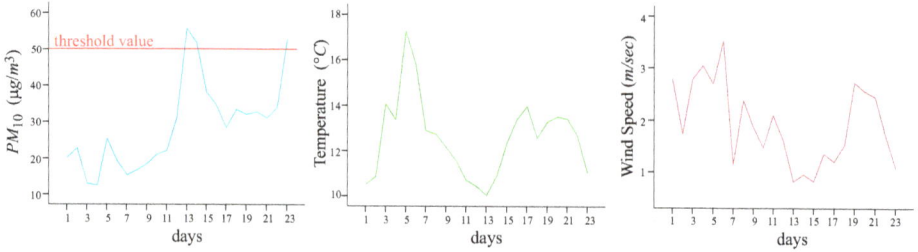

Figure 3. Time series plots of PM_{10}, Temperature and Wind Speed daily averages

been registered at all kinds of monitored stations, even at stations located at peripheral areas, likely for the transport effects caused by wind. As shown in Fig. 2(b), although the maximum PM_{10} values registered at stations close to heavy traffic area are greater than the threshold value, it is evident that these values are less than the ones measured at the other stations.

Spatio-temporal modeling and prediction techniques have been applied in order to assess PM_{10} risk pollution over the area of interest for the period 24-29 November 2009. In particular, the following aspects have been considered:

(1) estimating and modeling spatio-temporal correlation among the variables; in the fitting stage of a spatio-temporal *LCM*, the recent procedure [16] based on the nearly simultaneous diagonalization of several sample matrix variograms, has been applied and the product-sum variogram model [7] has been fitted to the basic components;

(2) spatio-temporal cokriging based on the estimated model, in order to obtain prediction maps for PM_{10} pollution levels during the period 24-29 November 2009 and indicator kriging [34] in order to construct risk maps related to the probability that predicted PM_{10} concentrations exceed the threshold value (50 $\mu g/m^3$) fixed by National Laws;

(3) generating and comparing, for two sites of interest (one close to an industrial area and the other one close to a heavy traffic area), the probability distributions that PM_{10} daily concentrations exceed some risk levels during the period 24-29 November 2009.

7.1. Modeling spatio-temporal *LCM*

Modeling the spatio-temporal correlation among the variables under study by using the spatio-temporal *LCM*, requires first to check the adequacy of such model. In particular, the symmetry assumption has been checked by

a) exploring the differences between the pseudo cross-variograms of the standardized variables (standardized by subtracting the mean value and dividing this difference by the standard deviation),

b) using the methodology proposed by [38].

As regards point a), the largest absolute difference has been equal to 0.135 and has been observed among the differences between the two pseudo cross-variograms concerning PM_{10} and Wind Speed standardized data. On the other hand, for the point b), three pairs composed by six stations, with consecutive daily average measurements have been selected for the test.

These pairs of monitoring stations have been picked to be approximately along the $SE - NW$ direction, since the prevalent wind direction over the area under study and during the analyzed period was $SE - NW$. Moreover, the temporal lag has been selected in connection with the largest empirical cross-correlations for all variable combinations; hence $h_t = 1$ has been considered for all testing lags. Three different variables and three pairs of stations generate 9 testing pairs in the symmetry test, and consequently the degree of freedom a for the symmetry test is equal to 9. The test statistic TS (25) has been equal to 0.34 with a corresponding p-value equal to 0.99. Hence, the results from both the procedures have highlighted that the spatio-temporal LCM is suitable for the data set under study.

By using the recent fitting procedure [16] based on the nearly simultaneous diagonalization of several sample matrix variograms computed for a selection of spatio-temporal lags, the basic independent components and the scales of spatio-temporal variability have been simply identified. In particular, the spatio-temporal surfaces of the variables under study have been computed for 7 and 5 user-chosen spatial and temporal lags, respectively (Fig. 4). Then, the 35 symmetric matrices of sample direct and cross-variograms have been nearly simultaneous diagonalization in order to detect the independent basic components. In this way, 3 scales of spatial and temporal variability have been identified: 10, 18 and 31.5 km in space, and 2.5, 3.5 and 6 days in time.

Thus, the following spatio-temporal LCM has been fitted to the observed data:

$$\Gamma(\mathbf{h}_s, h_t) = \mathbf{B}_1 \, g_1(\mathbf{h}_s, h_t) + \mathbf{B}_2 \, g_2(\mathbf{h}_s, h_t) + \mathbf{B}_3 \, g_3(\mathbf{h}_s, h_t), \tag{31}$$

where the spatio-temporal variograms $g_l(\mathbf{h}_s, h_t), l = 1, 2, 3$, are modelled as a generalized product-sum model, i.e.

$$g_l(\mathbf{h}_s, h_t) = \gamma_l(\mathbf{h}_s, 0) + \gamma_l(\mathbf{0}, h_t) - k_l \gamma_l(\mathbf{h}_s, 0) \, \gamma_l(\mathbf{0}, h_t). \tag{32}$$

The spatial and temporal marginal basic variogram models, $\gamma_l(\mathbf{h}_s, 0)$ and $\gamma_l(\mathbf{0}, h_t)$, respectively, and the coefficients $k_l, l = 1, 2, 3$, previously defined in (22), are shown below:

$$\gamma_1(\mathbf{h}_s, 0) = 86 \, \mathrm{Exp}(||\mathbf{h}_s||; 10), \qquad \gamma_1(\mathbf{0}, h_t) = 165 \, \mathrm{Exp}(|h_t|; 2.5), \qquad k_1 = 0.0057, \tag{33}$$

$$\gamma_2(\mathbf{h}_s, 0) = 0.95 \, \mathrm{Exp}(||\mathbf{h}_s||; 18), \qquad \gamma_2(\mathbf{0}, h_t) = 3.7 \, \mathrm{Exp}(|h_t|; 3.5), \qquad k_2 = 0.02418, \tag{34}$$

$$\gamma_3(\mathbf{h}_s, 0) = 0.29 \, \mathrm{Gau}(||\mathbf{h}_s||; 31.5), \qquad \gamma_3(\mathbf{0}, h_t) = 0.83 \, \mathrm{Exp}(|h_t|; 6), \qquad k_3 = 1.1633, \tag{35}$$

where $\mathrm{Exp}(\cdot; a)$ and $\mathrm{Gau}(\cdot; a)$ denote the well known exponential and Gaussian variogram models, with practical range a [19].

Finally, the matrices $\mathbf{B}_l, l = 1, 2, 3$, of the spatio-temporal LCM (31), computed by the procedure described in [15], are the following:

$$\mathbf{B}_1 = \begin{bmatrix} 0.982 & -0.044 & -0.024 \\ -0.044 & 0.018 & 0.006 \\ -0.024 & 0.006 & 0.007 \end{bmatrix}, \quad \mathbf{B}_2 = \begin{bmatrix} 43.421 & -2.632 & -1.550 \\ -2.632 & 0.395 & 0.118 \\ -1.550 & 0.118 & 0.105 \end{bmatrix}, \tag{36}$$

$$\mathbf{B}_3 = \begin{bmatrix} 71.429 & -10.119 & -4.167 \\ -10.119 & 1.726 & 0.414 \\ -4.167 & 0.414 & 0.357 \end{bmatrix}. \tag{37}$$

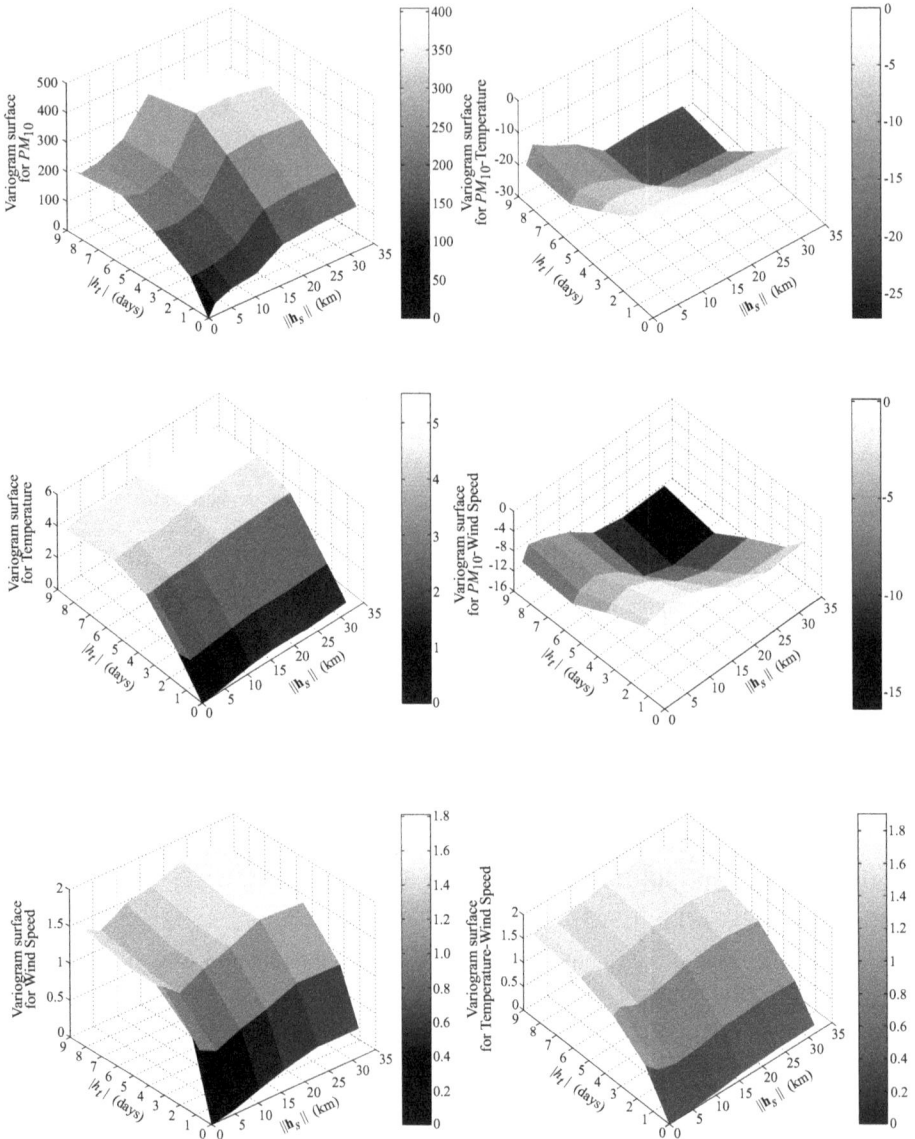

Figure 4. Spatio-temporal variogram and cross-variogram surfaces of PM_{10}, Temperature and Wind Speed daily averages

Fig. 5 shows the spatio-temporal variograms and cross-variograms fitted to the surfaces of PM_{10}, Temperature and Wind Speed daily averages. Then, the spatio-temporal LCM (31) has been used to produce prediction and risk assessment maps for PM_{10} daily concentrations, as discussed hereafter.

7.2. Prediction maps and risk assessment

In order to obtain the prediction maps for PM_{10} pollution levels for the period 24-29 November 2009, spatio-temporal cokriging has been applied, using the routine "COK2ST" [12]. Risk assessment maps have been associated to the prediction maps. Spatial indicator kriging has been applied to assess the probability that predicted PM_{10} daily concentrations exceed the PM_{10} threshold value fixed by National Laws for the human health protection (i.e. 50 $\mu g / m^3$), during the period of interest. Fig. 6 and Fig. 7 show contour maps of the predicted PM_{10} values and the corresponding risk maps, for the period 24-29 November 2009. The red points on the maps represent the monitoring stations.

It is important to highlight that the highest PM_{10} values are predicted in the Eastern part of the domain of interest: this area corresponds to the boundary between Lecce and Brindisi districts which is strictly close to an industrial site, such as the thermoelectric power station "Enel-Federico II", located in Cerano (Brindisi district). Moreover, in this area the probability that PM_{10} daily concentrations exceed 50 $\mu g / m^3$ is high during the predicted week. It is worth noting that on Saturday and Sunday the predicted values of PM_{10} daily concentrations show lower average levels than the ones estimated during the working days, when heavy traffic contributes to keep pollution concentrations high; consequently, the corresponding risk maps do not show hazardous PM_{10} conditions.

7.3. Probability distributions for different sites

After producing predicted maps of PM_{10} daily concentrations, it is also interesting to estimate the probability distribution that PM_{10} daily concentrations exceed some risk levels at sites characterized by sources of pollution.

Two different sites, one close to an industrialized area, located at Brindisi district, and the other one close to a heavy traffic area, located at Lecce district, have been considered in order to generate and compare the probability distributions that PM_{10} daily concentrations overcome several risk levels during the period 24-29 November 2009. Fig. 8 shows that, all over the industrialized area of interest, the probability that PM_{10} daily concentrations exceed the threshold fixed by National Laws (50 $\mu g / m^3$) is very high (greater than 80%) during the period 24-27 November 2009 (working days); on the other hand, during the weekend (28-29 November 2009) such a probability decreases at 40-42%.

Note that at the site close to a heavy traffic area, the probability that PM_{10} daily concentrations exceed the national law limit is very low during the analyzed period, except on the 25th of November; moreover, it drops off rapidly for values below the national threshold. This empirical evidence highlights that there is no critical PM_{10} exceeding for the selected traffic site, especially during the last 3 days of the week. As it is shown, this is a very powerful tool since any action of environmental protection might be adopted in advance by taking into account the actual likelihood of dangerous PM_{10} exceeding.

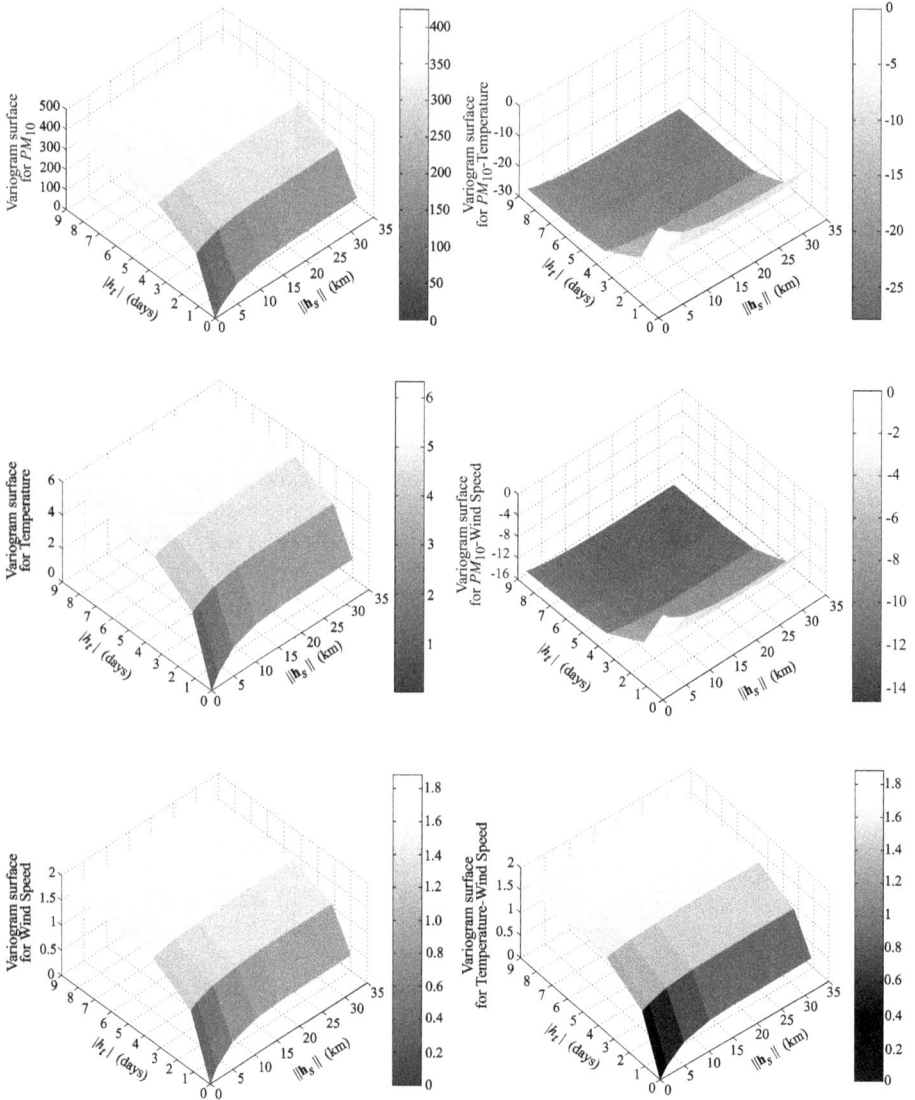

Figure 5. Spatio-temporal variograms and cross-variograms fitted to variogram and cross-variogram surfaces of PM_{10}, Temperature and Wind Speed daily averages

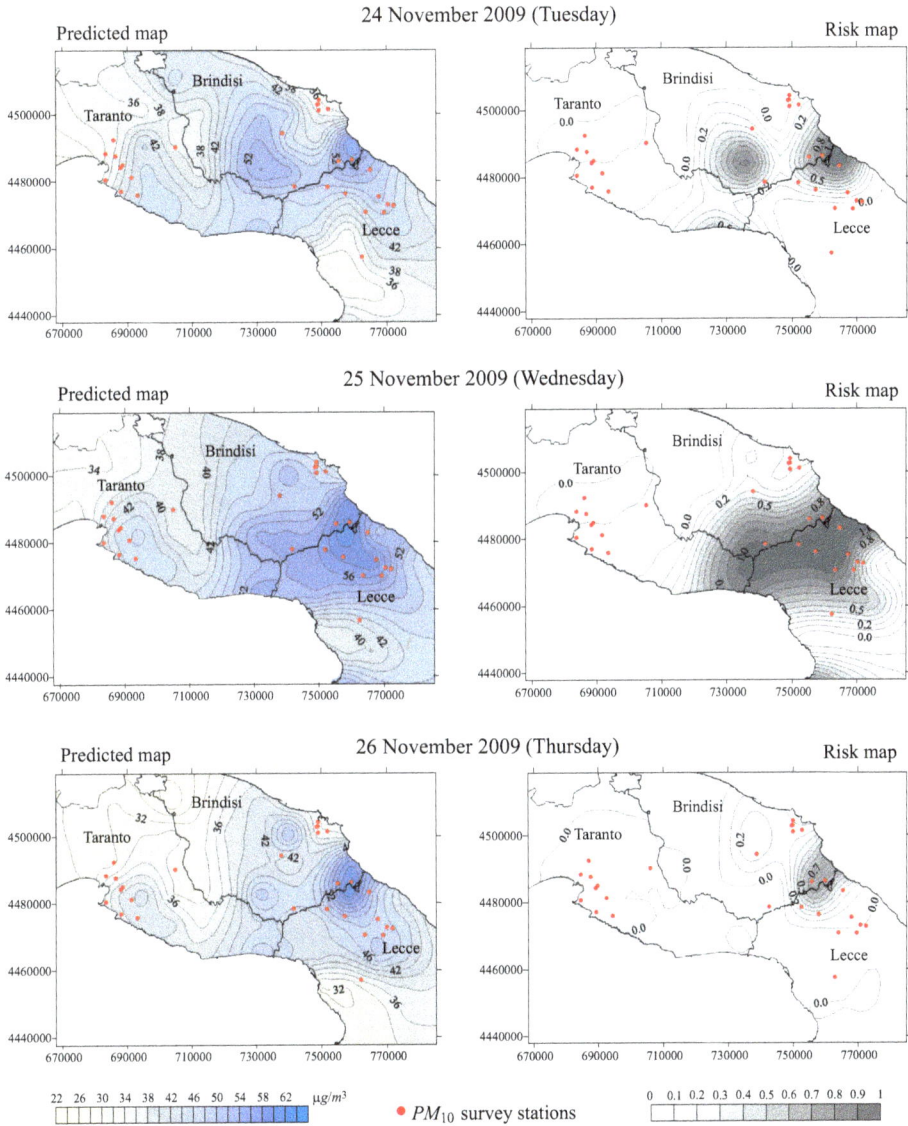

Figure 6. Prediction maps of PM_{10} daily concentrations and risk maps at the threshold fixed by National Laws, for the 24th, 25th and 26th of November 2009

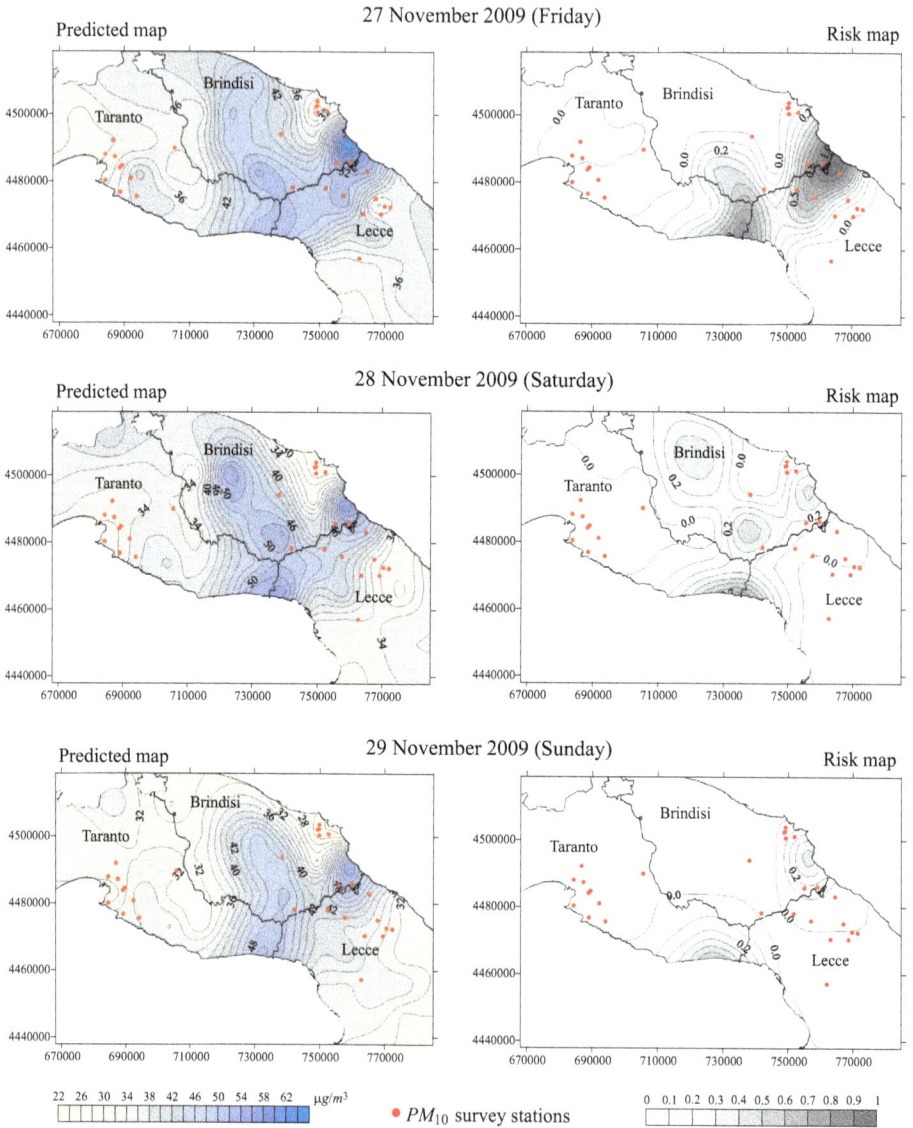

Figure 7. Prediction maps of PM_{10} daily concentrations and risk maps at the threshold fixed by National Laws, for the 27th, 28th and 29th of November 2009

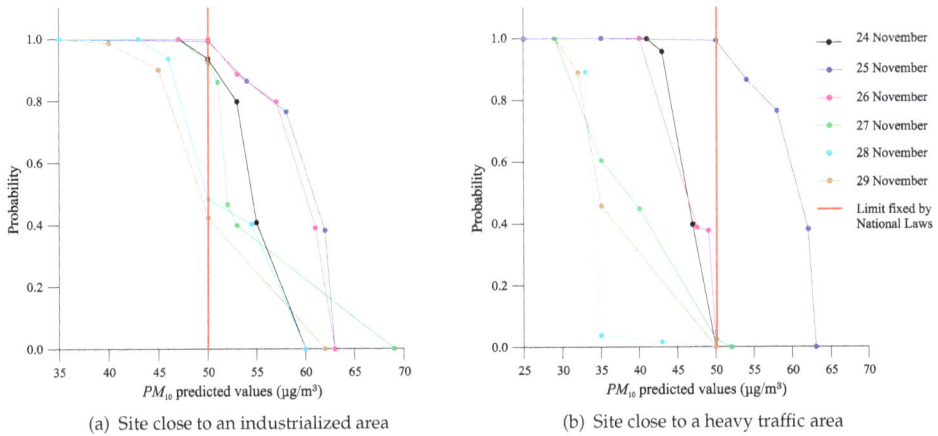

Figure 8. Probability distributions that PM_{10} daily concentrations overcome several risk levels during the period 24-29 November 2009

8. Conclusions

In this paper, some significant theoretical and practical aspects for multivariate geostatistical analysis have been discussed and some critical issues concerning sampling, modeling and computational aspects, which should be faced, have been pointed out. The proposed multivariate geostatistical techniques have been applied to a case study pertaining particle pollution (PM_{10}) and two atmospheric variables (Temperature and Wind Speed) in the South of Apulian region.

Further analysis regarding the integration of land use and possible sources of pollution through an appropriate geographical information system could be helpful to fully understand the dynamics of PM_{10}, which is still considered one of the most hazardous pollutant for human health.

Acknowledgments

The authors are grateful to the Editor and the reviewers, whose comments contribute to improve the present version of the paper. This research has been partially supported by the "5per1000" project (grant given by University of Salento in 2011).

Author details

S. De Iaco, S. Maggio, M. Palma and D. Posa
Università del Salento, DSE, Complesso Ecotekne, Italy

9. References

[1] Apanasovich, T.V. & Genton, M.G. (2010). Cross-covariance functions for multivariate random fields based on latent dimensions, *Biometrika* Vol. 97(No. 1): 15–30.

[2] Brown, P., Karesen, K. & Tonellato, G.O.R.S. (2000). Blur-generated nonseparable space-time models, *Journal of Royal Statistical Society, Series B* Vol. 62(Part 4): 847–860.

[3] Chilés, J. & Delfiner, P. (1999). *Geostatistics - Modeling spatial uncertainty*, Wiley.

[4] Choi, J., Fuentes, M., Reich, B.J. & Davis, J.M. (2009). Multivariate spatial-temporal modeling and prediction of speciated fine particles, *Journal of Statistical Theory and Practice* Vol. 3(No. 2): 407–418.

[5] Cressie, N. & Huang, H. (1999). Classes of nonseparable, spatial-temporal stationary covariance functions, *Journal of American Statistical Association* Vol. 94(No. 448): 1330–1340.

[6] De Iaco, S., (2011). A new space-time multivariate approach for environmental data analysis, *Journal of Applied Statistics* Vol. 38(No. 11): 2471–2483. 38, Issue 11 2471-2483.

[7] De Iaco, S., Myers, D.E. & Posa, D., (2001a). Space-time analysis using a general product-sum model, *Statistics and Probability Letters* Vol. 52(No. 1): 21–28.

[8] De Iaco, S., Myers, D.E. & Posa, D., (2001b). Total air pollution and space-time modeling. *GeoENV III. Geostatistics for Environmental Applications*, Eds. Monestiez, P., Allard, D. and Froidevaux, R., Kluwer Academic Publishers, Dordrecht: 45–56.

[9] De Iaco, S., Myers, D.E. & Posa, D. (2002). Nonseparable space-time covariance models: some parametric families, *Mathematical Geology* Vol. 34(No. 1): 23–42.

[10] De Iaco, S., Myers, D.E. & Posa, D. (2003). The linear coregionalization model and the product-sum space-time variogram, *Mathematical Geology* Vol. 35(No. 1): 25–38.

[11] De Iaco, S., Palma, M. & Posa, D. (2005). Modeling and prediction of multivariate space-time random fields, *Computational Statistics and Data Analysis* Vol. 48(No. 3): 525–547.

[12] De Iaco, S., Myers, D.E., Palma, M. & Posa, D. (2010). FORTRAN programs for space-time multivariate modeling and prediction, *Computers & Geosciences* Vol. 36(No. 5): 636–646.

[13] De Iaco, S., Myers, D.E., Posa, D. (2011a). Strict positive definiteness of a product of covariance functions, *Commun. in Statist. - Theory and Methods* Vol. 40(No. 24): 4400–4408.

[14] De Iaco, S., Myers, D.E. & Posa, D. (2011b). On strict positive definiteness of product and product-sum covariance models, *J. of Statist. Plan. and Inference* Vol. 141: 1132–1140.

[15] De Iaco S., Maggio S., Palma M. & Posa D. (2012a). Towards an automatic procedure for modeling multivariate space-time data, *Computers & Geosciences* Vol. 41: 1–11.

[16] De Iaco, S., Myers, D.E., Palma, M. & Posa, D. (2012b). Using simultaneous diagonalization to fit a space-time linear coregionalization model, *Mathematical Geosciences* Doi: 10.1007/s11004-012-9408-3.

[17] De Iaco, S. & Posa, D. (2012). Predicting spatio-temporal random fields: some computational aspects. *Computers & Geosciences*, Vol. 41: 12–24.

[18] de Luna, X. & Genton, M.G. (2005). Predictive spatio-temporal models for spatially sparse environmental data, *Statistica Sinica* Vol. 15(No. 2): 547–568.

[19] Deutsch, C.V. & Journel, A.G. (1998). *GSLib: Geostatistical Software Library and User's Guide*, Oxford University Press, New York.

[20] Dimitrakopoulos, R. & Luo, X. (1994). Spatiotemporal modeling: covariance and ordinary kriging systems, *in*: Dimitrakopoulos, R. (ed.), *Geostatistics for the next century*, Kluwer Academic Publishers, Dordrecht, pp. 88–93.

[21] Emery, X. (2010). Iterative algorithms for fitting a linear model of coregionalization, *Computers & Geosciences* Vol. 36(No. 9): 836–846.

[22] Fassò, A. & Finazzi, F. (2011). Maximum likelihood estimation of the dynamic coregionalization model with heterotopic data, *Environmentrics* Vol. 22: 735–748.

[23] Fuentes, M. (2006). Testing for separability of spatial-temporal covariance functions, *Journal of Statistical Planning and Inference* Vol. 136(No. 2): 447–466.

[24] Gaspari, G. & Cohn, S.E. (1999). Construction of correlation functions in two and three dimensions. *Quarterly Journal of the Royal Meteorological Society* Vol. 125(No. 554): 723–757.

[25] Gelfand, A.E., Schmidt, A.M., Banerjee, S. & Sirmans, C.F. (2004). Nonstationary multivariate process modeling through spatially varying coregionalization, *Sociedad de Estadystica e Investigacion Opertiva Test* Vol. 13: 263–312.

[26] Gelfand, A.E., Banerjee, S. & Gamerman, D. (2005). Spatial process modeling for univariate and multivariate dynamic spatial data, *Environmetrics* Vol. 16: 465–479.

[27] Gneiting, T. (2002). Nonseparable, stationary covariance functions for space-time data, *Journal of the American Statistical Association* Vol. 97(No. 458): 590–600.

[28] Gneiting, T., Kleiber, W. & Schlather, M. (2010). Matérn cross-covariance functions for multivariate random fields, *Journal of the American Statistical Association* Vol. 105(No. 491): 1167–1177.

[29] Goovaerts, P. (1997). *Geostatistics for natural resources evaluation*, Oxford University Press, New York.

[30] Goulard, M. (1989). Inference in a coregionalization model, *in* Armstrong, M. (ed.), *Geostatistics*, Kluwer Academic Publishers, Dordrecht, pp. 397–408.

[31] Goulard, M. & Voltz, M. (1992). Linear coregionalization model: tools for estimation and choice of cross-variogram matrix, *Mathematical Geology* Vol. 24(No. 3): 269–286.

[32] Guo, J.H. & Billard, L. (1998). Some inference results for causal autoregressive processes on a plane, *Journal of Time Series Analysis* Vol. 19(No. 6): 681–691.

[33] Journel, A.G. & Huijbregts, C.J. (1981). *Mining Geostatistics*, Academic Press, London.

[34] Journel, A.G. (1983). Non-parametric estimation of spatial distribution, *Mathematical Geology* Vol. 15(No. 3): 445–468.

[35] Kolovos, A., Christakos, G., Hristopulos, D.T. & Serre, M.L. (2004). Methods for generating non-separable spatiotemporal covariance models with potential environmental applications, *Advances in Water Resources* Vol. 27(No. 8): 815–830.

[36] Künsch, H.R., Papritz, A. & Bassi, F. (1997). Generalized cross-covariances and their estimation, *Mathematical Geology* Vol. 29(No. 6): 779–799.

[37] Lark, R.M. & Papritz, A. (2003). Fitting a linear model of coregionalization for soil properties using simulated annealing, *Geoderma* Vol. 115(No. 3): 245–260.

[38] Li, B., Genton, M.G. & Sherman, M. (2008). Testing the covariance structure of multivariate random fields, *Biometrika* Vol. 95(No. 4): 813–829.

[39] Lu, N. & Zimmerman, D.L. (2005a). The likelihood ratio test for a separable covariance matrix, *Statistics and Probability Letters* Vol. 73(No. 4): 449–457.

[40] Lu, N. & Zimmerman, D.L. (2005b). Testing for directional symmetry in spatial dependence using the periodogram, *Journal of Statistical Planning and Inference* Vol. 129(No. 1-2): 369–385.

[41] Ma, C. (2002). Spatio-temporal covariance functions generated by mixtures, *Mathematical Geolology* Vol. 34(No. 8): 965–975.

[42] Ma, C. (2005). Linear combinations of space-time covariance functions and variograms, *IEEE Transactions on Signal Processing* Vol. 53(No. 3): 489–501.

[43] Majumdar, A. & Gelfand, A.E. (2007). Multivariate spatial modeling using convolved covariance functions, *Mathematical Geolology* Vol. 39(No. 2): 225–245.

[44] Mardia, K.V. & Goodall, C.R. (1993). Spatial-temporal analysis of multivariate environmental monitoring data, *in* Multivariate environmental statistics, North-Holland Series in Statistics and Probability 6, Amsterdam, Olanda, pp. 347–386.

[45] Matheron, G. (1982). *Pour une analyse krigeante des données régionalisées* Rapport technique N732, Ecole Nationale Supérieure des Mines de Paris

[46] Mitchell, M.W., Genton, M.G. & Gumpertz, M.L. (2005). Testing for separability of space-time covariances, *Environmetrics* Vol. 16(No. 8): 819–831.

[47] Myers, D.E. (1991). Pseudo-cross variograms, positive definiteness, and cokriging, *Mathematical Geology*, Vol. 23(No. 6): 805–816.

[48] Pelletier, B., Dutilleul, P., Guillaume, L. & Fyles J.W. (2004). Fitting the linear model of coregionalization by generalized least squares, *Mathematical Geology* Vol. 36(No. 3): 323–343.

[49] Porcu, E., Mateu, J. & Saura, F. (2008). New classes of covariance and spectral density functions for spatio-temporal modeling, *Stochastic Environmental Research and Risk Assessment* Vol. 22, Supplement 1, 65–79.

[50] Scaccia, L. & Martin, R.J. (2005). Testing axial symmetry and separability of lattice processes, *Journal of Statistical Planning and Inference* Vol. 131(No. 1): 19–39.

[51] Shitan, M. & Brockwell, P. (1995). An asymptotic test for separability of a spatial autoregressive model, *Communication In Statistical Theory and Method* Vol. 24(No. 8): 2027–2040.

[52] Stein, M. (2005). Space-time covariance functions, *Journal of the American Statistical Association* Vol. 100(No. 469): 310–321.

[53] Thiebaux, H.J. (1990). Spatial objective analysis, *in* Encyclopedia of Physical Science and Technology, 1990 Yearbook. Academic Press, New York, pp. 535–540.

[54] Wackernagel, H. (2003). *Multivariate Geostatistics: an introduction with applications*, Springer, Berlin.

[55] Ver hoef, J.M. & Barry, R.P. (1998). Constructing and fitting models for cokriging and multivariable spatial prediction, *Journal of Statistical Planning and Inference* Vol. 69(No. 2): 275–294.

[56] Xie, T. & Myers, D.E. (1995). Fitting matrix-valued variogram models by simultaneous diagonalization (Part I: Theory), *Mathematical Geology* Vol. 27(No. 7): 867–876.

[57] Xie, T., Myers, D.E., Long, A.E. (1995). *Fitting matrix-valued variogram models by simultaneous diagonalization: (Part II: Applications)*, *Mathematical Geology* Vol. 27(No. 7): 877–888.

[58] Zhang, H. (2007). Maximum-likelihood estimation for multivariate spatial linear coregionalization models, *Environmetrics* Vol. 18, 125–139.

Permissions

The contributors of this book come from diverse backgrounds, making this book a truly international effort. This book will bring forth new frontiers with its revolutionizing research information and detailed analysis of the nascent developments around the world.

We would like to thank Budi Haryanto, for lending his expertise to make the book truly unique. He has played a crucial role in the development of this book. Without his invaluable contribution this book wouldn't have been possible. He has made vital efforts to compile up to date information on the varied aspects of this subject to make this book a valuable addition to the collection of many professionals and students.

This book was conceptualized with the vision of imparting up-to-date information and advanced data in this field. To ensure the same, a matchless editorial board was set up. Every individual on the board went through rigorous rounds of assessment to prove their worth. After which they invested a large part of their time researching and compiling the most relevant data for our readers. Conferences and sessions were held from time to time between the editorial board and the contributing authors to present the data in the most comprehensible form. The editorial team has worked tirelessly to provide valuable and valid information to help people across the globe.

Every chapter published in this book has been scrutinized by our experts. Their significance has been extensively debated. The topics covered herein carry significant findings which will fuel the growth of the discipline. They may even be implemented as practical applications or may be referred to as a beginning point for another development. Chapters in this book were first published by InTech; hereby published with permission under the Creative Commons Attribution License or equivalent.

The editorial board has been involved in producing this book since its inception. They have spent rigorous hours researching and exploring the diverse topics which have resulted in the successful publishing of this book. They have passed on their knowledge of decades through this book. To expedite this challenging task, the publisher supported the team at every step. A small team of assistant editors was also appointed to further simplify the editing procedure and attain best results for the readers.

Our editorial team has been hand-picked from every corner of the world. Their multi-ethnicity adds dynamic inputs to the discussions which result in innovative outcomes. These outcomes are then further discussed with the researchers and contributors who give their valuable feedback and opinion regarding the same. The feedback is then collaborated with the researches and they are edited in a comprehensive manner to aid the understanding of the subject.

Apart from the editorial board, the designing team has also invested a significant amount of their time in understanding the subject and creating the most relevant covers. They scrutinized every image to scout for the most suitable representation of the subject and create an appropriate cover for the book.

The publishing team has been involved in this book since its early stages. They were actively engaged in every process, be it collecting the data, connecting with the contributors or procuring relevant information. The team has been an ardent support to the editorial, designing and production team. Their endless efforts to recruit the best for this project, has resulted in the accomplishment of this book. They are a veteran in the field of academics and their pool of knowledge is as vast as their experience in printing. Their expertise and guidance has proved useful at every step. Their uncompromising quality standards have made this book an exceptional effort. Their encouragement from time to time has been an inspiration for everyone.

The publisher and the editorial board hope that this book will prove to be a valuable piece of knowledge for researchers, students, practitioners and scholars across the globe.

List of Contributors

Margherita Ferrante, Maria Fiore, Gea Oliveri Conti, Caterina Ledda, Roberto Fallico and Salvatore Sciacca
Department "G.F. Ingrassia", Sector of Hygiene and Public Health, Catania University, Italy

Selçuk Arslan and Ali Aybek
Department of Biosystems Engineering, Faculty of Agriculture, Kahramanmaraş Sütçü İmam University, Turkey

Helena Martins, Ana Miranda and Carlos Borrego
CESAM & Department of Environment and Planning, University of Aveiro, Aveiro, Portuga

Francisco A. Serrano-Bernardo and José L. Rosúa-Campos
Department of Civil Engineering. University of Granada, Spain

Luigi Bruzzi
Department of Physics. University of Bologna, Italy

Enrique H. Toscano
Joint Research Centre (JRC), European Commission, Karlsruhe, Germany

S. B. Nugroho, A. Fujiwara and J. Zhang
Transportation Engineering Laboratory, Graduate School for International Development and Cooperation, Hiroshima University, Japan, Kagamiyama, Higashi Hiroshima, Japan

An-Soo Jang
Division of Allergy and Respiratory Medicine, Department of Internal Medicine, Soonchunhayg University Hospital, Bucheon, Korea

Takao Matsumoto, Douyan Wang, Takao Namihira and Hidenori Akiyama
Kumamoto University, Japan

Parisa Shahmohamadi, Ulrich Cubasch and Sahar Sodoudi
Institut für Meteorologie, Freie Universität Berlin, Germany

A.I. Che-Ani
Department of Architecture, Faculty of Engineering and Built Environment, Universiti Kebangsaan Malaysia, Selangor, Malaysia

Wang-Kun Chen
Jinwen University of Science and Technology, Department of Environment and Property Management, Taiwan

Michael Hein and Manfred Kaiser
DMT GmbH and Co. KG, Cokemaking Technology Division, Essen, Germany

Hussein Ibrahim
TechnoCentre Éolien, Gaspé, QC, Canada

Adrian Ilinca
Université du Québec à Rimouski, Rimouski, QC, Canada

S. De Iaco, S. Maggio, M. Palma and D. Posa
Università del Salento, DSE, Complesso Ecotekne, Italy